Statistical Experiment Design and Interpretation

Statistical Experiment Design and Interpretation

An Introduction with Agricultural Examples

Claire A. Collins
ADAS Bridgets

and

Frances M. Seeney
United Kingdom Transplant Support Services Authority

JOHN WILEY & SONS, LTD
Chichester • New York • Weinheim • Brisbane • Singapore • Toronto

Copyright © 1999 by   John Wiley & Sons, Ltd,
Baffins Lane, Chichester,
West Sussex PO19 1UD, England

National   01243 779777
International (+44) 1243 779777
e-mail (for orders and customer service enquiries): cs-books@wiley.co.uk
Visit our Home Page on http://www.wiley.co.uk
or http://www.wiley.com

Other Wiley Editorial Offices

John Wiley & Sons, Inc., 605 Third Avenue,
New York, NY 10158-0012, USA

WILEY-VCH Verlag GmbH, Pappelallee 3,
D-69469 Weinheim, Germany

Jacaranda Wiley Ltd, 33 Park Road, Milton,
Queensland 4064, Australia

John Wiley & Sons (Asia) Pte Ltd, 2 Clementi Loop #02-01,
Jin Xing Distripark, Singapore 129809

John Wiley & Sons (Canada) Ltd, 22 Worcester Road,
Rexdale, Ontario M9W 1L1, Canada

British Library Cataloguing in Publication Data

A catalogue record for this book is available from the British Library

ISBN 0 471 96006 3

Typeset in 10/12pt Times by MHL Typesetting Limited
Printed and bound in Great Britain by Biddles Ltd, Guildford and King's Lynn
This book is printed on acid-free paper responsibly manufactured from sustainable forestry, in which at least two trees are planted for each one used for paper production.

Contents

Acknowledgements

We would like to thank ADAS for encouraging us to produce this book and Dr David Hughes for allowing us to use the data for our examples. Special thanks are due to Mr Brian Bastiman, Dr Howard Simmins and Dr Mike Proven who helped devise the training course this book is based on. Thanks are also due to our colleagues in statistics, especially Dr John Wafford, Mr Alan Heaven and Dr Douglas Wilson, for their patience in checking the content of this text.

Thanks are also due to all the staff at Sheffield Hallam University involved in the MSc. in Applied Statistics course that the authors attended. This course gave us the confidence to run our own in-house statistics training course and helped us identify some salient topics to be included. We are grateful to Addison-Wesley Longman for their permission to use definitions of statistical terms as published in Marriott's *A dictionary of statistical terms*.

We are also deeply indebted to our ADAS colleagues who attended our training course and provided much constructive advice and welcome feedback. Many of their comments have been taken on board in this book. We hope that you, the reader, will find this introductory book as useful as our colleagues did the course.

CHAPTER 1

Introduction

Much of our work at ADAS involves liaising with agriculture researchers involved in the design and management of arable and livestock experiments and who collect and analyse the data gathered. Although many of them produce statistical output from computer packages they are sometimes unsure why particular experiment designs are used and what the results actually mean. Part of our role is to train people in their understanding of experiment design and basic statistical techniques and interpretation. This was initially achieved by presenting a training course, which is further supported by general statistical support and consultancy. As part of this training course a set of lecture notes was produced for them to use as a reference document after training. The positive comments received from the researchers have indicated that the notes prove very useful in their work and we were encouraged to develop the notes into a book.

All the agriculture researchers had a good scientific background but statistical knowledge ranged from very little to years of practical experience. Most had had formal statistics training in academic situations so were familiar with basic statistical terminology but found it difficult to apply the theory to their own practical situations. The training course therefore concentrated on the practical application of statistics in agricultural research without delving too much into the theory behind the common statistical techniques used. Very little mathematical knowledge was assumed and formulae were kept to a minimum. It is important to realize that computers will do what you ask them even if it is nonsensical, so a user must know the correct techniques to use for their situation. The course concentrated on 'when' and 'why' rather than 'how', so computers and calculators were not used.

This text is based on the course notes. Some additional formulae have been presented and calculations behind some of the analysis of variance examples have been further detailed in the appendix for readers that are interested. This can be omitted without detracting from the earlier chapters. A section on computing packages used in our work has been incorporated. It is not a comprehensive coverage of statistical computing facilities. It covers the main pros and cons of some of the available packages from our point of view. You may disagree with our views but at least it highlights the importance of knowing what your statistical package is doing and how it may differ from another.

This text covers the basic concepts and theories underlying experiment design and the main statistical techniques that can be used to analyse and model data obtained from designed experiments. A theoretical discussion of statistical techniques is not intended and discussion concentrates on the practicalities and problems that may be encountered in agricultural research. A complete coverage of all statistical techniques encountered in research is beyond the scope of this book. This text concentrates on the analytical techniques of data exploration, analysis of variance, regression analysis and non-parametric techniques. The assumptions implicit in each technique are detailed, as are any restrictions on the type of data they can be used with. The main emphasis with each technique is on what it is testing and the basic conclusions from the test. Generally speaking, when techniques other than these were to be used with data the analysis was referred to a statistician so although other statistical techniques are mentioned it is only briefly as they extend beyond the original remit of the work. The reader is encouraged to read other texts that cover these techniques in more detail and reference is made throughout this text to a selection of publications the authors found useful.

Whilst it is recommended that you cover this text at least once in its entirety, it has been designed to allow you to dip into the separate techniques you require without needing to refer to numerous other sections. For this reason the assumptions, hypotheses and interpretation are included for each of the statistical tests covered in this text even when they repeat those for other tests, so cross-referencing other sections is minimalized. There are three main sections to the book: the first (Chapters 2–4) covers experiment design, the second (Chapters 5–7) discusses the interpretation of the statistical analysis of designed experiments, and the third (Chapter 8 and the appendices) contains relevant aspects of computing, formulae and a glossary of statistical terms used.

The 'correct' choice of experiment design is of primary importance in scientific experiments. If at the planning stage the design of an experiment is carefully and correctly thought through then the analysis will usually be straightforward and will have been decided upon with and by the objectives.

1.1   NOTATION

The following standard notation has also been used in the text:

df	degrees of freedom
H_0	null hypothesis
H_1	alternative hypothesis
LSD	least significant difference
ms	mean square
n	the number of observations
N (0,1)	a normal distribution with zero mean and variance of one
s	standard deviation of a sample
$s_{\bar{x}}$	standard error of the mean
$s_{\bar{x}_1 - \bar{x}_2}$	standard error of the difference between two mean values
SD	standard deviation of a sample
ss	sum of squares
vr	variance ratio
x_i	the value of the ith group or observation
$\bar{x}$	arithmetic mean of a sample
$|x|$	the modulus of x, i.e. the value regardless of the sign
α	significance level
μ	population mean
ρ	product moment correlation coefficient
σ	population standard deviation
$\sum_{i=1}^{n}$	sum over the n observations
υ	degrees of freedom

1.2 A LITTLE HISTORY

The need for statistics in agricultural experiments may not be immediately obvious; however, all such experiments involve collecting, analysing and interpreting numerical data, each stage of which involves statistics. If the experiments are robust and efficient they will provide the required information using the minimum resources in terms of money, time and human resources. The use of statistical principles helps in the design of such experiments.

R. A. Fisher worked as a statistician at, what is now, Rothamsted Agricultural Research Station and it was here that he founded statistical experiment design. He found that biological workers were shy of mathematics and only consulted a statistician after the data had been collected. He had been invited to work there to see what could be learnt from a statistical analysis of records of experiments and observational data collected over the previous 80 years. These data had been collected with no understanding of statistical planning and Fisher's analyses highlighted structural defects in the experiment layouts, from which grew the ideas of randomization and designed experiments. A full account of Fisher's work can be found in his biography, *The Life of a Scientist*, written by his daughter (Box, 1978).

1.3 POPULATIONS VERSUS SAMPLES

It is claimed that a new method of cultivating wheat produces a better crop. This would seem to be a simple claim to test, but how can one crop be judged to be better than another? Is the claim made about the crop yield, quality, disease resistance or some other criteria? Is it actually possible to measure the criteria, for example can disease resistance be measured or does it need to be qualified? Even if the criteria being measured have been defined, the statement as it stands is still not clear. Is the claim made about all varieties of wheat or just a selected few or just one? Already the simple claim has posed several questions.

If the claim is made more specific it might be stated that a new method of cultivating winter wheat variety X produces a higher yield than the old cultivation method. This is a clear specific statement, but how can it be tested? If wheat is grown by the two methods will the results only apply to the particular wheat crop that is grown? If this is the case there seems little point in any type of experiment as conclusions will only apply to the experiment material actually tested. Conclusions need to be drawn about future crops of this type but it is not possible to test all future crops; only a few can be tested. In other words, conclusions need to be applicable to a whole population but measurements are only taken on a sample. This concept underlies all experimentation: data are measured on a sample but conclusions are drawn about a population. Thus it should be clear at the outset of any investigation what population the conclusions should apply to and care taken that the sample actually represents this population (Steel and Torrie, 1980).

Data are measured on a sample but on their own are not particularly useful and need to be summarized in some way. Statistics can be used to summarize data and these summaries can be pictorially or numerically presented, or even presented in words.

The most common summary statistic is the sample mean. This is a measure of the location of the observations that is calculated as the sum of all of the observations divided by the number of observations. This statistic is an *estimate* of the population mean which is a value that represents the entire population. It is a good idea to know how precise the estimate of the population mean is. The sample variance provides information about how the individual observations cluster about their sample mean value.

These two statistics – the sample mean and the sample variance – provide useful information about the distribution of all individuals in the population. This is fundamental to many statistical methods.

REFERENCES

Box JF (1978) *RA Fisher: The Life of a Scientist*. John Wiley, New York.
Steel RGD and Torrie JH (1980) *Principles and Procedures of Statistics*, second edition. McGraw-Hill, New York, Chapter 2.

CHAPTER 2

Planning

Planning is the most important aspect of any experiment. Without this, the experiment has no direction, no meaning, no objectives and will therefore achieve nothing. An experiment will always start with an idea which has to be formulated into objectives. These objectives should be clear, concise, measurable and achievable.

Objectives are formulated into hypotheses to allow statistical analysis that will support any conclusions drawn. The population to which the conclusions of the experiment are to apply must be considered at this stage. Consideration should be given as to how representative samples are to be drawn from this population. The power and significance of a hypothesis test should be considered at the planning stage to give an idea of the degree of certainty associated with the conclusions. If treatment differences are expected, this can be considered at this stage and appropriate hypotheses formulated if prior knowledge of the anticipated differences is available.

2.1 FORMULATING THE IDEA

Scientists by their very nature are curious and therefore will always ask questions. Vague ideas can be developed or can lead to new ideas or old ideas being revamped, which in turn can provide the fundamental thinking behind an EXPERIMENT. The vague ideas soon become rephrased as more formal questions. Discussion with colleagues will quickly sort out whether the question remains an unanswered question or whether it will merit further investigation. Someone may already have an answer that will satisfy the curious scientist until the next idea springs to mind. If not, the scientist may start to develop their question and formulate a specific objective or series of specific objectives.

Arable example: wheat yield

How are crops best managed to obtain the best yield?
|
More specifically
↓
What fungicides and fertilizers give the best yield of a wheat crop in England?
|
Objective
↓
To determine the level of nitrogen fertilizer that will provide maximum wheat yield for the three varieties, A, B and C, when grown in England.

Livestock example: milk yield

Will changing the energy in a diet give better milk yields for cows?
|
More specifically
↓
Does increasing the energy concentration of a dairy cow's diet in mid-lactation improve its milk yield?
|
Objective
↓
To determine which of concentrations A, B or C, of high-energy diet in a cow's feed in mid-lactation gives the highest milk yields for breed Z dairy cows.

These two examples are used to illustrate the various points made throughout this chapter and the next regarding choice of variables and treatments, measurement methods and timing.

2.2 DEFINING OBJECTIVES

Objectives do not define themselves; it takes time and effort to produce useful objectives. Primarily, objectives should be realistic, measurable and achievable. For example, the main objective in the wheat yield experiment could have been 'to improve our ability to predict nitrogen fertilizer requirements', which could be achieved, but how is ability measured? If we cannot measure ability, how do we know if the objective has been met? The objectives should be expressed clearly and in an unambiguous manner. Many people will be involved in designing and carrying out the proposed experiment and it is extremely important that the team involved fully understands the purpose of the work at all stages. Waffled objectives could cause confusion and misunderstanding and may well be scanned rather than carefully read. This is easily avoided by keeping them short, simple and avoiding unnecessary jargon.

Finally, ensure that the objectives have not strayed from the original concept. It is a good idea to have an independent assessor to check the meaning and clarity. Do the objectives really say what was intended?

2.2.1 Main objective

Some experiments will only have one objective which obviously will be the main objective. Most experiments, however, have a series of specific objectives. It should be decided at the outset which one will be the main objective as this will affect all future decisions. If the experiment is to be successful nothing should detract from the main objective, regardless of any hurdles that may arise.

If no one objective can be considered as being the most important then consideration must be given to splitting it into smaller experiments or restating the objectives. You may be trying to get too much out of one experiment and risk losing everything. With smaller experiments operational management should be less complex, reducing the chance of anything going wrong.

2.2.2 Secondary objectives

It is all too easy to try and answer too many questions in one experiment, possibly resulting in ambiguous answers. Experiments are costly and time-consuming to run; it therefore makes sense to have additional objectives. Too few objectives will be wasteful of resources. The secondary objectives can provide much valuable information but should not be allowed to detract from the main objective. However, too many secondary objectives can cloud the issue so they must be chosen with care.

If disaster does strike and the main objective has to be abandoned, one of the secondary objectives can be promoted to the main objective and all is not lost. The experiment can be run bearing in mind the original objective has not been addressed. It may be necessary to reformulate the original objective. If it is not appropriate to redefine the main objective (for example, the majority of the experiment may have been completed) the secondary objectives may at least achieve something.

Arable example: wheat yield

A secondary objective of the wheat yield experiment first mentioned in Section 2.1 could be to determine which of four fungicides (C, T1, T2 and S2) best controls disease level at harvest in the wheat crops A, B and C.

This objective could provide additional information by investigating four fungicide applications to the three varieties at each level of nitrogen application used. If there were any problems with applying the nitrogen fertilizer treatments and no applications of fertilizer were made, this objective would still provide information at the zero nitrogen fertilizer level provided the four fungicides were applied to plots of the three varieties. If one of the wheat varieties became so diseased that the crop died completely before harvest then information on the fungicide comparisons would still be available for the other two varieties.

Livestock example: milk yield

Secondary objectives of the milk yield experiment first mentioned in Section 2.1 could be to determine the effect on breed Z's condition score of feeding with diet A, B or C; or to examine the effect on weight gain of feeding breed Z dairy cows with diet A, B or C.

It is expensive to run feeding studies on dairy cows and there are many different factors that will have an effect on milk yield. It costs relatively little to collect additional information and providing the taking of measurements does not affect the milk yield it makes sense to collect additional information. Measurements on animal weight and condition score could be informative in helping to explain the achieved milk yields.

2.2.3 Example: objectives

Objective: to determine the potassium efficiency of farm manure applications to sugar beet.

This objective indicates the variable that will be measured or calculated (i.e. potassium efficiency), the sort of treatments that will be used (i.e. farm manure applications), and the crop under investigation (i.e. sugar beet). The objective is phrased succinctly and the detail of how the potassium efficiency is to be derived and the exact treatments should be provided in an experiment protocol.

Objective: to define a yield response curve for wheat.

The main response variable for this objective is yield and the treatments will be varying rates of nitrogen in order to be able to produce a range of values suitable for a curve-fitting exercise. The actual nitrogen rates chosen will depend on how important it is to identify an optimum rate compared to identifying the general shape of the response curve.

Objective: to investigate the interaction.

There is no indication of the type of treatments in the trial or which interaction is being investigated. Nor is there any indication of the response variable. The objective is rather vague and a more explicit one would be 'to investigate the interaction between variety and rate of applied nitrogen on yield of winter wheat'.

Objective: to determine the effects on dairy cows and their calves of feeding high-energy rations during late pregnancy and early lactation.

This is a rather vague objective. Is late pregnancy and early lactation defined? What breeds are being investigated? What effects are being studied? There is no indication of response variables of interest. It looks as though effects on both the cow and the calf are of interest. Have two objectives been put together? It may be better to split the objective into one for the cows and one for the calves, particularly as it seems likely that the response variables of interest may be different for cows and calves.

Objective: to determine the effects of feed treatments on the performance of sows.

There are a number of aspects of performance that could be monitored, such as behaviour or meat quality. Which aspect of performance is the objective concerned with? What sort of feed treatments are being tested?

2.3   DEFINING THE POPULATION

Keypoints

- It is impractical to measure the whole population, therefore samples are used.
- Samples should be chosen to be representative of the population.

The objectives have defined the points of interest. It is now necessary to define the group of individuals about which the conclusions are to be drawn, i.e. the population of interest.

The objectives stated will cover a limited set of situations; they are not intended to cover all possible situations. For example, the milk yield experiment to consider the benefits of three feed rations for breed Z dairy cows would not cover the benefits of the feed rations for other breeds. Hence the proposed experiment has been restricted to a particular cattle population. It is impractical to include the entire breed Z cow population in an experiment, therefore a sample must be selected. This sample must be representative of the entire population of breed Z cows, so that valid conclusions can be drawn about the population, not just the sample (Steel and Torrie, 1980). For example, young and old cows must be included in the sample if it is intended to draw conclusions about young and old cows.

The sample will consist of a series of experiment units. An *experiment unit* is the smallest unit of experiment material to receive a single treatment. A plot of land or a cow or a group of cows or a shed of hens can all be experiment units. In group-fed experiments the group of cattle would be the experiment unit whereas in individual feeding trials each animal would be an experiment unit. It can be seen that an experiment unit can consist of an individual or a group of individuals, which must be clear from the design.

Example

Consider the wheat yield experiment in Section 2.1. The information should be applicable to farmers in England. Consequently sites with different climatic and soil conditions should be included in the experiment. Sites should be selected throughout the country to represent a wide range of soil types and weather conditions. The population of interest is arable land in England under a wheat crop of variety A, B or C. The experiment units will be single plots of land in the field growing the different wheat varieties.

2.4   FORMULATING HYPOTHESES

Keypoints

- A hypothesis states the variable, the statistic and the comparison of interest.
- Every objective has one null and one alternative hypothesis.

The objectives have been defined and agreed but have not yet been expressed in statistical terms. This will be required if statistical analysis is used to support conclusions from the experiment. A *hypothesis* is a restatement of the objective in statistical terms. This must state the variable of interest, the statistic under investigation and the comparison of interest. Consider the wheat yield experiment and a hypothesis of 'there is no difference in the mean yield of wheat varieties A, B and C'. It is a valid hypothesis; the variable of interest is yield, the comparison of interest is the three varieties of wheat and the statistic under investigation is the mean. A *statistic* is 'a summary value calculated from the observations in a sample' (Marriott, 1990). Variances, standard deviations and mean values are examples of statistics.

Expressing the objective as a statistical hypothesis allows formal statistical testing. If it cannot be expressed as a hypothesis the objective may have to be redefined but care should be taken not to stray from the original idea.

For every objective stated there will be a pair of hypotheses. These are known as the null hypothesis and the alternative hypothesis. A *null hypothesis* is the assertion that the statistic of interest takes a particular value. An *alternative hypothesis* expresses the way in which the value of the statistic of interest deviates from the value specified in the null hypothesis.

Example

Consider the secondary objective from the milk yield experiment: to examine the effect on weight gain of feeding breed Z dairy cows with diet A, B or C. The individual animals included in the experiment will be divided into three groups, one for each of the three diets being investigated. The objective will have the following hypotheses:

Null hypothesis   There is no difference in the mean weight gained by any of the three groups of cows.

$$H_0 : W_A = W \text{ and } W_B = W \text{ and } W_C = W$$

Or equivalently, the mean weight gained by the cows on diet A equals the mean weight gained by the cows on diet B which equals the mean weight gained by the cows on diet C which also equals the mean weight gained by all the cows.

$$H_0 : W_A = W_B = W_C = W$$

where

    W_A is the mean weight gained by cows under feed diet A,
    W_B is the mean weight gained by cows under feed diet B,
    W_C is the mean weight gained by cows under feed diet C,
    W is the overall population mean weight gain, and
    H_0 stands for the null hypothesis.

Alternative hypothesis   The mean weight gained by at least one of the groups of cows is different from the mean weight gained by all the cows.

$$H_1 : W_A \neq W \text{ and/or } W_B \neq W \text{ and/or } W_C \neq W$$

where

    W_A is the mean weight gained by cows under feed diet A,
    W_B is the mean weight gained by cows under feed diet B,
    W_C is the mean weight gained by cows under feed diet C,
    W is the overall population mean weight gain, and
    H_1 stands for the alternative hypothesis.

2.5   HYPOTHESIS TESTING

A *hypothesis test* is the name given to the statistical test used to test the objective. The principles of hypothesis testing hold regardless of the distribution of the data; the distribution could be normal, skewed or even unknown. Probability theory, although not discussed in detail, is the underlying principle used to indicate the chances of correctly accepting or rejecting a hypothesis by calculating Type I and Type II errors. Steel and Torrie (1980) and Sokal and Rohlf (1995) provide a more detailed coverage of test composition and probability. The salient points are identified in the following sections.

2.5.1   Probability

Probability is used to assign quantitative measures to situations where uncertainty occurs. This can be used to express the degree of belief, in numerical terms, in conclusions drawn from the test of the hypothesis. It is not possible to be completely sure (i.e. 100% sure) of any conclusion because future events may disprove them. After all we may well believe that all dairy cows are black and white and would be quite happy believing this until we find one that is not, which would of course disprove the theory. This leads to the concept of only ever being able to disprove a hypothesis and not being able to prove it.

    Given the above, it is possible to accept any conclusion when it is in fact false but it is desirable to minimize the chances of doing this. With any test procedure it is possible to make two types of error: Type I errors and Type II errors. A *Type I error* is made when the null hypothesis is rejected when it is in fact correct. The probability of making this type of error is called the *significance level* of the test. A *Type II error* is made when the null hypothesis is accepted when it is false, i.e. accepting the null hypothesis when the alternative hypothesis is true. The *power* of a test indicates the certainty that a difference will be detected given that it is real. In other words, it is the probability of correctly rejecting the null hypothesis and is calculated as one minus the probability of making a Type II error.

If the true state of affairs is that the null hypothesis is correct then the conclusion drawn from the sample can either be that the null hypothesis is true (with an associated probability of $1-\alpha$) or that the null hypothesis is false (with an associated probability of α). Similar conclusions can be drawn from the sample when the true state of affairs is that the null hypothesis is false. The conclusion from the sample can either be that the null hypothesis is true (with an associated probability of β) or that the null hypothesis is false (with an associated probability of $1-\beta$). In other words, regardless of the true state of affairs of the population there are two conclusions that can be drawn from the sample: either the null hypothesis is true or it is false. The actual state of affairs affects the probability with which these two conclusions can be drawn. This is summarized in Table 2.1.

Example

Consider the previous example in Section 2.2.2 examining the weight gain of cows. Working with the typical, but arbitrary, levels of $\alpha = 0.05$ and $\beta = 0.2$, the significance level is 0.05 (often quoted as 5%) and the power of the test is 0.8 (or 80%). The probabilities associated with each cell of the summary Table 2.1 are given in Table 2.2.

If the true state of affairs is that the null hypothesis is correct (there is no 'real' difference between the three diets) then the conclusion drawn from the sample can either be that the null hypothesis is true (with an associated probability of 0.95) or that the null hypothesis is false (with an associated probability of 0.05). Similar conclusions can be drawn from the sample when the true state of affairs is that the null hypothesis is false (there is a 'real' difference between at least two of the three diets). The conclusion from the sample can

Table 2.1 The outcomes of conclusions drawn from data in terms of power, significance, and Type I and II errors (after Steel and Torrie, 1980)

	Conclusion made from the sample data	
True state of the population	Accept the null hypothesis and say H_0 is correct	Reject the null hypothesis and say H_1 is correct
The null hypothesis, H_0, is correct	Conclusion is made with a probability of $(1-\alpha)$	A Type I error is made Conclusion is made with a probability of α, the significance level
The null hypothesis is false, H_1 is correct	A Type II error is made Conclusion is made with a probability of β	Conclusion is made with a probability of $(1-\beta)$, the power of the test

Table 2.2   Typical probabilities associated with power, significance, and Type I and II errors

	Conclusion made from the sample data	
True state of the population	Accept the null hypothesis and say H_0 is correct	Reject the null hypothesis and say H_1 is correct
The null hypothesis, H_0, is correct	0.95	0.05
The null hypothesis is false, H_1 is correct	0.20	0.80

either be that the null hypothesis is true (with an associated probability of 0.20) or that the null hypothesis is false (with an associated probability of 0.80).

2.5.2   Test composition

The aim of hypothesis testing is to use a set of data to decide whether to accept or reject the null hypothesis. This is done by calculating a test statistic from the data. The range of values which this statistic can take is divided into two regions by a critical value. These regions are the *acceptance region* where the value of the test statistic leads to the decision to accept the null hypothesis, and the *rejection, or critical, region* where the value of the test statistic leads to the decision to reject the null hypothesis. The actual critical value, and therefore the acceptance and rejection regions, are defined by the chosen significance level of the test. Under the null hypothesis the probability of rejecting the null hypothesis is chosen to be the significance level, α, of the hypothesis test and this will be the probability of being in the rejection region.

The concept is best illustrated by a diagram. Figure 2.1 shows the probability density function (pdf) of a normal distribution with a mean of zero and a variance of 1, N(0,1), the standardized normal distribution. When the probability density function (pdf) of a distribution is plotted against the possible values of the distribution the area under the resultant curve represents probabilities. The details of calculating pdf are beyond the scope of this text but the actual values are tabulated in normal tables for N(0,1) which can be found in statistical tables and texts such as Fisher and Yates (1963). A clear illustrated explanation of how to use normal tables can be found in Steel and Torrie (1980). As the area under the curve represents probability, the total area under the curve will equal one. The x-axis gives the range of values expected in a variable from such a distribution. With a N(0,1), very few observations are expected to be less than -3 or greater than $+3$. The critical value is the x-axis value that separates the rejection region from the acceptance region and for a N(0,1) distribution is often referred to as a Z-value which is the value formed by the rows and columns of the normal table. The shaded area gives the probability of an observation greater than the critical value occurring and it is these probabilities that are provided in the body of the normal table. In a hypothesis test the null hypothesis is rejected if the calculated test statistic is deemed to be 'too extreme' for it to be a genuine value under the null hypothesis. In other words, the null hypothesis is rejected if the test statistic exceeds a 'critical' value. As the probability of the test statistic exceeding a critical value is defined as the significance level, α, of the test, the shaded area under the probability density function

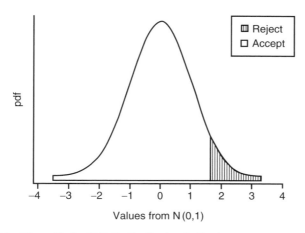

Figure 2.1   The pdf of a N(0,1) distribution indicating acceptance and rejection regions

represents α. Thus the critical value, acceptance and rejection regions are defined by the significance level of a test. These diagrams can be drawn for any sort of distribution and they will all have their own probability density function 'shape'. The pdf functions for normal distributions have a 'bell-shape', the width of which is determined by the distribution variance. The base of the 'bell' becomes wider and the height gets lower as the variance of the normal distribution increases.

The above argument can be extended. The critical region can be a range of possible values at the lower end of all possible values, or a range of values at the upper end of all possible values, or a combination of both of these. The different tests that arise from this are called lower one-tailed test, upper one-tailed test and two-tailed test respectively. One-sided tests are synonymous with one-tailed tests and they look for directional differences between the calculated test statistic and a specified value. They are only appropriate for situations where two entities are being compared. The lower one-tailed test looks to see if the calculated statistic is less than a specified value. The upper one-tailed test looks to see if the calculated statistic is greater than a specified value. Two-sided tests are also known as two-tailed tests which look to see if the calculated statistic is different from a specified value in either direction. The principle is still the same: the total area defined by the critical region represents the significance level of the test, α.

The above can be summarized with reference to Figure 2.2. If the calculated statistic lies within the area defined by the acceptance region (the non-shaded area in Figure 2.1) the null hypothesis is accepted. If the calculated statistic lies within the total area defined by the rejection region (the shaded area in Figure 2.1) the null hypothesis is rejected. The probability of the calculated statistic lying within the rejection region is equal to the significance level of the test. The areas defined by a significance level of 5% for a N(0,1) distribution are illustrated in Figure 2.2 for one- and two-tailed tests.

Upper one-tailed test

$$H_0 : W_i - W = 0 \qquad \text{for } i = 1, 2$$

$$H_1 : W_i - W > 0 \qquad \text{for } i = 1, 2$$

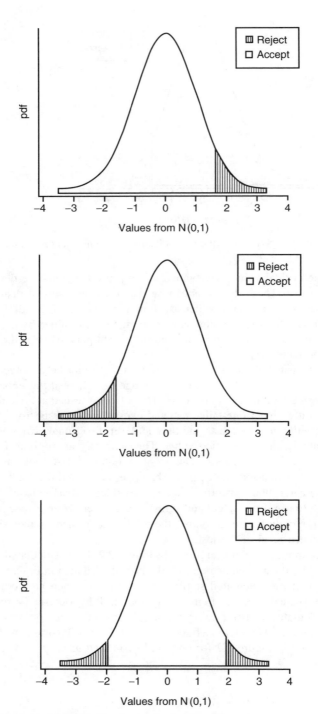

Figure 2.2 The pdf of a N(0,1) distribution indicating the acceptance and rejection regions for (a) an upper one-tailed test at 5%; (b) a lower one-tailed test at 5%, and (c) a two-tailed test at 5%

Prob (Reject H_0) $= \alpha$
$$= \text{Prob}(X > Z_\alpha = 1.645)$$
$$= 0.05 \text{ for } \alpha = 0.05$$

The probability represented by the shaded area in Figure 2.2(a) is 0.05.

Lower one-tailed test

$$H_0 : W_i - W = 0 \qquad \text{for } i = 1, 2$$
$$H_1 : W_i - W < 0 \qquad \text{for } i = 1, 2$$

Prob (Reject H_0) $= \alpha$
$$= \text{Prob}(X < Z_\alpha = -1.645)$$
$$= 0.05 \text{ for } \alpha = 0.05$$

The probability represented by the shaded area in Figure 2.2(b) is 0.05.

Two-tailed test

$$H_0 : W_i - W = 0 \qquad \text{for } i = 1, 2, \ldots, n$$
$$H_1 : W_i - W \neq 0 \qquad \text{for } i = 1, 2, \ldots, n$$

where

$n = $ total number of groups
Prob (Reject H_0) $= \alpha$
$$= \text{Prob}(X < Z_{\alpha/2} = -1.96) + \text{Prob}(X > Z_{\alpha/2} = 1.96) \text{ for } \alpha = 0.05$$
$$= 0.025 + 0.025 \text{ for } \alpha = 0.05$$

The two shaded areas in Figure 2.2(c) are equal in size so each one will represent half the probability required for the total significance level.

2.6   ANTICIPATING TREATMENT DIFFERENCES

Keypoint

- Formulate additional hypotheses to test for anticipated treatment differences when planning the experiment.

In many cases the conclusions drawn from the null hypothesis of equal treatment population means will not be particularly interesting. A conclusion that 'at least one of the population treatment means differs from the overall mean' is not particularly informative. However, the conclusion that 'there is no difference between population treatment means' will be of merit.

   Differences between specific treatments or groups of treatments may be of particular interest and these comparisons of interest can be derived from knowledge about the treatments included in an experiment. For example, crop disease levels on untreated control plots are expected to be different from those treated with a fungicide. In such cases it may be of interest to formulate additional null hypotheses at the *planning stage*. This will allow more detailed investigation of the anticipated differences in the treatment population means. For example, 'there is no difference between the mean disease level at harvest of plots treated with fungicide and those not treated with fungicide' may be an additional null hypothesis for the wheat yield experiment.

Additional hypotheses can be used to assess treatment differences for curvilinear effects if the treatments form a series of related treatments such as increasing dosages of a chemical or different amounts of feed additives fed to cattle. For example, is the effect of increasing nitrogen dosage on crop yield levels a straight-line effect or curved with yield increases gradually reducing as dosage increases? If it is curved, to what extent is it explained by a quadratic curve or some other higher order polynomial curve such as a cubic curve? These effects are covered in more detail in later sections (Sections 6.1.6.1 and 6.1.6.5).

If a hypothesis for a specific treatment comparison is formulated at the planning stage it will be impossible for the hypothesis to be biased by the data because they will not have been collected. The hypothesis formulation is based purely on knowledge of the treatments and not on the size of observed treatment differences. This means that the overall chance of rejecting the null hypothesis and stating that the treatment differences exist will be equal to the Type I error probability which will equal the significance level of the test.

These planned comparisons are known as contrasts and those designed to investigate curvilinear effects are known as polynomial contrasts. If more than one planned comparison is used they must all be mutually independent or, in other words, mutually orthogonal. The planned comparisons are then referred to as orthogonal contrasts. The number of planned comparisons must be no more than the number of degrees of freedom available. The *degrees of freedom* of a system is the total number of observations in that system minus the number of mathematical constraints imposed within that system; these vary according to the complexity of the system. Hence the size of the degrees of freedom reflects the number of 'observations' that are free to vary after certain restrictions have been placed on the data. For example, if there are n observations and the mean value is known, $n-1$ observations can take any value but the value of the nth observation would be defined by the values of the $n-1$ observations and that of the mean.

Example

Consider the secondary objective of the wheat yield experiment: to determine which of four fungicides (C, T1, T2 and S2) best controls disease level at harvest in the wheat crops A and B. Let the four fungicide programmes be no spray (C), a two-spray programme (S2) and a single-spray programme with different application times, (T1) and (T2). The null hypothesis is that there is no difference between the mean disease levels at harvest from any of the fungicide programmes. The alternative hypothesis is that at least one of the mean disease levels differs between fungicides.

Prior to running the experiment it has been planned to test

- whether the two single-spray programmes differ in mean disease level (T1 versus T2);
- whether the two-spray programme disease level is different from the mean disease level of the single-spray programmes (S2 versus the mean of T1 and T2);
- whether, on average, applying fungicide controls disease levels (C versus the mean of S2, T1 and T2).

These are three mutually independent contrasts (questions). It should be verified that these are independent. It is not possible to make any further comparisons as all the degrees of freedom available have been used; there are four treatments and therefore three degrees of freedom.

Suppose that instead of the above orthogonal comparisons it had been planned to compare every treatment in turn against the control treatment (C versus S2, C versus T1 and C versus T2). This still involves only three comparisons. However, they are no longer independent since a comparison is made against the same treatment (the control) each time. This cannot be analysed using contrasts, though a special test is available and is detailed in Section 6.1.6.2.

The treatment degrees of freedom give the *maximum* number of comparisons that can be made. This does not mean that they should all be used. Only meaningful comparisons should be tested, thus it would very rarely be sensible to test all pairwise comparisons.

REFERENCES

Fisher RA and Yates F (1963) *Statistical Tables for Biological, Agricultural and Medical Research*, sixth edition. Oliver and Boyd, Edinburgh.

Marriott FHC (1990) *A Dictionary of Statistical Terms*, fifth edition. Addison-Wesley Longman.

Steel RGD and Torrie JH (1980) *Principles and Procedures of Statistics*, second edition. McGraw-Hill, New York, Chapters 2, 3 and 5.

Sokal RR and Rohlf FJ (1995) *Biometry*, third edition. Freeman, New York, Chapter 1.

CHAPTER 3

Design

At the planning stage of any experiment it should be decided which variables will be measured, how the measurements will be made and what method will be used to record them. Treatments should be chosen to satisfy the experiment objectives bearing in mind any physical constraints on resources.

One of the basic underlying principles of experimentation is replication. The number of replicates required for any experiment influences the ability of the experiment to recognize treatment mean differences if they exist. The number of replicates must be determined when planning the experiment.

Blocking is a technique used to group experiment units together into homogeneous sets whilst allowing the experiment to use heterogeneous material. This allows the experiment conclusions to be applicable in more situations. The number of blocks used in an experiment should be determined by the possible number of homogeneous groups the experiment material forms and not by the number of replicates required.

The allocation of treatments to experiment units could bias the results in favour of one treatment over another if not done objectively. This is avoided by randomizing the allocation of the treatments to the units. Randomization also validates the statistical analysis of the experiment.

3.1 VARIABLES

The terms variates and variables are frequently used synonomously although there is a distinction between the two. In this text the term variable is used to refer to the collection of data points measured for a single characteristic.

Any experiment will measure variables in one form or another. Variables of the most interest are those that are affected by the treatments applied in an experiment. These are called *response variables* and their values may vary from one experiment unit to the next, for example, weight gain in a feed study and yield in a variety study. Other variables may also be measured; these could be background variables or covariates.

3.1.1 What to measure

Keypoints

- Decide when measurements are to be taken.
- Define the end-point of the trial.
- Consider the use of covariates.

It is necessary to decide at the outset what will be the main response variable. This will be determined by the main objective. This main variable must be capable of answering the original question and may be derived from a series of measured variables. Further response variables will be measured to meet secondary objectives. Again these can be derived variables. Variables that are used as part of a set of calculations may not merit separate analysis since they could be meaningless on their own or would only duplicate the results from the derived variables.

After choosing the variables to be measured, careful consideration should be given to determining the frequency and timing of measurements. The measurements should be taken so that the results will be as widely applicable as possible. The objectives often determine whether measurements may be taken at intervals throughout the experiment or only at the end-point or both. In some cases the timing and frequency of the measurements is determined by the variable itself.

In some experiments it may not be clear when the end-point has been reached. The end-point should be clearly defined in the experiment design. Practical limitations may sometimes determine the end-point, and this should be stated at the outset of the experiment. For example, slaughter may not take place at an animal's optimum weight and condition but will be determined by when the animal can be sold for slaughter. In experiments where the end-point changes for the different experiment units, consideration must be given as to whether this will affect the remaining units. Problems such as this must be addressed at the planning stage of the experiment. The method used to overcome the problem will depend on the particular situation and will often be determined by the objectives of the study. For example, all animals could be slaughtered at the same time whatever their condition scores. However, this would not be suitable if the objective of the experiment is to assess the length of time taken to reach optimum score but may be appropriate if weight gain in a given time period is the objective.

Some variables will be measured to provide additional background information to assist interpretation. These variables may not be directly affected by the treatments in the study. In designing experiments these variables should be carefully chosen so as not to detract from the response variables or waste resources. Consideration should be given as to whether the information could be gained from another source; for example, rainfall data could be obtained from the local Meteorological Office.

Further information may sometimes be available about the individual experiment units in the form of the values taken by some additional variable believed to be related in some way to the response variable. Such a variable is called a covariate. The use of one or more covariate(s) in the analysis may help remove some of the variation between experiment units. See Section 3.7 for further details.

Arable example: wheat yield

Yield at 85% dry matter would be the main response variable to meet the main objective. It is not measured directly but is derived from the variables: plot yield, dry matter content and plot area. Plot area is only measured to calculate the main variable and is meaningless as an analysed variable. Yield in tonnes per hectare provides the same information as yield in kilograms per metre squared so the derived main variable would be measured and analysed in one or other of the units, not both. Another variable that needs to be measured to meet the secondary objective is disease level. Weather conditions such as amount of rainfall and temperature could be measured or obtained from the Meteorological Office. They will not be affected by the experiment treatments but they may help explain the onset of any disease epidemics that may occur. Soil moisture at the time of sowing may influence crop yield to the extent that plots of land that are dry at the beginning of the growing season may result in stunted development compared with plots that are more moist. At sowing time this variable will not have been affected by the treatments and so using it as a covariate may help explain some of the variability between plots.

Deciding when to measure the variables is not as straightforward as it may first seem. Yield can only be measured at harvest time, the end-point of the experiment, but the experiment involves three different varieties of wheat. Should the varieties be harvested at the same time or should each be harvested when it is mature? The main objective is concerned with obtaining maximum yield for applied fertilizer, not with when the crop matures, so the former choice may be the more appropriate one. Disease in a crop can develop at any point of crop growth. This experiment is only interested in disease at harvest but it could have been measured and assessed at the different growth stages of the crop.

Livestock example: milk yield

The main variable of interest is milk yield and it will be measured at each milking session. The milk yield of a cow changes during a lactation, therefore each cow should enter the experiment at the same point in their lactation period. This experiment is concerned with mid-lactation so cows enter the experiment six weeks post-calving. As the cow's entry into the experiment depends on when it calves, the starting point of the experiment will not be the same for all cows. Cows could be starting on the experiment over a period of several weeks, with a subsequent 'knock on' effect when the experiment finishes, and it could be that some cows are just starting the experiment as others finish. Each animal will stay in the study for six weeks, which is sufficient for the diet to take effect. Condition score and weight gain need to be measured in this experiment to satisfy the secondary objectives. It should be decided at the outset how often these measurements need to be taken. This will depend on whether the experiment is concerned with overall weight change or changes during the course of the experiment. Weight gain is derived from initial weight, which can only be measured at the start of the experiment, and current weight, which can be measured at any time during the experiment's lifetime. For this experiment it would be measured weekly throughout the experiment to observe the effects of the different diets, A, B and C, and how weight gain may affect average daily milk yield during a cow's mid-lactation.

The cows will be housed in cubicles with individual feeding stations which allow daily intake to be measured for each cow. Since all animals are not starting the study together there may be only a few animals initially so these cows will have more room and cubicle space per cow than those entering the experiment later on. As new animals are introduced to the group this could cause problems such as stress for the animal which has entered most recently, due to dominant behaviour and bullying from other animals, which in turn may affect the milk yield and other measured variables for the new entrant. Since these environmental changes could affect the milk yield and other results, the response variables from each animal's first week on the experiment will be measured and used as covariates. Therefore an animal's weight in the first week will be used as a covariate when analysing weight gain, and average daily milk yield in the first week will be used as a covariate when analysing average daily milk yield, etc.

3.1.2   Method of measurement

Keypoints

- Measurement techniques should be consistent across all sites and years of an experiment.
- Potential observer effects should be taken into account through the experiment design.

At the planning stage of an experiment the measurement methods used must be decided upon and any effect they may have on the response variable considered. For example, very different results will be produced when weed assessments are made by assessing percentage cover in a quadrat or counting the number of quadrat subdivisions in which the weed is present. A quadrat is a frame (usually a metre square) that is placed on the ground over the plot area. Frequency of occurrence of plants within the frame can be counted or the frame can be sectioned into small squares and an estimate of the area covered by the weeds made. The sampling techniques differ according to the variables being measured and the limitations of recording devices. The physical size of the sample required to take a measurement and the scale of measurement may be determined by the sampling technique used. For example, a minimum amount of material will be required to take a dry matter sample and the weight of a cow can only be measured to within the limitations of the scales used. In multi-site trials all measurement techniques for a variable should be consistent across all sites throughout the complete trial. In the wheat yield experiment disease assessments could be made for whole plots by estimating the percentage of crop affected by the disease, or made for samples taken from the plants within a plot by assessing the percentage leaf area covered by the disease. All sites in England involved in the experiment must use the same type of measurement throughout the complete trial.

Assessments can be objective or subjective. All objective assessments should be measured to the same standard regardless of who performs the assessments. With subjective assessments observer effects are inevitable and cannot be ignored. Observer effects can occur when measurements are taken at different times by the same person or when different people are involved in taking measurements thereby influencing the results. Whenever possible, subjective assessment should be avoided due to these observer effects. If subjective measures have to be used, the possible observer effects should be taken into account. This could be done by either one assessor performing all assessments for a particular variable or by ensuring that the observer effect is accounted for in the experiment design structure.

3.1.3   Recording the results

Keypoints

- Do not exaggerate the precision of results.
- Round results after calculations, not before.
- Record all missing values with an appropriate code such as an asterisk (*).

There are a number of points that should be considered before recording any results. Three very important aspects are the degree of precision to which the data are recorded, the overall precision of the data and the accuracy of the measurements.

Assessments should be recorded to the degree of precision (decimal places or significant figures) appropriate to the equipment used. The converse should also be considered; the measuring equipment needs to be capable of measuring to the appropriate level of precision. For example, the scales used to measure the weight of dairy cows in the milk yield experiment need to be capable of measuring in grams as well as kilograms. The results of any calculations should also be quoted to a consistent degree of precision. All decimal places should be used until the final result to minimize rounding errors. When two variables are included in a calculation the final result is reported to the same number of decimal places or significant figures as the variable with the least degree of precision (decimal places or significant figures).

Both accuracy and precision should be considered when recording the data and these refer, respectively, to the average values and the variability of the measurements. If a measurement is accurate its value will be close to the true value and the mean of the sample of measurements will be a close estimate of the true population mean. The sample is then said to be unbiased. For measurements to be precise they need to be repeatable, producing a similar value for each measurement taken. The variance of the sample of measurements will then be relatively small indicating the measurements are all clustered close to the sample mean. These ideas are also explored by Cochran and Cox (1957) and illustrated in Figure 3.1.

Figure 3.1 illustrates different ways in which data points (represented by vertical lines) may fall around the true population mean value (represented by the upward pointing arrow). Diagram (a) illustrates data that are accurate (the lines centre around the arrow) and precise (the lines are close together); diagram (b) illustrates data that are accurate (the lines centre around the arrow) and imprecise (the lines are quite scattered). Data in diagram (c) are precise (the lines are close together) but not accurate (the lines do not centre around the arrow) and data in diagram (d) are both imprecise (the lines are quite scattered) and inaccurate (the lines do not centre around the arrow).

Results can be recorded in electronic format or paper format. Electronic data capture is the more reliable method of data entry though suitable back-up is required. The units used

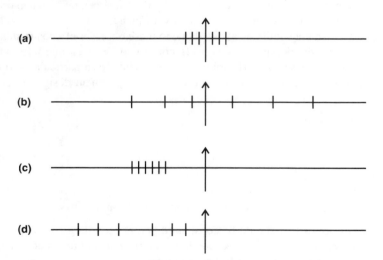

Figure 3.1   Illustration of (a) accuracy and precision, (b) accuracy and imprecision, (c) precision and inaccuracy and (d) imprecision and inaccuracy, where the arrow is the target and the vertical lines are the data

for the different assessments should always be stated on the appropriate forms when results are written and all records should be legible. When results are recorded electronically a sensible degree of precision should be used and the units used should be stated. For example, a plot yield of 16.73591756578 kg would be meaningless and would more sensibly be recorded as 16.74 kg. Quoting too many decimal places can give a misleading impression of the precision of the data and care should be taken to avoid this by using a sensible number of decimal places that is determined by practicality or by the measuring device. For example, if visual assessments are made of ground cover there is little point in recording a value such as 54.2% as most people would only be able to make the assessment to the nearest 5 or 10%.

Missing values can occur by virtue of the experiment design, for example, in a multi-sowing date trial some plant emergence assessments, such as height, taken before the final sowing date will be missing. They can also occur because sampling is not possible, or by accident when taking the samples or taking the measurements. They could occur in the wheat yield experiment if the crop dies on a plot so disease assessments and yield measurements cannot be taken. In the milk yield experiment, missing weight and milk yields would arise if a cow dies. They would also occur if an individual cow gets mastitis so that the milk yield data have to be discarded; they can still be weighed for weight gain assessments and their condition scores can be obtained but they will no longer have any valid milk yield results. A missing value should have a unique code; generally this will be recorded as an asterisk (*) although different computer packages may use different codes or symbols. All other data cells should be completed with a data value. If blank cells are left in a data file this could mean that no attempt has been made to enter the data, that the data are still outstanding or that the result was a missing observation. However, it should be noted that some computer packages use a blank cell or a full-stop as a missing value.

Care should be taken to avoid missing values due to human or equipment error though missing values may occur due to circumstances outside the experimenter's control.

Consider the livestock milk yield experiment. A sub-sample of a typical data set for this experiment may look as shown in Table 3.1. Week 1 is the covariate week for each animal, but is not necessarily the same actual week due to the staggered entry into the study. CS is condition score and LWT is the live weight of the animal.

3.2 CHOOSING THE TREATMENTS

The objective of many experiments is to compare how the response variable differs for a variety of treatments. These could be unrelated treatments where there is no natural order to the treatments, such as a list of different fungicides that will be used. Alternatively the treatments could be related, with an inherent ordering to the treatments, such as a single fungicide applied at different rates. A combination of these two situations could also be used. In this case a treatment would be the combination of fungicide and rate.

A *treatment* is 'anything which is capable of controlled application according to the requirements of the experiment' (Marriott, 1990). It is any combination of treatment factors and treatment levels. A *treatment factor* defines a method or type of application (e.g. crop variety, or nitrogen fertilizer) and is therefore a term that refers collectively to a list of actual treatments. The individual treatments included in a treatment factor can be referred to as treatment factor levels, the level being the specific treatment applied. It follows from this that all treatment factors will have a set of treatment factor levels (Steel and Torrie, 1980).

Table 3.1   Typical data set for six dairy cows (hypothetical data)

Cow ID	Block	Treat	Week 1 Milk yield (kg/day)	Week 1 CS	Week 1 LWT (kg)	Week 1 Feed intake (kg DM/day)	Week 2 Milk yield (kg/day)	Week 2 CS	Week 2 LWT (kg)	Week 2 Feed intake (kg DM/day)
178	1	1	24.5	3.5	684	14.5	28.2	3.5	666	28.3
222	1	2	32.2	3.5	655	15.9	*	3.5	669	18.6
524	1	3	29.6	3.0	707	15.8	37.1	2.5	717	17.0
137	2	1	33.7	2.0	594	16.1	34.1	2.5	717	17.0
129	2	2	26.3	2.5	588	15.4	35.9	2.0	579	17.1
537	2	3	35.9	3.0	643	15.5	42.5	3.0	650	16.5
⋮	⋮	⋮	⋮	⋮	⋮	⋮	⋮	⋮	⋮	⋮

Applying this to the three examples cited in the first paragraph, in the first example, the unrelated fungicide treatments can be referred to collectively as fungicides and thus the treatment factor is 'fungicide'. The types of fungicide to be used form the individual elements of the collective group and are the treatment factor levels (e.g. fungicide A, fungicide B or fungicide R). In the second, the collective name for the treatment is fungicide rate and the treatment factor could be called this or shortened to just 'rate' because the same fungicide is being used for all treatments. The individual elements within 'rate' are the actual rates of fungicide application (e.g. 20 kg/ha, 50 kg/ha or 70 kg/ha) and these are the treatment factor levels.

The third example can be considered in one of two ways: as a set of unrelated treatments or as a set of structured (related) treatments. In the first instance there is a collective term, 'fungicide and rate', representing the treatment factor. This comprises all the combinations of different fungicides and the rates at which they are to be applied (e.g. fungicide A at 20 kg/ha, fungicide B at 40 kg/ha, and fungicide B at 50 kg/ha) and these combinations are the treatment factor levels, the actual treatments to be applied. If, however, all the fungicides are to be applied at all the rates, although the treatments to be applied may be the same the actual structure of the treatments could be regarded slightly differently. In this second approach there are two treatment factors, 'fungicide' and 'rate', each with their own factor levels, i.e. type of fungicide and application rate, respectively. The actual applied treatments are the combinations of the fungicide and rate treatments. Selection of treatment factors and factor levels will depend on the experiment objectives. All treatment factor combinations may not always be practical. Where two or more treatment factors are used, the interactions between their levels may be of interest. The final choice of treatment factors and factor levels will take into consideration the most appropriate statistical design, agricultural practicalities and interpretation.

The factor levels selected depend on the objectives. If the experiment objective is a straightforward comparison of treatments, the factor levels used will differ from those selected for an experiment examining a relationship between a response variable and the factor levels of a treatment factor. In the first instance the levels chosen will depend on the comparisons of interest. In the second, the levels chosen will have to provide sufficient points in an appropriate range for a modelling exercise to determine the shape of the response or optimum rates. When the treatment levels in a treatment factor represent a series of increasing rates, doses or applications, it is possible to investigate the presence of a trend in the response to increasing treatment levels. If just two treatment levels are used, only straight-line responses can be examined. With a minimum of three levels the existence of curvilinear effects or trends can be investigated although the number of treatment levels used restricts the complexity of any trends that can be investigated. To identify quadratic trends a minimum of three treatment levels is required.

A standard or null treatment (control treatment) may be included in the design if the comparisons between it and the experiment treatments are of interest, or it can be included to provide useful background information such as the background level of disease (Cochran and Cox, 1957). In some cases special provision for this will be needed in the design although in most cases it will be included as an extra treatment or treatments. If the control treatment is used for background information then it does not have to be analysed with the data from the other treatments; it would be used purely to assist in interpreting the results. The amount of prior knowledge of the control treatments, the objectives of the experiment and practical considerations will determine the number of control treatments to include.

It may be appropriate at this stage to consider emergency plans should naturally occurring phenomena prevent treatment application. In some drastic instances it may be necessary to change the design of the experiment.

Arable example: wheat yield

This experiment has three treatment factors: fungicide type, wheat variety and nitrogen fertilizer. There are four treatment factor levels for fungicide type and these are the fungicide product, C, T1, T2 and S2. The secondary objective of this experiment is a comparison of the fungicide products, rather than rates of application, so each should be applied at a standard rate. Wheat variety has three treatment factor levels, A, B and C. Treatment factor levels for the nitrogen treatment are the actual rates of nitrogen fertilizer applied to the crop. The experiment's main aim is to determine the optimum rate of nitrogen fertilizer to apply to achieve the maximum wheat yield. Therefore the treatment factor levels need to be a series of nitrogen rates clustered around the expected optimum, with a few more rates spread out above and below the cluster and one at the zero level. Such an arrangement will help pinpoint the optimum more closely, provided it occurs roughly where expected, and provide general information about the shape of the curve that will help confirm the accuracy of the expected optimum. If nothing is known in advance about expected optimum levels then equispaced nitrogen levels between zero and a maximum should be used to provide detailed information on the overall shape of the response curve. Biological growth curves tend to be exponential in their character. A bare minimum of four nitrogen rates would be required to provide estimates of model parameters for the simplest model but it is recommended that more rates be used. The reader is referred to Section 6.1.7.2 for further advice on the number of points to use when fitting curves.

The control fungicide treatment of no fungicide application may be included in the experiment analysis as a direct treatment comparison with the different fungicide programmes. Alternatively it could be used to provide background information on the disease levels generally experienced in the season, helping to explain potential large differences in overall disease levels that may occur between the different sites throughout England.

Contingency plans for the experiment could be the delayed spraying of the fungicide treatments if the weather is too stormy to apply the treatment on the original spray dates. This may involve changing the two-spray programme to a single one if there is then insufficient time to apply the two sprays at the correct time interval.

Livestock example: milk yield

This experiment has a single treatment factor, diet, with three treatment factor levels, A, B and C, which are different concentrations of high-energy feed. The high-energy feed is mixed with a silage to form the complete diet. To ensure the treatment effects are not confounded with a silage effect the silage for all the animals' feed needs to come from the same source. Thus the number of treatment factor levels in a dietary experiment needs to be kept low to ensure there is an adequate supply of the same silage. Using three different high-energy concentrations will enable an investigation of possible linear trends in the response variables.

Milk yields of dairy cows can be very variable. Cows enter the experiment over a period of weeks depending on when they are six weeks post-calving. The actual animals used will be selected from an available pool of animals; the animals selected will be those which can form homogeneous blocks and are considered sufficiently healthy to complete the experiment. As well as blocking, covariate measurements will be taken to account for the variability. Each response variable will be measured in the first week of the study and used as a covariate measure for that variable, so that average daily milk yield in the first week will be used as a covariate when analysing all subsequent average daily milk yield data.

Selecting cows for the experiment from an available pool has the benefit of ensuring there will be sufficient animals for the study. Should an animal be removed from the experiment, possibly due to illness or death, then they can be replaced with another animal from the pool, provided there is another animal that can form a homogeneous block. However, it would have to be clear that the cause of the removal was due to some external factor and not due to the treatments.

3.3   CONSTRAINTS

Throughout the design process consideration must be given to the resources available. Sometimes compromises must be reached in order to run the experiment with limited resources. If the objectives cannot be met or the statistical tests will not be powerful enough to detect real differences within the constraints the experiment should be abandoned.

It could be that for a particular experiment there may be insufficient experiment units available to run the trial with the required number of replicates to detect a useful difference in mean values. The trial would have to be redesigned or abandoned.

Consideration should be given to human and financial resources, the time available before results are required (the agricultural timetable), and the equipment and materials required.

3.4   REPLICATION

Keypoints

- Replication is one of the basic principles underlying experiments.
- Replication is the application of a single treatment to one or more experiment units.

- Replication is needed to estimate experimental error.
- Replication increases the power of an experiment to detect differences.

One of the basic underlying principles of experimentation is replication. In statistical terms replication is usually defined as the application of each treatment of interest to one or more experiment units, which can sometimes cause confusion as it can be argued that there must be two of anything to have a replicate. Usually no two experiment units will produce the same result even when the same treatment is applied to each. Biological data are variable so for the experiment to be scientifically and statistically valid, replication must be employed.

It is important to note that replicate experiment units for each treatment are required, rather than repeated measurements from a single experiment unit. Recording an observation from sub-samples from a single experiment unit simply provides repeated observations for that experiment unit. Replicate observations can only be achieved by applying the treatment to more than one experiment unit within an experiment and recording an observation from a single sample from each of those units.

In some cases treatments may be unequally replicated though it is usual to replicate all treatments equally. Unequal replication can involve the use of complex statistical designs resulting in a more difficult analysis and interpretation. These designs are sometimes necessary, particularly when resources are limited. If they are carefully planned and executed the analysis (and mathematics behind it) may be more complex but the interpretation can be straightforward.

Occasionally single replication trials are carried out. In these cases the effect of a treatment must be compared against some estimate of background variability. This measure of variability has to be defined at the outset and is usually made up of some treatment combination that is not expected to show any effect, e.g. a high-order interaction (Cochran and Cox, 1957). These experiments usually take the form of large multiple-factor factorial experiments in which high-order interactions are pooled to give an estimate of experiment error. In practice, high-order interactions are not usually significant and have no practical meaning so it is justified to pool the interactions and use them as the error term. The technique of pooling interaction terms to form an error term can also be used when there are few residual degrees of freedom though it must be decided which interactions are to be pooled before any analysis takes place. A problem with this technique is that if an interaction term is significant and it has been pooled into the error term then the significant interaction result will not be found. An advantage of the technique is that more treatment factors can be used without needing a lot more experiment units; a single replication experiment can be conducted using pooled unimportant interaction terms for the error term with the treatment effects of interest being examined.

Replication has two vital functions in experimentation (Montgomery, 1991). Since the majority of experiments cannot measure the whole population, a sample is taken and the sample used to estimate various characteristics of the population, e.g. the treatment population mean will be estimated by the sample mean. The estimate will improve as more observations are taken, i.e. a more precise estimate of the population mean will be achieved by taking a larger sample. In other words, the experiment will increase in power, in terms of detecting treatment differences, if the number of replicates is increased, all other things being equal.

The natural variation between experiment units means that two similar experiment units that receive the same treatment will not give the same response. This natural variation is known as experimental error or background variability. A measure of this natural variation can be obtained as a variance called the error variance or residual variance. The experimental error or background variability is compared against the treatment variability to determine if there is a treatment effect. It is not possible to measure the experimental error without using replication and so replication is required to test for treatment effects.

3.4.1   The number of replicates

Keypoints

- The more variable the data, the more replicates required.
- The smaller the treatment difference to be detected, the more replicates required.

An experiment that has insufficient power to detect treatment differences that are of practical significance will waste resources and achieve nothing. It is possible for differences to be statistically significant but still be too small to be of any practical worth. If an experiment has insufficient replicates to detect differences of a predetermined size it would be more realistic to conclude that there were not enough replicates rather than that there was no significant difference because a real difference may exist but not be detected.

The number of replicates of a treatment is one of the factors that will determine whether a treatment difference will be detected if it exists. Too few replicates will lead to an inability to detect the required differences, providing the differences actually exist. Using too many replicates is wasteful of resources. The number of replicates required is determined by the required power of the test, the size of difference between treatment mean values to be detected, the significance level of the test used and the background variability of the data.

The relationship between the power of the test and the number of replicates used is not linear, and doubling the number of replicates does *not* mean that the ability to detect a real difference has been doubled. For a given significance level, error variance and size of treatment difference to be detected, the power of a test increases as the number of replicates increases. For a specified significance level, power and difference between treatment sample mean values of a response variable, the number of replicates required to detect the difference will depend on the variability of the response variable. The greater the variability, the larger the number of replicates required. In practice, the population variability will not be known and will have to be estimated. The variance estimate will be constant for a given response variable. The number of replicates required to detect a specified difference for a given significance and power and level of variability will increase as the size of difference to be detected decreases. However, the relationship between size of difference and number of replicates is not linear. If the number of replicates required is doubled, the size of difference that can be detected is not halved.

However, a compromise may have to be reached between the size of the difference to be detected, the statistical power and level of significance of the test and the number of treatments to be tested in order to obtain adequate replication. For example, if treatments are expensive or only a few animals are available or there is limited land space then a compromise would have to be reached.

It is important to get as good an estimate of the error variance as possible. If the error variance obtained from the actual experiment differs greatly from the estimate used to obtain the required number of replicates then the actual experiment could either have insufficient replication or too much replication. For a given set of circumstances (significance level, power, background variability and size of treatment difference) there will be a minimum number of replicates required to detect the difference in the treatment sample mean values. There will also be a number of replicates above which there is no practical or statistical benefit to be gained. The chosen number of replicates should lie between these two extremes for an experiment to be efficient both statistically and economically.

Although many variables may be measured in an experiment it is usually the main response variable that will be used to determine the number of replicates required in the experiment. In cases where a number of variables are of equal interest it is the most variable variable that is used to determine the number of replicates required. The error variance, or experimental error, of a response variable is used as a reference against which to test for differences between sample means and it is this variance that is used to calculate the number of replicates required.

3.4.2   Information needed

Keypoints

- The size of treatment difference to be detected, usually dictated by common sense and the measuring equipment.
- An estimate of error variability, which can be found from various sources.
- Power and significance levels of the test, which must be decided.

'How many replicates do I need?' There is never just one answer to this question. It depends on many aspects of the particular trial in question. There is no right or wrong answer. Compromises have to be reached and a value judgement made to arrive at one answer that will meet all requirements. Often the information required to calculate the number of replicates is available from a variety of sources. A little thought will be needed to produce some of the information required but if this is done at the planning stages then any problems may be resolved before the experiment starts.

Before any attempt can be made to answer the question a decision must be made on what size of difference in response between treatment mean values is of interest. The size of difference is determined by practical considerations and must be of practical significance. For example, there is no practical value in finding differences of 0.01 t/ha in wheat yields. There are no rules for determining this, only common sense, but it can be limited by the accuracy of the recording devices to be used. For example, with a ruler capable of measuring only in centimetres it is pointless to look for differences in millimetres, regardless of the number of replicates taken.

Background variability, such as that due to differences in wheat yield obtained from two equal-sized plots of land even when no treatments have been applied, is another consideration and there are many ways of expressing this. The most common measure of variability is the sample variance which provides information on how the individual sample observations cluster about their sample mean. A reliable estimate of the error variance for the response variable should be available. The *error variance* 'measures the variability' between observations 'due to unexplained causes or experimental error' (Marriott, 1990) and it can also be called the residual variance.

Estimates of error variability can be obtained from numerous sources, including literature and other historical sources. Often a previous experiment will have considered the response variable under similar conditions before, and the estimate of error variability from that experiment can be used. In the few situations where there is no literature and no records of previous experimentation a pilot study may be carried out to provide this information (Chatfield, 1995). A *pilot study* is a small-scale run of the experiment to gain information about variability and to test any novel application methods prior to the running of the main experiment. If it is not possible to run a pilot study and there is no information available on background variability you may decide to go ahead with the experiment anyway. In that instance use as many replicates as possible, financially and practicably, and then estimate the background variability after the experiment has been run. It is then possible to check retrospectively whether or not sufficient replication was used and either draw the conclusion of insufficient replication or sufficient replication with the appropriate interpretation. This estimate of background variability will then form the beginning of past historical records for future experiments.

Two further pieces of information are required before the question of how many replicates are needed can be answered. These are the significance of the test (the probability of saying there is evidence that a difference exists when one does not exist) and the power of the test (the probability of correctly saying there is evidence that a difference does exist). Once these have been decided upon, an appropriate test statistic can be selected from statistical tables which can then be used to determine sample size.

The choice of significance levels and power deals with the risk involved in rejecting the null hypothesis when it is in fact correct and the ability of the test to correctly recognize when the null hypothesis is false. The significance level is often chosen to be 5% as this is the risk people are prepared to take in most situations but there is no statistical reason why other values cannot be chosen. The power is usually taken to be 80%, though there is no statistical reason for choosing this value.

There are a variety of methods that can be used to calculate the number of replicates required for an experiment. All the methods are based on the same principles and give similar results. Montgomery (1991) covers a variety of different techniques and Cochran and Cox (1957) and Steel and Torrie (1980) cover this topic in slightly more detail. All of these texts provide various formulae, illustrated with examples, for calculating the number of replicates required. All of the methods use a test statistic which is found from tables, the value of the statistic depending on the chosen power and significance level required. The following formula is one taken from Steel and Torrie (1980) that can be used when there are more than two treatments involved in an experiment:

$$r = 2(Z_{\alpha/2} + Z_\beta)^2 \left(\frac{\sigma}{d}\right)^2$$

where

r	is the number of replicates,
α	is the probability of making a Type I error (the significance level of the test),
β	is the probability of making a Type II error $(1 - \text{power of the test})$,
Z	is the area in the tail of a standardized normal distribution (see Section 2.5.2),
σ	is the error standard deviation, and
d	is the size of treatment difference in sample mean values to be detected.

It is necessary to adjust the value for *r* calculated from this equation to account for the use of Z-values, as Z-values assume the error variance is actually known rather than estimated. Details are provided in Steel and Torrie (1980). Cochran and Cox (1957) recommend using t-values instead of Z-values to get round this problem.

3.4.3 Example: replication

Consider the milk yield livestock experiment. The number of replicates per diet will be determined by the variability of milk yield as this is the main response variable. The known information is as follows. A difference between the milk yields of approximately 4 to 5 kg/ day is considered to be of practical importance when comparing the three diets. From previous experiments conducted under similar conditions the standard deviation of milk yield was estimated as 3.5 to 5.5 kg/day. It was decided that the two extreme values for each statistic would be used to produce a range of number of replicates required. Calculations were carried out at two different power and significance levels to cover a range of situations and choices.

From Table 3.2 it can be seen that with a power of 80% and a significance level of 5% it should be possible to detect a difference in the diet treatment mean values for average daily milk yield of at least 4 kg/day with 12 cows per diet if the error standard deviation is 3.5 kg/ day. However, to detect the same size difference with an increased power of 90% and reduced significance level of 1% an additional 11 cows per diet would be required.

As you can see, compromises may have to be made and drawing up tables like these can help in assessing the various options and combinations available. Also such a table indicates the importance of obtaining accurate and precise estimates of the overall mean and error variability. The higher the variability, the greater the number of replicates required for a constant power, significance level and difference.

Table 3.2   Number of replicates required per diet treatment for the milk yield livestock experiment

Milk yield				
Standard deviation	Difference (kg/day)	Power (%)	Significance level (%)	No. of cows per diet
3.5	4	80	5	12
5.5	4	80	5	30
3.5	5	80	5	8
5.5	5	80	5	19
3.5	4	90	1	23
5.5	4	90	1	56
3.5	5	90	1	15
5.5	5	90	1	36

3.5   BLOCKING

Keypoints

- Blocking creates homogeneous groups of experiment units.
- Blocking increases the precision of the experiment if the experiment material is heterogeneous.
- Blocking gives the experiment wider coverage by using diverse experiment units.

Experiment units that respond similarly to a treatment are described as being homogeneous. If units are homogeneous then the variation between units will be small. There may be variation within the experiment material in a unit itself, but a group of units are homogeneous if this element of variation within the unit is the same for all units in the group (Clarke, 1994). However, homogeneous units are rare. For example, it would be unusual for a field or flock of sheep to be totally homogeneous; the field may be boggy at one end or the sheep may be different weights. The conclusions from an experiment, though, need to apply to heterogeneous populations otherwise they would not apply to 'real life'. There is a conflict between wanting homogeneous units to keep the experimental error small, which helps to detect treatment differences if they exist, and wanting heterogeneous units so the experiment results are useful for 'real life' heterogeneous populations. This conflict is resolved by the technique of blocking.

A *block* is a group of homogeneous experiment units. The idea behind blocking is to form groups of units that are as alike as possible but the blocks themselves can be heterogeneous. The experimental error is estimated from within a block (Steel and Torrie, 1980) so will be kept small but since heterogeneous units have been included the experiment results will be applicable to the heterogeneous population represented by the blocks. In many experiments a block will consist of a single replicate of all treatments, this being the simplest case. It is also possible to run experiments with more than one complete replicate in a block or with only partial replication in a block.

Example

Consider a field in which the soil type changes from sandy soil at one end to sandy loam at the other. If a single experiment were to use plots (experiment units) from both ends of the field then the plots would not be as homogeneous as a group of plots selected from just one side of the field from one soil type. If the experiment conclusions are to be applicable to only sandy soil then the experiment units should only come from this soil type. However, if it is desirable for conclusions to be applicable to both soil types then plots from both soil types must be included in the experiment. Each treatment would need to be applied to each soil type at least once. The way to minimize the experimental error would be to group the experiment units into blocks, there being as many blocks as there are replicates required. Each block would comprise units of the same soil type but obviously the blocks themselves would be from the different soil types. The plots within a block would be homogeneous but the blocks themselves would be heterogeneous. A method of laying out this trial is shown in Figure 3.2. The conclusions of this study would be applicable to both sandy soils and sandy loam soils and not just one or the other.

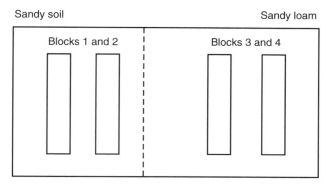

Figure 3.2   Illustration of an appropriate direction in which to position blocks in a field with varying soil type

3.5.1   How to block

Each block usually contains a whole number of replicates of all the treatments. It is not necessary for each block to contain only one replicate although within a block experiment material must be homogeneous. Treatments must be randomly assigned to the experiment units for each block separately.

Before deciding how to block experiment units, all known sources of variation that may affect the measured responses must be identified, e.g. birth weight, condition score, previous cropping, soil type. Within each block the known sources of variability should be kept as constant as possible. For example, if drainage in a field varies due to the slope of the field then the experiment units should be grouped such that the units within a block all have similar drainage characteristics. This is usually achieved by having blocks running along the slope rather than down the slope, as indicated in Figure 3.3.

If the experiment units are plots of land the blocks need not necessarily be a regular shape, or even adjacent, provided that they are homogeneous. If an individual animal is an experiment unit then a block may consist of animals of a similar weight or blocking may be performed using a single blocking factor or two or more combined factors. For example, condition score could be used or condition score and parity could be combined to generate a single blocking factor.

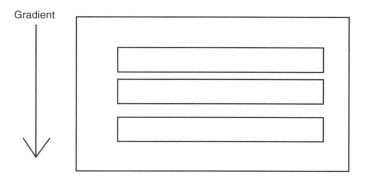

Figure 3.3   Positioning blocks in a sloping field

Arable example: wheat yield

It is proposed that the experiment will be carried out using three replicates, it having been previously determined that this will be sufficient to detect the treatment differences required for wheat yield, the main response variable. Each replicate will contain 60 plots as there are 60 treatment combinations: three varieties × four fungicide programmes × five nitrogen rates. The only field available to run the trial is quite heterogeneous due to the site of an old river bed which will probably have different drainage characteristics from the rest of the field. It is also thought that the soil types on either side of the old river are different. These three distinct areas of the field are large enough to hold 60 plots each, with one area being large enough to hold 120. The field is therefore able to cater for four blocks, each containing one replicate.

The diagram in Figure 3.4 shows the blocking carried out with total disregard for the soil types and drainage. Three blocks were apparently chosen because there were three replicates of each treatment and they have not been designed to minimize experimental error.

The diagram in Figure 3.5 illustrates a method of blocking that takes the drainage and different soil types in the field into consideration. Note that the blocks are not necessarily a regular shape. The plots in each of these blocks are more homogeneous than those in the blocks in Figure 3.4. The experimental error variance for the layout in Figure 3.5 should be smaller than that in Figure 3.4 assuming the correct blocking factor has been used.

Livestock example: milk yield

From the calculations of the number of replicates required in Section 3.4.3 it is proposed to carry out an experiment with 12 replicates. Since there are three diet treatments, each replicate will contain three breed Z cows so a total of 36 cows are required. These will be selected from a pool of 54 available breed Z cows. Each cow is an experiment unit. The number of previous lactations for these 54 cows ranges from one to five so the available

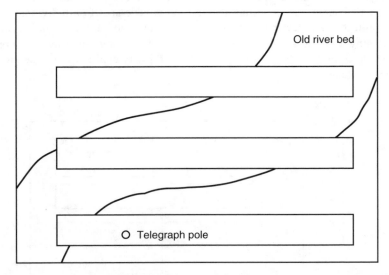

Figure 3.4   An example of blocking that ignores any knowledge of the field that would lead to variation within a block

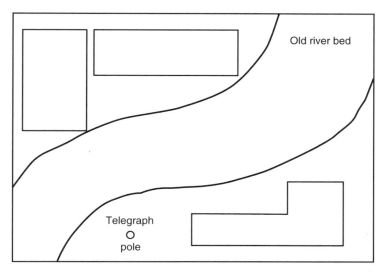

Figure 3.5   An example of blocking taking into account knowledge of the field that would lead to variation within a block

cows form a heterogeneous group. Blocking is therefore required. The frequency table for the number of previous lactations is shown in Table 3.3.

Equal numbers of cows per block are required. The majority of animals had two, three or four previous lactations so only animals from these previous lactation groups will be used. The group with four previous lactations can provide four replicates. Twelve of the animals in the group with three previous lactations can provide four replicates. There are 19 animals in the group with two previous lactations that can provide six replicates. The 12 replicates will comprise three from each of the groups with four and three previous lactations and six from the group with two previous lactations. Average daily milk yield from the week prior to the study is known for all cows so it is possible to further divide these three groups of animals according to their previous average daily milk yield. Within each group formed by the number of previous lactations, the animals can be divided into sub-groups of three cows with similar previous average daily milk yields. This should have the effect of reducing the variability in the milk yield response variable during the experiment. The overall effect has been to produce 12 groups of three cows. These 12 groups form 12 blocks for the experiment and there are sufficient cows in each block to cater for a single replicate. The experiment has been blocked using two blocking factors. Within each block the cows are as homogeneous as possible; the three animals have had the same number of previous lactations and produced similar milk yields immediately prior to the start of the experiment. Results would only be applicable to breed Z dairy cows with two, three or four previous lactations.

Table 3.3   Frequency table of lactations in breed Z dairy cows (hypothetical data)

Number of previous lactations	1	2	3	4	5
Frequency	5	19	14	12	4

3.6   RANDOMIZATION

Keypoints

- Randomization avoids accidental bias.
- Randomization ensures experiment errors are random variables.

Treatments must be allocated to the experiment units in an unbiased way to ensure that a particular treatment is not continually handicapped or favoured in successive replicates. Suppose that an investigation was being conducted into the pigmentation of egg yolks from two different hen housing systems and the yolk pigments from barn eggs were being compared with the yolk pigments from battery eggs. If the assessments were made on all of the battery eggs first and then on all of the barn eggs, any difference in pigmentation may be due to changes in light intensity rather than differences in the housing systems. For example, if daylight was fading before all assessments were made, the second batch may have been examined under artificial light and the first in natural daylight. This could bias the results of the trial as colours can look different in different types of light. A way of avoiding such bias is randomization (Cochran and Cox, 1957). *Randomization* is the allocation of treatments to experiment units in an unbiased manner. In the example, if the eggs had been examined in a random order then it would be unlikely that all the battery eggs were examined first and all the barn eggs examined last. Neither treatment, barn nor battery, would then be disadvantaged or advantaged over the other and accidental bias would be avoided.

A treatment randomization which is produced following the correct procedure may appear non-random but this is not a problem in the analysis (Cochran and Cox, 1957). Therefore it is *not* necessary to re-randomize treatments if it appears that one treatment may be favoured. Treatments should *not* be swapped around after the randomization process as this may well introduce bias. If the randomization produced causes major practical problems in the operation of the experiment then the whole trial should be re-randomized following the correct procedure.

3.6.1   Reasons for randomization

Randomization offers protection against accidental biases (Snedecor and Cochran, 1989). If an experiment has been correctly randomized then the experimenter cannot bias the results in any way. Consider the wheat yield experiment where one of the treatment factors was wheat variety. If all varieties were to be harvested on the same day and the plots of variety A were harvested first, followed by those of variety B and then those of variety C, there would be a danger of introducing bias. This could be due to a deterioration in weather, or operator fatigue or any other less obvious reasons (e.g. a breakdown of equipment). If the harvesting sequence had been randomized across the varieties this bias may have been avoided by ensuring that each variety has an equal chance of being tested under more favourable conditions.

Randomization is, perhaps most importantly, used to ensure that the statistical analysis of the data is valid (Montgomery, 1991). Most statistical analyses make various assumptions about the data. The one assumption that is made by almost every statistical analysis (and all that are covered in detail in this text) is that the experiment errors are independent. The only way in which this assumption can be met is by randomly allocating the treatments to experiment units. If the correct randomization procedure is not used then the random error assumption cannot be made and the statistical analysis will be invalidated.

3.6.2 Practical aspects

Truly random selection is not easy to achieve and, as Steel and Torrie (1980) point out, choosing at random in the sense of an individual selecting which units receive which treatments is unsatisfactory and should never be considered. The individual will probably introduce personal bias by avoiding placing certain treatments together or favouring a particular placing of treatments. For example, they may favour placing a control plot that requires different husbandry methods at the end of a block rather than in the middle.

Various pseudo-random devices, such as tossing coins, throwing a dice, using a telephone directory and so on, are sometimes advocated but these are liable to introduce non-random elements into the selection process. Any reliable method uses a statistical table of random numbers or a computer or calculator, and all numbers generated must be used. As many random numbers as there are experiment units should be generated. A random number table is a table in which any of the possible entries is equally likely to occur in any given position. They can be found in all books of statistical tables. To generate a sequence of random numbers the starting point in the table is randomly selected and one of the four possible directions (left, right, up or down) is also randomly selected. The sequence of numbers is generated by taking consecutive numbers from the tables in the selected direction. If the edge of the table is reached a new direction is selected and consecutive numbers used until no more random numbers are required.

The sequence of random numbers generated is used to allocate treatments to experiment units. A random number is allocated to each experiment unit. The basic principle is that these numbers are ranked and a treatment code assigned to the rank. This method guarantees that the treatments will be truly randomized to the experiment units and thus allows one of the assumptions of most statistical analyses to be met.

3.7 COVARIATES

Keypoints

- A covariate must be unaffected by the treatments.
- Using a covariate can help increase the precision of an experiment.
- A covariate is related to the response variable.
- The covariate must be measured on all experiment units.

The basic idea behind a covariate is that some of the variation in the response variable is due not to the treatments or background error variability but to the variation in the covariate. If the variation due to the covariate is removed before estimating the treatment effect of the response variable then the estimate of the error variance made to test the treatment effect will be reduced and so the experiment will be more precise. If the covariate is to have any effect on the response variable then the two must be related and it is usually assumed that it is a straight-line relationship. If this is found to be a statistically significant relationship then the response variable is adjusted for the effect of the covariate and then analysed in the usual way for identifying treatment differences. Analysis of covariance is the standard technique for taking account of covariates in an analysis using multiple treatments. The analysis of covariance is a combination of analysis of variance (Chapter 4 and Sections 6.1.3 and 6.1.5) and linear regression (Section 6.1.7).

Covariates can be used with any of the experiment designs mentioned in Chapter 4 but the interpretation of the resultant analysis can be difficult with the multi-treatment factor designs. It can also be assumed that the relationship between the covariate and the response variable is curvilinear, such as quadratic. The analysis will proceed in exactly the same way as if a straight-line relationship had been used but with a different regression model. Multiple covariates can also be used.

It is important that the covariate is not affected by the treatments. Consequently covariates tend to be measured prior to any treatments being applied. For example, the previous yield of each plot could be used to reflect plot to plot differences in plot fertility if the same plots were to be used again for the wheat yield experiment. As the previous yield is from a previous crop it cannot be affected by the treatments applied to the current crop. Plots with low fertility would tend to produce a low yielding crop whereas plots with high fertility would tend to produce a higher yielding crop so yield differences in the current crop may be due in some part to the fertility level of the plot rather than a treatment effect. Milk yield measured the week before a cow entered an experiment could be used in a dietary study for dairy cows. An animal which has a high milk yield in one week tends to have a high yield throughout a lactation. Since initial milk yield is measured before the treatment diet is offered it cannot be influenced by the treatment. The effect of initial milk yield on overall milk yield will be removed.

Montgomery (1991) gives a clear presentation of the formulae and calculations involved in covariance analysis with a worked example. Both Steel and Torrie (1980) and Sokal and Rohlf (1995) provide more detailed coverage including assumptions made in the analysis. Many other texts give a thorough coverage of this topic but the interested reader should find the above references informative and instructive.

3.8   CONFOUNDING

Keypoints

- Confounding can be designed into an experiment to reduce block size.
- Confounding can be useful for controlling extraneous sources of variability.
- Care must be taken to ensure confounding is not used inadvertently.

Confounding occurs 'when certain effects can be estimated only for treatments in combination and not for separate treatments' (Marriott, 1990). If each of the three different wheat varieties in the arable wheat yield experiment was sown in a separate block then block and variety would be confounded and it would be impossible to distinguish between the block effect and the variety effect. If identifying a variety effect was one of the experiment objectives it would not be possible to meet the objective because if an effect was found it would not be known whether it was due to differences in varieties, differences in the blocks or differences in both. Even if there was no evidence of any effect this could just be due to any block differences cancelling out any variety differences.

Confounding may sometimes occur inadvertently. Great care must be taken in the design and running of an experiment to avoid this as important information may be lost. In the same experiment the fungicide programmes, T1, T2 and S2, should all be applied in a similar manner; they should either be hand-sprayed or be mechanically sprayed. Otherwise a perceived treatment effect may not be due to a true effect of the different fungicides but may be due to the type of application. A more even spread of fungicide can be obtained with a mechanical sprayer than with a hand-sprayer and it could be this that is affecting the variable assessment rather than the fungicide.

Confounding is a technique that can be useful for controlling sources of variability that are known to exist but are of no interest in the experiment. If they can be confounded with some other experiment factor that is also of no interest then the source of variability has been confined to a part of the experiment that is of no interest and it has no influence on the factors of interest in the experiment. In an arable experiment the block effects themselves are not usually of interest but they are identified. If it is required to take samples from the plots or to make some instant assessments on the plots, such as percentage covered by disease, and it is not possible for one person to make the assessment on every plot, two or more people will have to be involved. There could be potential observer effects due to the different people taking the samples or making the assessment, which would inflate the variability in the data. If each block in the experiment was dealt with by a single assessor then any potential observer effects will be confounded with block effects. As the block effects are not of interest this does not matter. The treatment effects that are of interest can still be estimated.

A further important use of confounding is in helping to reduce block sizes, particularly in factorial experiments (see Sections 4.3 and 5.1.5 for details of factorial designs). The idea behind blocking is to produce groups of homogeneous plots but as individual plots will never be identical there will always be an element of variability between the plots in a block. With only a few plots per block it is likely that the variability between the individual plots will be less than if there were a large number of plots in the block. In other words, smaller block sizes tend to be more homogeneous than larger ones so an experiment with a smaller number of plots per block could be more precise than the same experiment with a larger number of plots per block (Cochran and Cox, 1957).

This can be a problem in experiments with a large number of treatments. In order to put a complete replicate of all treatments in a block the block size gets big with a possible resultant increase in the background error term which is being used to test for treatment effects. By confounding some of the treatments with blocks it is possible to reduce the block size in an experiment. The price paid for the increased experiment precision is the inability to estimate the effects of the treatments that were confounded with the blocks. It is usual to confound higher-order interactions in a factorial experiment which may be deemed to be unimportant or of little practical consequence or which may be expected to be non-significant. The remaining main effects and lower-order interactions can then be estimated with greater precision than would otherwise have been possible (Chatfield, 1995).

REFERENCES

Chatfield C (1995) *Problem Solving: A Statistician's Guide*, second edition. Chapman and Hall, London, Appendix III.

Clarke GM (1994) *Statistics and Experimental Design: An Introduction for Biologists and Biochemists*, third edition. Edward Arnold, London, Chapters 15 and 16.

Cochran WG and Cox GN (1957) *Experimental Designs*, second edition. John Wiley, New York, Chapters 1, 2, 4 and 6.

Marriott FHC (1990) *A Dictionary of Statistical Terms*, fifth edition. Addison-Wesley Longman.

Montgomery DG (1991) *Design and Analysis of Experiments*, third edition. John Wiley, New York, Chapters 1 and 17.

Sokal RR and Rohlf FJ (1995) *Biometry*, third edition. Freeman, New York, Chapters 1 and 14.
Snedecor GW and Cochran WG (1989) *Statistical Methods*, eighth edition. Iowa State University Press, Ames, Chapter 6.
Steel RGD and Torrie JH (1980) *Principles and Procedures of Statistics*, second edition. McGraw-Hill, New York, Chapters 2, 5, 6, 9, 15 and 17.

CHAPTER 4

Trial Structure*

All decisions about the main factors, the main and secondary response variables, and the amount of replication have been made. The final decision is the exact trial plan or design. It should take account of the number of available experiment units and any other constraints, such as the size of machinery.

This plan should detail the treatment randomization to be used with the treatment codes and labels and descriptions to be used throughout the whole experiment. Experiment units should be uniquely identified, usually with a number, so there will be no ambiguity about which treatment or treatment combination should be applied to each unit. Where applicable, a diagram of the physical layout, for example, site plans or the position of pens, etc., should be included.

In 'real' experiments there is usually more than one objective and there will be many variables. However, to keep things simple all the examples in this section only consider one objective and one variable of interest. The variable used is not necessarily the main variable of interest. All the data and designs used in the examples in this section arose from actual experiments.

Note that the site plans shown throughout this document show the position of the blocks in the trial and other features of note. Plot numbers are given but treatment codes have not been included on the site plan for ease of clarification. The treatment lists can be found in the randomization tables above each site plan. In practice, if treatment codes are not included in the site plan a separate list should be attached detailing the plot number, treatment code and treatment name. In all cases if treatment codes are used on the site plan then the treatment names should be attached.

4.1 CONSIDERATIONS

The experiment design should be kept as simple as possible. This will ensure that the statistical analysis is straightforward which will help make the interpretation clear. Practical mistakes made in performing the experiment or interpreting instructions will be reduced by keeping the experiment design and assessment methods simple.

There are sometimes cases where more complex design may be very useful and necessary to achieve the desired result. This is most common when resources are limited. For example, the physical size of an experiment must be considered during its design; a trial requiring 144 plots, each measuring 20×24 m, would require a very large site. There are designs which can overcome physical limitations (field size, number of animals, amount of equipment); such as balanced incomplete blocks designs, partial squares, or designs using confounding.

---

* Note: ADAS kindly gave permission to use data from some of their experiments. These experiments have been described in this chapter in very simplistic terms with all reference to product names removed.

A selection of these designs are discussed in more detail in Section 4.4. If the designs are carefully thought out and planned before any practical work is started they can be run successfully and be problem free, and although the actual analysis may be more mathematically complex the interpretation can be as easy as that for the more straightforward designs. These are discussed in more detail at the end of this chapter. In multi-site experiments the minimum physical requirement must be available at all sites and *all* sites must use the same design if there is to be any cross-site analysis.

4.2   SINGLE-TREATMENT FACTOR DESIGNS

There are three basic experiment designs and these form the basis of many other experiment designs. The designs have only one treatment factor and use the techniques of randomization, replication and blocking. The basic designs are the completely randomized design, the randomized blocks design and the Latin square design. Multi-treatment factor designs are considered in Sections 4.3 and 4.4. For each design the actual design is discussed and the appropriate statistical model is given along with the hypotheses that are being tested. Details are also given about the correct randomization procedure and the advantages and disadvantages of the model presented. After each model a detailed example of its simplest, basic use is given.

The treatments to be used in an experiment can be selected in different ways, which will influence the inferences that can be drawn from any hypothesis testing. If the experimenter selects a specific set of treatments (unstructured or structured) the model is termed a 'fixed effects' model. Any inferences drawn from the resultant analysis will apply only to that set of treatments. If the experimenter randomly selects treatments to represent a wider population of treatments then the model is termed a 'random effects' model. Any inferences drawn from such a model apply to the wider treatment population from which the treatments actually used were drawn. It is also possible to have combined models with both fixed and random effects. This text considers only fixed effects models so formulae and conclusions presented only apply to these models. Further information on fixed and random effects and how they influence the analysis of variance models can be found in Montgomery (1991).

A designed experiment is used to compare treatments by fitting a model to the data collected. This simply expresses the observations in terms of random variables that represent the error in the model (there will always be an element of observation variability that cannot be explained) and parameters that represent the different treatment effects. A *parameter* is a 'quantity which may vary over a certain set of values' (Marriott, 1990). The usual terminology for the general observation is y_{ij}, and this represents the response obtained for the *j*th unit receiving treatment *i*, that is, the *j*th replicate of treatment *i*.

The aim when fitting the statistical model to data from an experiment designed to compare treatments is to estimate the model population parameters. Since only a sample of the population is measured then population parameters are estimated by sample parameters. For example, the mean value of a sample of observations for a single treatment is the treatment sample mean, and this is an estimate of the population parameter called the treatment population mean.

Many texts dealing with experiment design provide details of the designs described in this section. Model formulae and hypotheses are discussed in Montgomery (1991). Steel and Torrie (1980) provide clear instructions for allocating treatments randomly to the experiment units and detail the general interpretation of the F-tests used to test hypotheses in these designs. Further detail about the advantages and disadvantages of each design are given in Steel and Torrie (1980) and Cochran and Cox (1957).

4.2.1 Completely randomized designs

Keypoints

- Completely randomized designs compare t treatments.
- All experiment units must be homogeneous.
- Unequal treatment replication is possible.
- Coverage of results is limited.
- No blocking is required.
- There is no need to estimate missing data.

4.2.1.1 Design

The COMPLETELY RANDOMIZED design is the simplest of all designs. This design is only appropriate if the experiment units that are used are homogeneous, otherwise this design should not be used. This means all units in the experiment must respond in a consistent way; they should produce similar results if they all receive the same treatment. This will mean that the experiment error and hence the residual variance is as small as possible, which will assist in the detection of treatment differences if they exist.

The purpose of the design is to compare the effects of a set of treatments on a response variable, y. If the various treatments to be compared are of equal interest then they should all be replicated equally often. This means that if there are t treatments in the experiment and each is to be replicated r times then a total of $n = rt$ experiment units will be needed. Unequal replication of treatments is possible. In this case the ith treatment will be replicated r_i times so for t treatments the number of experiment units needed will be the sum of the r_i's for $i = 1 \ldots t$. It is essential that all N experiment units are randomly assigned to the various treatments.

The appropriate statistical model for this design is as follows:

$$y_{ij} = \mu + T_i + e_{ij} \qquad \text{for } i = 1 \ldots t; j = 1 \ldots r_i$$

where μ represents the overall population mean response, T_i represents the mean effect of treatment i, r_i represents the number of replicates of treatment i, e_{ij} represents the random experimental error, and y_{ij} is the observation from the jth unit receiving treatment i. μ and T_i are the model parameters.

4.2.1.2 Hypotheses

$H_0 : T_i = \mu_i - \mu = 0 \qquad \text{for } i = 1 \ldots t$

The null hypothesis is that all the treatment population mean values are equal.

There is only one possible alternative hypothesis:

$H_1 : T_i = \mu_i - \mu \neq 0 \qquad \text{for at least one } i$

At least one of the treatment population mean values is not equal to the overall mean.

where T_i is the treatment effect, μ is a constant representing the overall population mean, μ_i is the ith treatment population mean and t is the number of treatments.

4.2.1.3   Randomization

1. Assign a unique number to identify each experiment unit (the unit identifier).
2. Generate as many random numbers as there are experiment units.
3. Assign the random numbers in turn to the experiment units.
4. Rank the random numbers in ascending (or descending) order but keeping the same unit identifier with the random numbers.
5. The r_i replicates of the first treatment are allocated to the first i ranked random numbers, the r_i replicates of the second treatment are allocated to the second i ranked random numbers. This continues until all t treatments have been allocated to the random numbers.
6. The experiment unit with the unit identifier associated with the random number then receives the treatment that was allocated to the random number.

Example   In an experiment in which three treatments, A, B and C, are each replicated four times, a total of 12 experiment units are needed. To randomly allocate these treatments to the units each unit must be assigned a unique number and 12 random numbers generated and assigned to the experiment units.

Step 1	Unit	1	2	3	4	5	6	7	8	9	10	11	12
Steps 2 and 3	Random no.	425	865	908	692	800	199	321	971	269	036	275	019

Rank the random numbers in ascending order.

	Unit	1	2	3	4	5	6	7	8	9	10	11	12
	Random no.	425	865	908	692	800	199	321	971	269	036	275	019
Step 4	Rank	7	10	11	8	9	3	6	12	4	2	5	1

Experiment units with ranks 1 to 4 will receive treatment A, those with ranks 5 to 8 will receive treatment B and those with ranks 9 to 12 will receive treatment C.

	Unit	1	2	3	4	5	6	7	8	9	10	11	12
Steps 5 and 6	Treatment	B	C	C	B	C	A	B	C	A	A	B	A

4.2.1.4   Advantages

This design can have any number of treatments and any number of replicates, and treatments do not need to be equally replicated. The analysis for this design is straightforward. Missing values cause no problems in the design since unequal replication is possible.

4.2.1.5 Disadvantages

Homogeneous experiment units are needed for this design. Variability between the units, other than that caused by the treatments, is assigned to the experimental error term which can make the design inefficient if the units are not particularly homogeneous. In practice this means that completely randomized designs are only used on small experiments due to the difficulty in finding a large number of homogeneous experiment units. Any conclusions from such an experiment must be limited to populations which only contain homogeneous units such as those included in the design.

4.2.1.6 Example: a completely randomized design

Description of experiment The experiment is investigating the effect of different stocking densities of hens on various aspects of egg production and hen behaviour. A total of 3000 hens are available for the trial and it is decided that colonies of 300 birds will be used so different stocking densities will be obtained by varying the pen size for the colony. A colony is considered to be an experiment unit because the variables measured will be totals or averages for the colony. Thus there are 10 experiment units available for the trial. Four stocking densities, A, B, C and D, are under consideration, with the densities increasing from A (the lowest) to D (the highest). Given that there should be a minimum of two replicates of any treatment in an experiment, all four treatments are to be replicated twice. This leaves two spare experiment units. There is no reason to believe that there is any consistent source of variation throughout the 10 pens so the pens are considered to form a homogeneous group of experiment units. Consequently the two spare pens can be utilized by including an extra replicate of the lowest and highest stocking densities as a completely randomized design can cope with unequal replication. The design is therefore a completely randomized design with four treatments; the treatment factor is stocking density and the treatment factor levels are the different densities, A, B, C and D.

Specific objective The aim of this study is to investigate the effect of stocking density of hens on the weight of eggs produced.

Hypotheses The null hypothesis is that there are no differences in the mean egg weights for the four stocking densities.
  The alternative hypothesis is that at least one of the stocking density mean egg weights is different from the others.

Randomization The treatment randomization plan is given in Table 4.1.

Table 4.1 Treatment randomization plan for the hen stocking density experiment, a completely randomized design

	Pen									
	1	2	3	4	5	6	7	8	9	10
Random no.	645	284	428	618	968	227	866	519	551	375
Rank	8	2	4	7	10	1	9	5	6	3
Treatment	D	A	B	C	D	A	D	B	C	A

Site plan   The 10 pens are in a large room containing other pens used in other experiments. There is no special requirement for a site plan as the randomization plan gives all the necessary detail.

4.2.2   Randomized blocks designs

Keypoints

- Randomized blocks designs compare t treatments.
- Experiment units are grouped into homogeneous groups.
- Designs can include heterogeneous blocks.
- This design can give wide coverage of results depending on the heterogeneity between blocks.
- Missing data need to be estimated to maintain the balance of the design.

4.2.2.1   Design

The design that is used most often is the RANDOMIZED BLOCKS design. It is appropriate when differences between groups of experiment units can be identified, i.e. when the experiment units are not homogeneous but can be grouped so that units within a group are homogeneous but are not necessarily homogeneous between groups. For example, in an arable trial a field might be known to have a high weed population at one end of the field compared with another, which could well affect the way in which the plot responds to any given treatment. If such differences between experiment units can be identified and the units grouped into blocks then a completely randomized design would not be appropriate and could lead to treatment differences being masked by the variability between the units. By using a randomized blocks design the variability between experiment units can be taken into account so that the size of the experiment error and hence the residual variance is as small as possible. In some instances it may be advantageous to include plots which are very different if the population of interest is also diverse.

   The purpose of the design is to compare the effects of a set of treatments on a response variable, y. If there are t treatments in the experiment and each is to be replicated b times then a total of $N = bt$ experiment units will be needed. In the simplest case of this design each block will contain one complete replicate of the treatments and therefore there will be b blocks with t units in each. It is possible to include more than one complete replicate of all treatments in each block in these designs but all blocks must contain the same number of replicates. It is also possible to include unequal replication of the treatments in each block, provided this inequality is the same for each block. Throughout this text all the designs using randomized blocks will be considered with one complete replicate in each block but the two situations just mentioned are straightforward extensions of the design.

   It is essential that the treatments are randomly assigned to the experiments units within each block.

   The appropriate statistical model for this design is as follows:

$$y_{hi} = \mu + T_i + Bl_h + e_{hi} \qquad \text{for } i = 1 \ldots t; h = 1 \ldots b$$

where μ represents the overall population mean response, T_i represents the mean effects of treatment i, Bl_h represents the effect of block j, e_{hi} represents the random experimental error and y_{hi} is the observation from the experiment unit in the hth block receiving treatment i. μ, Bl_h and T_i are the model parameters.

Example   Most fields in which arable experiments are performed have some source of known variation, such as drainage, slope, previous cropping history, weed population, or soil characteristics. The sources of variation that affect the response variable should be identified and taken into account in the design. If an experiment is to be run in a field that is on a slope it is likely that plots lower down the slope will produce a better yield than those at the top of the slope since the soil will probably have more moisture and better fertility lower down the slope. If blocks of homogeneous plots are to be formed then a block should include plots with the same gradient.

Plots within a block should be perpendicular to the direction of the variability. This will mean that the material may not be homogeneous within a plot but all plots within a block will differ in the same way. Figure 4.1 illustrates a randomized blocks design with two replicates of five treatments. If the slope is imagined to be down the page then the blocks should run across the page to make them as homogeneous as possible. However, the plots within a block are placed to run down the page. This will mean that one end of plot 1 will be different from the other end (the end with the plot identifier is higher than the bottom end), but this is true for all plots within block 1. In this way there may be variation within a plot but it will be similar for all of the plots within any block.

4.2.2.2   *Hypotheses*

$H_0 : T_i = \mu_i - \mu = 0$ for $i = 1 \ldots t$

The null hypothesis is that all the treatment population mean values are equal.

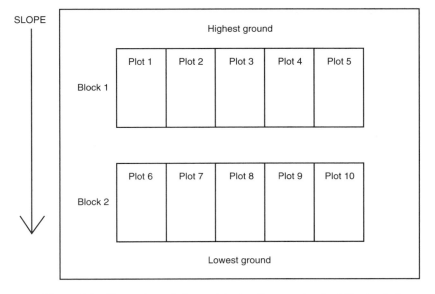

Figure 4.1   Illustration of plot positioning for a randomized blocks design

There is only one possible alternative hypothesis:

$H_1 : T_i = \mu_i - \mu \neq 0$ for at least one i

At least one of the treatment population mean values is not equal to the overall mean.

where T_i is the treatment effect, μ is a constant representing the overall population mean, μ_i is the ith treatment population mean and t is the number of treatments.

4.2.2.3 Randomization

1. Assign a unique number to identify each experiment unit (the unit identifier).
2. Generate as many random numbers as there are experiment units.
3. Assign the random numbers in turn to the experiment units.
4. For each block separately, rank the random numbers in ascending (or descending) order, keeping the associated unit identifier with the random number.
5. For each block separately
   - when there is equal replication of all treatments the first treatment is allocated to the first ranked random number, the second treatment is allocated to the second ranked random number and so on until all treatments have been assigned to a ranked random number;
   - when there is unequal treatment replication the r_i replicates of the first treatment are allocated to the first i ranked numbers, the r_i replicates of the second treatment are allocated to the second i ranked numbers and so on until all treatments have been allocated to a ranked random number.
6. The experiment unit with the unit identifier associated with the random number then receives the treatment that was allocated to the ranked random number.

Example   In an experiment, six treatments, A, B, C, D, E and F, are to be replicated four times, so a total of 24 experiment units are needed. The treatments are to be compared in a randomized blocks design with one replicate per block so there will be four blocks of six plots in the design.

To randomly allocate the treatments to the experiment units, assign a unique number to each unit and generate 24 random numbers. Associate each random number with an experiment unit.

		Block 1						Block 2					
Step 1	Unit	1	2	3	4	5	6	7	8	9	10	11	12
Steps 2 and 3	Random no.	527	500	330	021	071	118	339	266	093	347	208	361

		Block 3						Block 4					
Step 1	Unit	13	14	15	16	17	18	19	20	21	22	23	24
Steps 2 and 3	Random no.	284	713	530	875	019	572	072	982	905	351	594	572

Within each block rank the random numbers in ascending order.

		Block 1						Block 2					
	Unit	1	2	3	4	5	6	7	8	9	10	11	12
	Random no.	527	500	330	021	071	118	339	266	093	347	208	361
Step 4	Rank	6	5	4	1	2	3	4	3	1	5	2	6

		Block 3						Block 4					
	Unit	13	14	15	16	17	18	19	20	21	22	23	24
	Random no.	284	713	530	875	019	572	072	982	905	351	594	572
Step 4	Rank	2	5	3	6	1	4	1	6	5	2	4	3

Experiment units with rank 1 in each block will receive treatment A, those with rank 2 will receive treatment B and so on until all treatments are allocated.

		Block 1						Block 2					
Steps 5 and 6	Treatment	F	E	D	A	B	C	D	C	A	E	B	F

		Block 3						Block 4					
Steps 5 and 6	Treatment	B	E	C	F	A	D	A	F	E	B	D	C

4.2.2.4   Advantages

This design can have any number of treatments replicated any number of times, the only constraint being that each block must contain a whole number of replicates and each block must contain the same number of replicates. The analysis and interpretation of the design is straightforward.

   The design has the advantage that if results are not obtained for a whole block the block can be omitted and the analysis be run with one block less. Balance of the design is retained and any heterogeneity between remaining blocks can still be taken into account. It is for this reason that it is sometimes recommended to perform a randomized blocks design in preference to a completely randomized design, even if there is no obvious blocking factor. If the number of blocks is relatively small compared with the total number of experiment units then little power will be lost by employing this design even if the blocks are homogeneous.

4.2.2.5   Disadvantages

If there is a large number of treatments in the design it may be difficult to find a block of homogeneous units large enough for a single replicate of treatments. If the units within a block are not homogeneous the experiment error may be large, making treatment differences more difficult to detect. Any missing values in the design need to be estimated.

4.2.2.6   Example: a randomized blocks design

Description of experiment   The experiment is a winter wheat field trial investigating the effectiveness of nine fungicides in controlling crop disease compared with an untreated control and how they may influence crop yield and other aspects of crop quality. The plots (experiment units) will need harvesting so a minimum plot size of 24×2 m will be required. The available field has a hedge running along the top of the field and it slopes away from the hedge. All the plots in the field will, therefore, not be homogeneous so blocking will be required. If the blocks are laid out so the length of a block runs parallel to the hedge and the length of the plots within the block run perpendicular to the hedge then each block will, in theory, be as homogeneous as possible. Using the minimum plot size the field is wide enough to take 12 experiment units. It makes sense, therefore, to use one replicate of all treatments in a block. From previous experience the background variability is expected to be such that four replicates of each treatment will be required in order to detect a difference in thousand grain weight of, at least, 2.5 g when corrected to 85% dry matter. Four blocks, all running parallel to the hedge, will therefore be laid out, one under the other. Since there is only one treatment factor, fungicide, with 10 levels, a single treatment factor randomized blocks design is appropriate. The treatment factor levels are the different fungicides and have been coded from A to J inclusive, with treatment A being the untreated control treatment.

Specific objective   The aim of this study is to investigate the effect of nine fungicide treatments and an untreated control on the grain size of variety W winter wheat.

Hypotheses   The null hypothesis is that there are no differences in the 10 treatment mean thousand grain weights of the harvested ears of wheat.
    The alternative hypothesis is that at least one of the 10 treatments mean thousand grain weights is different from the overall mean.

Randomization   The randomization is shown for blocks 1 and 4 in Table 4.2. Blocks 2 and 3 are randomized in a similar manner.

Site plan   The site plan is given in Figure 4.2.

4.2.3   Latin square designs

Keypoints

- Latin square designs compare *t* treatments.
- Experiment units are grouped using simultaneous classification factors.
- One replicate of each treatment is required in every row and column.
- No row and column interaction terms are estimated.
- Missing data need to be estimated to maintain the balance of the design.

Table 4.2   Treatment randomization plan for the fungicide experiment, a randomized blocks design

Block 1

	Unit									
	1	2	3	4	5	6	7	8	9	10
Random no.	410	922	778	767	337	216	964	665	400	987
Rank	4	8	7	6	2	1	9	5	3	10
Treatment	D	H	G	F	B	A	I	E	C	J

Block 4

	Unit									
	31	32	33	34	35	36	37	38	39	40
Random no.	102	928	422	683	968	537	817	15	777	121
Rank	2	9	4	6	10	5	8	1	7	3
Treatment	B	I	D	F	J	E	H	A	G	C

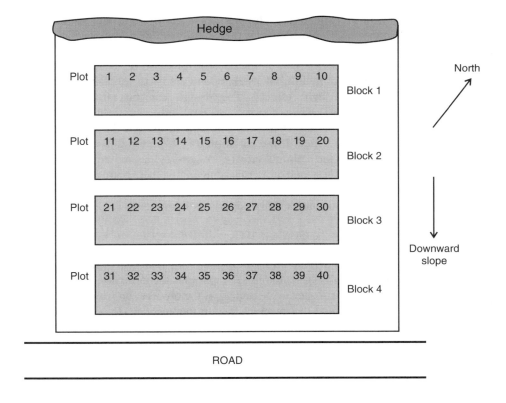

Figure 4.2   Outline site plan for the randomized block experiment described in Section 4.2.2.1

4.2.3.1 Design

The randomized blocks design reduces experiment error and hence the residual variance by taking account of a known source of variation occurring in one direction. The LATIN SQUARE design extends this idea by removing two sources of variation. The first source can be thought of as the rows of a square and the second source of variation as the columns of a square. The square is constructed in such a way that each treatment occurs exactly once in each column and in each row, i.e. every column forms a complete replicate and every row forms a complete replicate of the treatments. So the design must contain as many replicates as it has rows and columns and this equals the number of treatments. The idea of accounting for different sources of variation can be extended to other designs with more than two sources of variability but these designs are beyond the scope of this text.

The purpose of the design is to compare the effect of a set of treatments on a response variable, y. If there are t treatments in the experiment there will need to be t rows and t columns in the design which will have t replicates, and therefore a total of $N = t^2$ experiment units will be needed. If more replicates are needed for the experiment to have sufficient power to detect treatment differences then more than one square will be needed (this is discussed briefly in Section 4.4).

A Latin square can be useful in livestock experiments when an animal is to receive all treatments but this will only be appropriate if there are no carry-over effects from one treatment to the next. In this situation each animal will be different and so the animals themselves provide one source of variation. If the experiment has to be conducted over a number of days then as an animal will be expected to give different results on different days the second source of variation will be the days. It may also be useful in greenhouse experiments where there could be differences along and across the greenhouse caused by variation in ventilation and light effects. The design can be effective in arable experiments if there are two known sources of variation in a field which are not coincident or if it is known that the field is not homogeneous but the true direction of variation cannot be clearly identified. If the variation runs from top to bottom of the field it will be accounted for by the column effects, if it is from side to side it will be accounted for by the row effects and if it runs diagonally then both row and column effects take account of this.

The appropriate statistical model for this design is as follows:

$$y_{ijk} = \mu + R_i + C_j + T_k + e_{ijk} \qquad \text{for } i = 1 \ldots t; j = 1 \ldots t; k = 1 \ldots t$$

where μ represents the overall population mean response, R_i represents the mean effect of row i, C_j represents the mean effect of column j, T_k represents the mean effect of treatment k, e_{ijk} represents the random experimental error, and y_{ijk} is the observation from the experiment unit in the ith row and jth column that receives the kth treatment. μ, R_i, C_j, and T_k are the model parameters.

Example   Some examples of Latin squares are given below. The first is a 4×4 Latin square with treatments A, B, C and D. The total number of observations is $N = 4^2 = 16$.

	Columns			
Rows	1	2	3	4
1	C	B	D	A
2	A	D	B	C
3	D	C	A	B
4	B	A	C	D

The second example is a 5×5 Latin square with treatments A, B, C, D and E. The total number of observations is $N = 5^2 = 25$.

	Columns				
Rows	1	2	3	4	5
1	E	B	A	D	C
2	D	A	E	C	B
3	B	D	C	A	E
4	A	C	B	E	D
5	C	E	D	B	A

Below is a 6×6 Latin square with treatments A, B, C, D, E and F. The total number of observations is $N = 6^2 = 36$.

	Columns					
Rows	1	2	3	4	5	6
1	B	F	E	C	D	A
2	E	A	D	F	B	C
3	A	B	F	D	C	E
4	D	C	A	B	E	F
5	C	D	B	A	F	B
6	F	E	C	E	A	D

4.2.3.2   Hypotheses

$H_0: T_k = \mu_k - \mu = 0$ for $k = 1 \ldots t$

The null hypothesis is that all the treatment population mean values are equal.

There is only one possible alternative hypothesis:

$H_1: T_k = \mu_k - \mu \neq 0$ for at least one k

At least one of the treatment population mean values is not equal to the overall mean.

Where T_k is the treatment effect, μ is a constant representing the overall population mean, μ_k is the kth treatment population mean and t is the number of treatments.

4.2.3.3   Randomization

As with all designs, a Latin square must be randomized so that the statistical assumptions underlying the analysis are met. Standard Latin squares are provided in statistical tables but they must also be randomized. The actual randomization procedure to use depends on the size of the square. For this reason no example of randomization will be given. Guidance is always provided in the statistical tables on the randomization procedure to use for each square. See, for example, Cochran and Cox (1957) and Fisher and Yates (1963).

4.2.3.4   Advantages

The Latin square design controls two sources of variability so, in situations where variability can be identified in two directions, the design will be more powerful than previous designs discussed since the experimental error will be smaller. The design can have any number of treatments.

4.2.3.5   Disadvantages

The number of rows, columns and treatments must be equal and must equal the number of replicates. This constraint means that the number of experiment units needed can get large; for example, with eight treatments, 64 experiment units will be necessary. However, small Latin squares have the problem that there are few residual degrees of freedom. The total number of degrees of freedom is divided amongst the rows, columns and treatments, so few are left for the residual; a 3×3 Latin square has two residual degrees of freedom and a 4×4 Latin square has only six residual degrees of freedom. It is for these reasons that most Latin square designs using single squares are no smaller than 5×5 and no larger than 12×12.

It should be noted in these designs that the rows and columns are blocking factors and interactions between them, or between them and the treatment, cannot be estimated.

4.2.3.6   Example: a Latin square design

The following example of a Latin square design is a 4×4 square which is being used for illustrative purposes. The actual data are those from a single square in an experiment using repeated 4×4 Latin squares.

Description of experiment   This is a nutritional study with sheep. The experiment investigates the digestibility of dry matter in different feeds made up of grass hay and a forage combined in different ratios. It is proposed individually to feed each sheep maintenance diets and for any refusals to be measured. The digestibility of the diets will be determined for each animal by measuring the dry matter disappearance (DMD). There are four different grass hay to forage ratios in the study, hence four treatments, A, B, C and D, in the trial. The animals will be acclimatized to the diet for a period of 14 days. The trial period will last for seven days so the total study time is three weeks. If all treatments can be applied once in a three-week period, then the three-week period can be used as a blocking factor. An individual animal can only receive a single diet at any one time so in any one three-week period there must be four animals, one for each treatment. After the first three-week period all four animals will change diets and another three-week period commences. The whole trial will be completed in a minimum of 12 weeks during which each animal will have been on each diet. The four animals will not be exactly alike although they are the same breed and

Table 4.3 Treatment randomization plan for the nutritional study, a Latin square design

	Sheep 1	Sheep 2	Sheep 3	Sheep 4
Period 1	B	D	C	A
Period 2	D	B	A	C
Period 3	A	C	B	D
Period 4	C	A	D	B

roughly the same weight so animal can also be considered as a blocking factor. Thus there are two simultaneous blocking (or classification) factors, period and animal, in this trial so a Latin square design is employed. A 4×4 square provides only six residual degrees of freedom so a repeated Latin square design with two squares is proposed to provide additional degrees of freedom. Only a single Latin square is illustrated below; the treatment factor is diet and the factor levels are the grass hay to forage ratios, A, B, C and D.

Specific objective To study the digestibility, as measured by dry matter disappearance (DMD), of grass hay with varying proportions of forage.

Hypotheses The null hypothesis is that there are no differences in mean DMD for the four diets.
   The alternative hypothesis is that at least one of the four mean DMD values is different from the rest.

Randomization The treatment randomization plan is given in Table 4.3.

Site plan The sheep will be housed in individual pens to allow them to be individually fed. The randomization plan provides all the information needed so a separate site plan is not required.

4.2.4 Summary of single-treatment factor designs

- *Completely randomized*: homogeneous units; limited coverage of results; no blocks.
- *Randomized blocks*: homogeneous groups of units; heterogeneous blocks; wide coverage of results; missing values need to be estimated.
- *Latin square*: two simultaneous blocking factors; no row and column interactions estimated; estimate missing values.

4.3 MULTI-TREATMENT FACTOR DESIGNS

Keypoints

- Multi-treatment factor designs use a blocking structure and a treatment structure.
- They can be used to investigate interaction effects.

The designs mentioned so far have considered a single treatment factor and have compared several levels of that one treatment factor. To study the effect of two or more factors on a variable a series of experiments could be set up. The levels of one factor would be varied whilst keeping the other factors all at a fixed level. This approach has a number of drawbacks. As only one factor is allowed to vary in any one experiment, information is only provided on how the variable measured varies with the levels of one factor for a specific combination of levels of the other treatment factors. No information is provided on the way in which the variable values measured differ as the levels of *all* factors change. This approach may also use more resources than is necessary.

Factorial experiments overcome this problem by combining the treatment factors to give a single set of treatments which are tested in one experiment. This allows interaction effects to be investigated which would not have been possible had separate experiments been run for each factor. *Interactions* occur between two factors when the differences between the response to the levels of one factor differ at each level of another factor.

The same basic designs used for single treatment factor designs are used for designs with more than one treatment factor, thus these can be completely randomized, or can have a blocking structure such as the randomized blocks or Latin square designs. Therefore, multi-treatment factor designs have a treatment structure in addition to having a block structure.

Factorial designs are, therefore, ones in which the treatments to be investigated are generated by combining the levels of two or more factors of interest. Each of the factors can be qualitative or quantitative. They are the only designs which can investigate the interactions of various factors. There are two forms of multi-treatment factor designs: FACTORIAL and SPLIT-PLOT designs. In both designs, each treatment factor combination is applied to one or more experiment units. It is the way in which the treatments are laid out that determines whether a split-plot design or a factorial design is employed.

Factorial designs are ones in which the treatment factors can be combined into a single treatment factor combination. Each treatment factor combination is randomly assigned to an experiment unit.

The basic idea behind split-plot designs is that the treatment factors are split into main-plot treatment factors and sub-plot treatment factors. The experiment material is divided into main-plots and each main-plot is split into sub-plots. Each main-plot receives a single main-plot factor level and receives all sub-plot factor levels. This means that comparisons between all sub-plot factor levels can be made within each main-plot.

It is usually for practical reasons that treatments are applied to the sub-plot or to the main-plots, however the implications of this must be considered since there will usually be more variation between main-plots than between sub-plots. This is because the sub-plots use a smaller amount of experiment material and so will tend to be more homogeneous than the main-plots, which means that the sub-plot background error term will be expected to be smaller than that for the main-plots. This means that any effects estimated at the sub-plot level will be estimated with more precision than those at the main-plot level. It would therefore be preferable to put the treatment of most interest on the sub-plots, if this is practicably possible. Consequently it is important to determine at the outset which are the most important treatment factors (Snedecor and Cochran, 1989).

As with the single treatment factor designs (Section 4.2), multi-treatment factor designs can be formed from either fixed or random effects models or they can be mixed models containing both fixed and random effects. This text covers only fixed effects models so the formulae, hypotheses and conclusions presented in this section apply to fixed models. For each two factor design the actual design is discussed and the appropriate statistical model is given along with the hypotheses that are being tested. Details are also given about the correct randomization procedure and the advantages and disadvantages of the models presented. After each model a detailed example of its simplest, basic use is given. Designs with three or more factors are an extension of the two factor designs and are discussed with reference to three factor designs.

There are other designs that include many treatment factors, each with only two or three treatment levels. These designs are particularly useful in the exploratory stages of experiments to indicate which treatment factors are having an effect. Each factor being used would usually be included at high and low levels. These experiments can get large very quickly, for example if there were five factors at two levels there would be 32 treatment combinations, and so single replicates of these experiments are often run. High-order interactions are rarely found to be significant so when a single replicate is used the higher-order interactions are pooled and used as the estimate of experimental error. However, it must always be decided in advance which interactions are to be used for this. These designs are not covered in any further detail in this text but a comprehensive description of this type of design can be found in Montgomery (1991).

4.3.1 Two-way factorial designs

Keypoints

- Two-way factorial designs investigate two treatment factors at different levels.
- They investigate the presence of interaction effects between the two treatment factors.
- Any of the blocking structures can be used.
- All possible treatment factor combinations need to be present and equally replicated.
- Missing values need to be estimated to maintain the balance of the design.

4.3.1.1 Design

The simplest factorial experiment is a TWO-WAY FACTORIAL COMPLETELY RANDOMIZED design. The same restrictions apply as for a single treatment completely randomized design, i.e. for the design to be sensible all of the experiment units must be homogeneous. If the units can be classified into some form of group then a more complicated factorial design with blocking structure must be used. All treatment factor combinations should be present and equally replicated. The problem of missing values will be considered later.

As with all of the other designs considered, the purpose of the design is to compare the effects of the treatments on some response variable, y. This design has two treatment factors, treatment A with a levels and treatment B with b levels, therefore the number of treatments, t, in this experiment is ab. In most cases each treatment is equally replicated r times so the total number of experiment units, N, in the experiment will be abr. Any of the models used so far will not be appropriate for this situation since they do not allow for any interaction terms. A parameter representing this interaction term must be introduced into the model. The interaction is represented by the term $(AB)_{ij}$; this term is not a product but represents a single parameter.

The appropriate statistical model for the design is as follows:

$$y_{ijk} = \mu + A_i + B_j + (AB)_{ij} + e_{ijk} \qquad \text{for } i = 1 \ldots a; \ j = 1 \ldots b; \ k = 1 \ldots r$$

where μ represents the overall treatment mean response, A_i represents the mean effect of treatment A at level i, B_j represents the mean effect of treatment B at level j, $(AB)_{ij}$ represents the mean effect of the interaction between treatment A at level i and treatment B at level j,

e_{ijk} represents the random experimental error and y_{ijk} is the observation from the kth replicate receiving the ith treatment A and the jth treatment B. μ, A_i, B_j, and $(AB)_{ij}$ are the model parameters.

This design tests whether there is an interaction between treatments A and B, i.e. whether the response in the different levels of treatment A is not the same for all levels of treatment B. It also tests whether there is a main effect of each of the two treatment factors when averaged over the levels of the other treatment factor, i.e. whether there is a difference between the levels of A when they are averaged over all levels of B and whether there is a difference in the levels of B when they are averaged over all levels of A.

The factorial arrangement of treatments can be used with the blocking structures of the randomized blocks and Latin squares. The principles are exactly the same as described above for the completely randomized layout but use the underlying blocking structure from the appropriate design.

In a two-way factorial randomized blocks design all treatment factor combinations should be randomized within each block. The model will separate the treatment effects into the main effects and interactions shown for the completely randomized design but will additionally contain a parameter representing the block effect.

A Latin square design can also be used to run a factorial experiment. The number of rows and columns in the square must equal the number of treatment combinations generated. This fact makes most Latin square factorial designs impractical due to the very large Latin square that would be needed. If such a design is ever used the statistical model would be built in the same way as before. The treatment main effects and interactions would be represented, as would parameters for the row and column effects of the Latin square.

4.3.1.2   Hypotheses

There are three null hypotheses to be tested simultaneously:

$H_0 : A_i = \mu_i - \mu = 0$ for $i = 1 \ldots a$

$H_0 : B_j = \mu_j - \mu = 0$ for $j = 1 \ldots b$

$H_0 : AB_{ij} = \mu_{ij} - \mu = 0$ for $i = 1 \ldots a, j = 1 \ldots b$

The null hypotheses are as follows:

All treatment factor A population mean values are equal.
All treatment factor B population mean values are equal.
All AB interaction population mean values are equal.

There is only one possible alternative hypothesis for each null hypothesis:

$H_1 : A_i = \mu_i - \mu \neq 0$ for at least one i

$H_1 : B_j = \mu_j - \mu \neq 0$ for at least one j

$H_1 : AB_{ij} = \mu_{ij} - \mu \neq 0$ for at least one i or j

The alternative hypotheses are as follows:

At least one of the treatment factor A population mean values is not equal to the overall mean.

At least one of the treatment factor B population mean values is not equal to the overall mean.

At least one of the interaction population mean values is not equal to the overall mean.

where A_i is the factor A treatment effect, B_j is the factor B treatment effect, AB_{ij} is the interaction effect, μ is a constant representing the overall population mean, μ_i is the ith factor A treatment population mean, μ_j is the jth factor B treatment population mean and a and b are the number of treatments in factors A and B respectively.

4.3.1.3 Randomization

The method of randomization depends on the blocking structure of the experiment. Each design will be randomized as previously described in the single treatment factor section as the appropriate method is determined by the blocking structure. Treatment factor combinations will be randomized to the experiment units rather than individual treatments.

Example   Consider an experiment with two treatment factors, A with two levels (a_1 and a_2) and B with three levels (b_1, b_2 and b_3) which are to be replicated three times. A total of 18 experiment units are needed. The treatments are to be compared in a randomized blocks design with one replicate per block, therefore there will be three blocks of six plots in the design.

To randomly allocate the treatments to the experiment units, assign a unique number to each unit, generate 18 random numbers and associate each random number with an experiment unit:

		Block 1						Block 2						Block 3					
Step 1	Unit	1	2	3	4	5	6	7	8	9	10	11	12	13	14	15	16	17	18
Steps 2 and 3	Random no.	559	617	298	826	577	782	255	774	745	584	600	67	147	77	496	713	415	391

Within each block, rank the random numbers in ascending order:

	Block 1						Block 2						Block 3					
Unit	1	2	3	4	5	6	7	8	9	10	11	12	13	14	15	16	17	18
Random no.	559	617	298	826	577	782	255	774	745	584	600	67	147	77	496	713	415	391
Step 4	2	4	1	6	3	5	2	6	5	3	4	1	2	1	5	6	4	3

Experiment units with rank 1 in each block will receive treatment $a_1 b_1$, those with rank 2 will receive treatment $a_1 b_2$ and so on until all treatments are allocated:

		Block 1						Block 2						Block 3					
Unit		1	2	3	4	5	6	7	8	9	10	11	12	13	14	15	16	17	18
Steps 5 and 6	Treatment	$a_1 b_2$	$a_2 b_1$	$a_1 b_1$	$a_2 b_3$	$a_1 b_3$	$a_2 b_2$	$a_1 b_2$	$a_2 b_3$	$a_2 b_2$	$a_1 b_3$	$a_2 b_1$	$a_1 b_1$	$a_1 b_2$	$a_1 b_1$	$a_2 b_2$	$a_2 b_3$	$a_2 b_1$	$a_1 b_3$

4.3.1.4 Advantages

A factorial design is a design in which interactions between factors can be investigated. The design is flexible in that any of the blocking structures can be used.

4.3.1.5 Disadvantages

These designs can become very large: even with only a few factors at a few levels, many treatment combinations are generated. For example, a design with four levels of one treatment and six levels of another will have 24 treatment combinations. This means that 24 experiment units will be needed for one replicate and it may be difficult to find sufficient homogeneous experiment units to run such a design.

The structure of a factorial experiment will be destroyed if there are any missing values. Estimation techniques can be used if there are relatively few missing values.

These designs are given a good clear coverage in Montgomery (1991). Models and hypotheses are given and discussed in detail. Steel and Torrie (1980) provide a more detailed discussion of the advantages and disadvantages of the designs. Cochran and Cox (1957) and Snedecor and Cochran (1989) also give a thorough coverage of these designs.

4.3.1.6 Example: a two-way factorial design

Description of experiment   The experiment involves two treatment factors, diet and housing. Each factor has only two levels, a low and a high concentrate diet (D1 and D2) and two types of housing environment (H1 and H2). Thirty-six heifers are available for the trial which allows nine replicates of all four treatment combinations. The weight of the animal can have an influence on many of the assessments made in livestock trials so it makes sense to block on the animal's weight before entering the trial. This is done so that the four animals in each block are of as similar a weight as possible. The four treatment combinations are randomly allocated to the four animals in each block. All animals are group fed to appetite.

Objective   To determine the effect on milk quality of heifers, measured by mean milk protein content, of different diets and housing environments.

Hypotheses   The three null hypotheses are as follows:

There is no difference between the high and low feed concentrate mean milk protein mean values.
There is no difference between the two housing mean milk protein mean values.
There are no differences in the four interaction mean milk protein mean values.

The three alternative hypotheses are as follows:

The high and low concentrate mean milk protein means are different.
The two types of housing environment mean milk protein means are different.
At least one of the interaction mean milk protein mean values is different from the rest.

Randomization   The treatment combinations are ordered and allocated to the treatment numbers as follows: treatment 1 is diet 1, housing 1; treatment 2 is diet 1, housing 2; and treatments 3 and 4 are housing 1 and 2 respectively for diet 2. The final randomization is shown in Table 4.4.

Table 4.4 Treatment randomization plan for the milk quality trial, a two-way factorial in randomized blocks design

Heifer	Block	Random no.	Treatment	Heifer	Block	Random no.	Treatment
1	1	83	4	19	5	56	2
2	1	52	3	20	5	33	1
3	1	51	2	21	6	74	3
4	1	14	1	22	6	84	4
5	2	47	2	23	6	47	2
6	2	56	3	24	6	4	1
7	2	91	4	25	7	99	4
8	2	29	1	26	7	83	3
9	3	80	4	27	7	81	2
10	3	44	1	28	7	74	1
11	3	63	2	29	8	77	4
12	3	74	3	30	8	8	2
13	4	40	2	31	8	7	1
14	4	53	3	32	8	23	3
15	4	79	4	33	9	14	2
16	4	9	1	34	9	1	1
17	5	62	3	35	9	50	4
18	5	82	4	36	9	49	3

Site plan No site plan is required for this experiment as the animals are group fed. It is not necessary to isolate the animals in each block from animals in another block as the blocking factor is the animal's weight which cannot be influenced by any other animal.

4.3.2 Multi-way factorial designs

Keypoints

- Multi-way factorial designs can be used with any of the blocking structures.
- Many treatment factors can be included in the same experiment.
- High-order interactions are difficult to interpret.
- All possible treatment factor combinations need to be present and equally replicated.
- Missing values have to be estimated to maintain the balance of the design.

4.3.2.1 Design

The ideas for two-way factorial experiments can be easily extended to experiments with more than two treatment factors, each at several levels. These can all be combined to give a series of treatment combinations.

As before, the treatment factors can be used with any of the blocking structures, completely randomized, randomized blocks or Latin square design. It is important that all treatment factor combinations are present and equally replicated. These designs should be used with good purpose and not as a way of including many treatment factors in one experiment since high-order interactions are difficult to interpret and often have no practical meaning. Main effects and lower-order interactions can become meaningless if they are included in any significant interactions. This will be considered further in the interpretation of such designs. The order of an interaction is one less than the number of treatment factors involved in the interaction. So an interaction between two treatment factors is a first-order interaction whilst one between four treatment factors is a third-order interaction.

Designs can have as many treatment factors as required but the implications of significant interaction effects should be considered at the design stage. If high-order interactions are expected to be significant but the objectives only consider main effects or low-order interactions then a different design might be more appropriate. For example, if the objective of a two-way factorial experiment is concerned with main effects only then the interaction term (the first-order interaction) complicates the issue and a simpler design should be chosen. However, investigating interactions can be a source of much useful information when attempting to assess the combined effects of multiple treatment factors.

4.3.2.2   Hypotheses

There is one null hypothesis for each of the main effects and one for each of the interaction effects. All these will be tested simultaneously. Examples of hypotheses are shown below. Main effect null hypotheses will be of the form:

$H_0: C_k = \mu_k - \mu = 0$ for $k = 1 \ldots c$

All treatment factor C population mean values are equal.

First-order interaction effect null hypotheses will be of the form:

$H_0: BC_{jk} = \mu_{jk} - \mu = 0$ for $j = 1 \ldots b; \ k = 1 \ldots c$

All BC interaction population mean values are equal.

Second-order interaction effect null hypotheses will be of the form:

$H_0: BCD_{jkl} = \mu_{jkl} - \mu = 0$ for $j = 1 \ldots b; \ k = 1 \ldots c; \ l = 1 \ldots d$

All BCD interaction population mean values are equal.

There is only one possible alternative hypothesis for each null hypothesis. For those illustrated they will be as follows:

$H_1: C_k = \mu_k - \mu \neq 0$ for at least one k

$H_1: BC_{jk} = \mu_{jk} - \mu \neq 0$ for at least one j or k

$H_1: BCD_{jkl} = \mu_{jkl} - \mu \neq 0$ for at least one j or k or l

At least one of the treatment factor C population mean values is not equal to the overall mean.
At least one of the BC interaction population mean values is not equal to the overall mean.
At least one of the BCD interaction population mean values is not equal to the overall mean.

Higher-order interactions will each have their own null hypothesis and alternative hypothesis which will be in similar formats.

4.3.2.3 Randomization

As with two-way factorial designs, the treatment factor combinations are randomized to the experiment units in accordance with the design used; completely randomized, randomized blocks or Latin square.

4.3.2.4 Advantages

Many main effects can be investigated in an experiment provided no interaction terms are significant. Multiple interactions between the many treatment factors can be investigated. Single replicates of the design can be used with high-order interactions used as estimates of experiment error.

4.3.2.5 Disadvantages

Experiments can become very large and complex. High-order interactions are difficult to interpret. Equal replication of all treatment factor combinations is required though estimation techniques can be used if there are only a few missing values.

These designs are frequently mentioned simply as an extension of the two-way factorial designs. Montgomery (1991) does provide details of the model and both Montgomery (1991) and Steel and Torrie (1980) give clear detailed descriptions of the design with complete worked examples.

4.3.2.6 Example: a three-way factorial design

Description of experiment   This experiment is an investigation into the quality of a number of oat varieties and how sowing date and treating the crop with a standard nitrogen application may affect the quality of the crop. Five varieties of oats, A, B, C, D and E, are considered for this experiment with all varieties to be sown at an early and a late sowing date, D1 and D2 respectively. A standard optimum nitrogen application, N1, is usually given to oat crops, however with the general increased awareness of environmental considerations one of the experiment aims is to assess the effect of not using a nitrogen treatment at all, N2. Drilling equipment at the farm includes an øjord drill which is a small plot drill that can be used to sow narrow (2 m) strips; conventional sized drills cannot sow such narrow areas. Consequently it is possible to fully randomize the variety and sowing date treatments keeping the trial within the limited area available. The nitrogen treatments can be applied by hand-held sprays so this treatment can also be applied to the small plots. Three replicates are considered to be sufficient for the requirements of the trial. The available field for the experiment runs alongside a long tall farm building which could have a shading effect on any crop grown near to the building. As space is limited, it is not possible to keep all of the 60 plots required out of the area affected by shading. Blocking can be used to counteract this and the length of the randomized blocks run parallel to the farm building. The lengths of the plots within each block run perpendicular to the building. Each of the 20 treatment combinations are to be applied to a single plot in each block. The design is a three-way factorial in three randomized blocks. The three treatment factors are variety, sowing date and nitrogen rate. Treatment factor levels are the different oat varieties, A, B, C, D and E, the two sowing dates, D1 and D2, and the optimum and nil nitrogen treatments, N1 and N2 respectively.

Specific objective   To determine whether sowing date and nitrogen application affects the quality of five varieties of oats as measured by oven dry matter recorded at 40% moisture.

Hypotheses   The three main effect null hypotheses are as follows:

There are no differences in the five variety mean oven dry matter values.
There is no difference in the two sowing date mean oven dry matter values.
There is no difference in the two nitrogen rate mean oven dry matter values.

The three first-order interaction effect null hypotheses are as follows:

There are no differences in the 10 variety/sowing date interaction mean oven dry matter values.
There are no differences in the 10 variety/nitrogen rate interaction mean oven dry matter values.
There are no differences in the four sowing date/nitrogen rate interaction mean oven dry matter values.

The second-order interaction null hypothesis is as follows:

There are no differences in the 20 variety/sowing date/nitrogen rate interaction mean oven dry matter values.

The alternative hypotheses are as follows:

At least one of the five variety mean oven dry matters is different from the rest.
The two sowing date mean oven dry matters are different.
The two nitrogen rate mean oven dry matters are different.
At least one of the 10 variety/sowing date interaction mean oven dry matters is different from the rest.
At least one of the 10 variety/nitrogen rate interaction mean oven dry matters is different from the rest.
At least one of the four sowing date/nitrogen rate interaction mean oven dry matters is different from the rest.
At least one of the 20 variety/sowing date/nitrogen rate interaction mean oven dry matters is different from the rest.

Randomization   The treatment combinations are allocated numbers 1 to 20 with the factor levels being ordered within variety and then sowing date within variety and nitrogen rate within sowing date. Thus treatment 1 is variety A, sowing date 1 and nitrogen rate 1, and treatment 20 is variety E, sowing date 2 and nitrogen rate 2. The final treatment randomization is shown in Table 4.5.

Site plan   The outline site plan is shown in Figure 4.3.

Table 4.5   Treatment randomization plan for the oat experiment, a three-way factorial in randomized design

	Block 1				Block 2				Block 3		
Plot	Random no.	Rank	Treatment	Plot	Random no.	Rank	Treatment	Plot	Random no.	Rank	Treatment
1	261	9	Cd1n1	21	591	11	Cd2n1	41	265	5	Bd1n1
2	875	16	Dd2n2	22	992	20	Ed2n2	42	746	14	Dd1n2
3	120	4	Ad2n2	23	807	17	Ed1n1	43	115	3	Ad2n1
4	939	19	Ed2n1	24	988	19	Ed2n1	44	73	1	Ad1n1
5	125	5	Bd1n1	25	676	14	Dd1n2	45	426	8	Bd2n2
6	80	3	Ad2n1	26	416	6	Bd1n2	46	880	16	Dd2n2
7	436	12	Cd2n2	27	672	13	Dd1n1	47	809	15	Dd2n1
8	664	14	Dd1n2	28	360	4	Ad2n2	48	109	2	Ad1n2
9	912	17	Ed1n1	29	477	9	Cd1n1	49	387	7	Bd2n1
10	856	15	Dd2n1	30	382	5	Bd1n1	50	615	11	Cd2n1
11	930	18	Ed1n2	31	449	7	Bd2n1	51	714	13	Dd1n1
12	384	11	Cd2n1	32	9	1	Ad1n1	52	247	4	Ad2n2
13	40	1	Ad1n1	33	526	10	Cd1n2	53	586	10	Cd1n2
14	954	20	Ed2n2	34	844	18	Ed1n2	54	295	6	Bd1n2
15	155	7	Bd2n1	35	785	16	Dd2n2	55	992	20	Ed2n2
16	150	6	Bd1n2	36	623	12	Cd2n2	56	700	12	Cd2n2
17	72	2	Ad1n2	37	687	15	Dd2n1	57	906	17	Ed1n1
18	285	10	Cd1n2	38	167	2	Ad1n2	58	918	19	Ed2n1
19	474	13	Dd1n1	39	454	8	Bd2n2	59	915	18	Ed1n2
20	159	8	Bd2n2	40	328	3	Ad2n1	60	505	9	Cd1n1

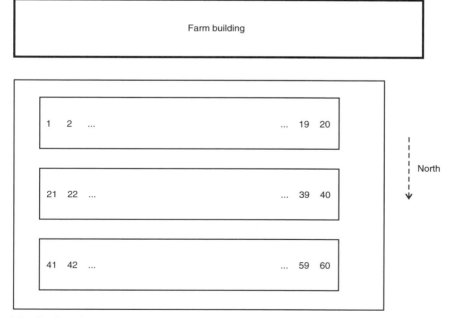

Figure 4.3   Outline site plan for the three-way factorial in randomized blocks experiment described in Section 4.3.2.6

4.3.3   Two treatment factor split-plot designs

Keypoints

- Two-factor split-plot designs investigate two treatment factors at different levels.
- They investigate the presence of interaction effects between the two treatment factors.
- They estimate treatment effects of factors on main-plots less precisely than those on sub-plots.
- Treatment factors using different amounts of experiment material can be tested together.
- Any of the blocking structures can be used.
- All possible treatment factor combinations need to be present and equally replicated.
- Missing values need to be estimated to maintain the balance of the design.

4.3.3.1   Design

The SPLIT-PLOT design is a special case of the factorial design. In split-plot designs the experiment material is divided into main-plots and each main-plot is split into sub-plots. One treatment factor is applied to the main-plots, the other randomized to the sub-plots. The simplest split-plot experiment is a TWO-FACTOR SPLIT-PLOT COMPLETELY RANDOMIZED design. The same restrictions apply as for factorial completely randomized design, i.e. for the design to be sensible all of the experiment units must be homogeneous. If the units can be classified into some form of group then a more complicated split-plot design with a different blocking structure must be used. It is important that all of the treatment factor combinations are equally replicated. The problem of missing values will be considered later.

The design involves just two treatment factors, treatment A with a levels and treatment B with b levels. The chief practical advantage of the split-plot arrangement of treatments is that it enables treatment factors that require relatively large amounts of experiment material and treatment factors that require small amounts to be combined in the same experiment. With a split-plot experiment the treatment factor applied to the sub-plots and the interaction between the two treatment factors will be estimated more precisely than the treatment factor effect applied to the main-plots since the error mean square is usually larger for the main-plots than for the sub-plots. If treatment factors A and B both require the same amount of experiment material but for good practical reasons a split-plot design is proposed, consideration must be given to which treatment factor is of the most interest. If factor B is of most interest then factor A will be applied to each main-plot and factors A and B can be termed the main-plot and sub-plot factors, respectively. The main-plots will be split into as many sub-plots as there are levels of factor B and factor B will be applied to these sub-plots. Each level of factor A will be randomly allocated to the main-plots and each level of factor B is randomly allocated to each sub-plot. Thus each main-plot is tested at only one level of factor A but at each level of factor B. Thus the split-plot design is most advantageous if one of the treatment factors and the interaction is of greater interest or if one of the treatment factors requires a relatively large amount of experiment material.

The purpose of the design is to compare the effects of a set of treatments on some response variable, y. The design has two treatment factors, treatment A with a levels and treatment B with b levels, so the number of treatment combinations, t, is ab. Each factor A level is randomly assigned to r equally replicated main-plots. Thus there will be ar main-plots. Within each main-plot there are b sub-plots, one for each level of factor B, hence there will be abr sub-plots. From this it follows that factor B will be replicated b times more than factor A. Thus the effect of factor B will be estimated with greater power, as will its interaction with factor A. The effect of factor B and the interaction effect will be tested at the sub-plot level and the effect of factor A will be tested at the main-plot level. It must, therefore, be carefully planned at the outset which factor levels should be applied to main-plots and which to sub-plots. For practical reasons it may not always be possible to do this but the implications on the analysis outcome should be considered. For example, if the main objective can no longer be met with the required power then consideration should be given to other designs.

The appropriate statistical model for the design is as follows:

$$y_{ijk} = \mu + A_i + em_{ki} + B_j + (AB)_{ij} + e_{ijk} \qquad \text{for } i = 1 \ldots a; \; j = 1 \ldots b; \; k = 1 \ldots r$$

where μ represents the overall treatment mean response, A_i represents the mean effect of treatment A at level i, B_j represents the mean effect of treatment B at level j, $(AB)_{ij}$ represents the mean effect of the interaction between treatment A at level i and treatment B at level j, em_{ki} represents the main-plot error and e_{ijk} the sub-plot error. μ, A_i, B_j, and $(AB)_{ij}$ are the model parameters.

This design tests whether there is an interaction between treatments A and B, i.e. whether the response in the different levels of treatment A is not the same for all levels of treatment B. It also tests whether there is a main effect of each of the two treatment factors when averaged over the levels of the other treatment factor, i.e. whether there is a difference between the levels of A when they are averaged over all levels of B and whether there is a difference in the levels of B when they are averaged over all levels of A.

As with a simple factorial arrangement of treatments, the split-plot experiment can be used with any of the other designs. The blocking structure underlying the other designs is used to allocate experiment units to the main-plot factors. For example, in a RANDOMIZED BLOCKS TWO-FACTOR SPLIT-PLOT design each block would be split into main-plots with a complete replicate of the main-plot factor levels. The levels of the main-plot treatment factor would be randomly allocated to the main-plots in each block and the levels of the sub-plot treatment factor are randomly allocated to the sub-plots in each main-plot.

4.3.3.2 Hypotheses

There are three null hypotheses to be tested simultaneously:

$H_0 : A_i = \mu_i - \mu = 0$	for $i = 1 \ldots a$
$H_0 : B_j = \mu_j - \mu = 0$	for $j = 1 \ldots b$
$H_0 : AB_{ij} = \mu_{ij} - \mu = 0$	for $i = 1 \ldots a; \; j = 1 \ldots b$

The null hypotheses are as follows:

All treatment factor A population mean values are equal.
All treatment factor B population mean values are equal.
All AB interaction population mean values are equal.

There is only one possible alternative hypothesis for each null hypothesis:

$H_1 : A_i = \mu_i - \mu \neq 0$	for at least one i
$H_1 : B_j = \mu_j - \mu \neq 0$	for at least one j
$H_1 : AB_{ij} = \mu_{ij} - \mu \neq 0$	for at least one i or j

The alternative hypotheses are as follows:

At least one of the treatment factor A population mean values is not equal to the overall mean.
At least one of the treatment factor B population mean values is not equal to the overall mean.
At least one of the interaction population mean values is not equal to the overall mean.

where A_i is the factor A treatment effect, B_j is the factor B treatment effect, AB_{ij} is the interaction effect, μ is a constant representing the overall population mean, μ_i is the ith factor A treatment population mean, μ_j is the jth factor B treatment population mean and a and b are the number of treatments in factors A and B respectively.

4.3.3.3 Randomization

The method of randomization depends on the design of the experiment. The main-plots for each design will be randomized as previously described in the single treatment factor section as the appropriate method is determined by the blocking structure. Within each main-plot, the sub-plot allocations are randomly allocated to the sub-plots.

Example Consider a two-factor split-plot in three randomized blocks with one replicate of all treatment combinations in each block. Factor A has two levels, $a1$ and $a2$, and is the main-plot factor. Factor B is the sub-plot factor with three levels, $b1$, $b2$ and $b3$.

The procedure for randomly allocating the treatments to the experiment units is shown for the first block. All other blocks are randomized in a similar way.

Assign a unique number to each main-plot and sub-plot and generate as many random numbers are there are sub-plots and main-plots. Associate a random number with each main-plot and one with each sub-plot:

Block 1

Step 1 for main-plots	Main-plot identifier		1			2	
Step 1 for sub-plots	Sub-plot identifier	1	2	3	1	2	3
	Plot number	1	2	3	4	5	6
Steps 2 and 3 for main-plots	Main-plot random number		502			902	
Steps 2 and 3 for sub-plots	Sub-plot random number	495	920	324	11	936	743

Rank the random numbers associated with the main-plots in ascending order (or descending order) within each block:

Block 1

	Main-plot identifier		1			2	
	Sub-plot identifier	1	2	3	1	2	3
	Plot number	1	2	3	4	5	6
	Main-plot random number		502			902	
			2			2	
Steps 4 for main-plots	Main-plot random number rank		1			2	

The main-plot with rank 1 in each block will receive treatment a_1, those with rank 2 will receive treatment a_2.

Block 1

	Main-plot identifier	1			2		
	Sub-plot identifier	1	2	3	1	2	3
	Plot number	1	2	3	4	5	6
Steps 5 and 6 for main-plots	Main-plot treatment	a_1			a_2		

For each main-plot, rank the random numbers assigned to the sub-plots in ascending (or descending) order:

Block 1

	Main-plot identifier	1			2		
	Sub-plot identifier	1	2	3	1	2	3
	Plot number	1	2	3	4	5	6
Step 4 for sub-plots	Sub-plot random number	495	920	324	11	936	743
	Sub-plot random number rank	2	3	1	1	3	2

The sub-plots with rank 1 will receive treatment b_1, those with rank 2 will receive treatment b_2, and so on.

Block 1

	Main-plot identifier	1			2		
	Sub-plot identifier	1	2	3	1	2	3
	Plot number	1	2	3	4	5	6
Steps 5 and 6 for main-plots	Sub-plot treatment	b_2	b_3	b_1	b_1	b_3	b_2

So the final treatment randomization is as follows:

Block 1

	Main-plot identifier	1			2		
	Sub-plot identifier	1	2	3	1	2	3
	Plot number	1	2	3	4	5	6
	Main-plot treatments	a_1			a_2		
	Sub-plot treatments	b_2	b_3	b_1	b_1	b_3	b_2

4.3.3.4 Advantages

Treatment factors that require different amounts of experiment material can be tested in the same experiment. Interaction terms can be estimated.

4.3.3.5 Disadvantages

Main-plot effects are estimated with less precision than if a simple factorial design had been employed. Equal replication of all treatment factor combinations is essential. Estimation techniques can be used if there are relatively few missing values.

Good clear descriptions of these designs can be found in Cochran and Cox (1957) and Steel and Torrie (1980). Both of them detail the randomization of the treatments to the experiment units. Montgomery (1991) presents the model formula and a basic description of the designs can be found in Snedecor and Cochran (1990).

4.3.3.6 Example: a two-factor split-plot design

Description of experiment   This experiment investigates the effects of seven nitrogen treatments, N1 to N7, and four fallow treatments, A, B, C and D, on the specific weight of a winter wheat crop grown on the site. One of the experiment aims is to model the crop response to increased applications of nitrogen fertilizer for the four fallow treatments. Consequently the nitrogen treatments are rates of nitrogen increasing in equal increments from zero, $N1$, to a maximum rate, $N7$, that exceeds the expected optimum value and hence catches any likely turning point in the response curve. It has been decided from past experimentation that four replicates of each treatment should be sufficient to detect the treatment differences of interest. However, the application of the four fallow treatments poses a problem to the extent that the machinery required to apply them is quite large and there is insufficient space in the field to position 112 plots large enough to take the machinery. The nitrogen treatments can be applied by hand so the plots can be much smaller for these treatments. It is therefore decided to make a plot receiving a fallow treatment sufficiently large, not only to cope with the fallow application machinery but also to be split into seven smaller plots which could be used for the nitrogen treatments. This allows the experiment to be conducted in a much smaller area than otherwise possible but it is at the expense of losing precision on the fallow treatments. The experiment is therefore designed as a two-factor split-plot experiment with fallow treatments on the main-plots and nitrogen treatments on the sub-plots; one treatment factor is fallow with four factor levels (A, B, C and D), the second treatment factor is nitrogen, with the seven rates of application (N1 to N7), being the treatment factor levels. There are four blocks, each with a complete replicate of all 28 treatment combinations.

Specific objective   To determine the effects of fallowing on the specific weight of a winter wheat and how this may affect response to different rates of nitrogen application.

Hypotheses   The three null hypotheses are

   There are no differences in four fallow mean specific weight values.
   There are no differences in seven nitrogen mean specific weight values.
   There are no differences in the interaction mean specific weight values.

The three alternative hypotheses are

   At least one of the four fallow mean specific weight values is different from the others.
   At least one of the seven nitrogen mean specific weight values is different from the others.
   At least one of the 28 interaction mean specific weight values is different from the others.

Table 4.6   Treatment randomization plan for the winter wheat experiment, a two-factor split-plot design

Block 1	Main-plot				Sub-plot			
Units	Plot no.	Random no.	Rank	Treatment	Plot no.	Random no.	Rank	Treatment
1					1	749	4	D
2					2	927	7	G
3					3	342	2	B
4	1	386	3	C	4	290	1	A
5					5	355	3	C
6					6	803	5	E
7					7	880	6	F
8					1	912	7	G
9					2	500	4	D
10					3	457	3	C
11	2	303	1	A	4	643	5	E
12					5	277	1	A
13					6	696	6	F
14					7	413	2	B
15					1	299	1	A
16					2	674	7	G
17					3	331	2	B
18	3	383	2	B	4	477	5	E
19					5	366	3	C
20					6	499	6	F
21					7	375	4	D
22					1	203	2	B
23					2	856	6	F
24					3	439	4	D
25	4	760	4	D	4	412	3	B
26					5	122	1	A
27					6	752	5	E
28					7	915	7	G

Randomization   The randomization for block 1 is shown in Table 4.6. Blocks 2, 3 and 4 are randomized in a similar manner.

Site plan   The outline site plan is given in Figure 4.4.

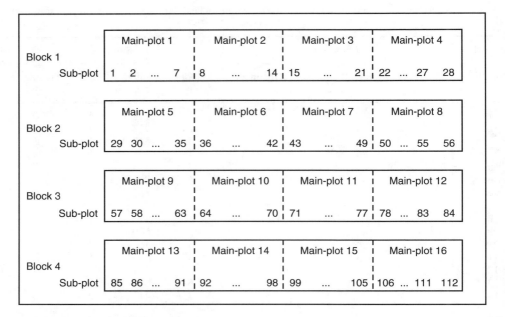

Figure 4.4   Outline site plan for the two-factor split-plot in randomized blocks experiment described in Section 4.3.3.6

4.3.4   Multi-treatment factor split-plot designs

Keypoints

- In multi-treatment factor split-plot designs, the effects of main-plot treatments will be estimated less precisely than the effects of sub-plot treatments.
- The effects of sub-plot treatments will be estimated less precisely than the effects of sub-sub-plot treatments and so on.
- High-order interactions are difficult to interpret.
- All possible treatment factor combinations need to be present and equally replicated.
- Multi-treatment factor split-plot designs can be used with any of the blocking structures.
- Missing values have to be estimated to maintain the balance of the design.

4.3.4.1   Design

The basic ideas and principles of split-plot designs discussed in the previous section can be very easily extended to more complex split-plot designs. It is important that the design is balanced and that all treatment factor combinations are equally replicated. In general, a split-plot design can include any number of main-plot factors and any number of sub-plot factors. Therefore there will be a factorial arrangement of treatment factors on either the main-plots, the sub-plots or both.

Consider an example where three treatment factors are being tested as a split-plot design with one main-plot factor, *A*, and two sub-plot factors, *B* and *C*. As with the two-factor split-plot design, factor *A* is allocated to the main-plots. The treatment factor combination of *B*

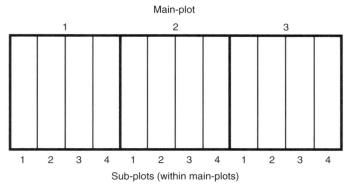

Main-plot

Figure 4.5   Schematic layout for a split-plot design with one treatment factor on the main-plots and two sub-plot treatment factors

and C will be randomly allocated to the sub-plots. Each main-plot is split into as many sub-plots as there are treatment combinations of B and C. The effect of factor A is tested at the main-plot level and the effects of factors B and C will be tested at the sub-plot level. The effects of the interactions between A and the other two factors, including the third-order interaction, are tested at the sub-plot level, as is the interaction between factor B and factor C. For example, if factor A has three levels and factors B and C have two levels there would be three main-plots and four (2 × 2) sub-plots in each main-plot, as illustrated in Figure 4.5. The thick lines indicate the perimeters of the main-plots and the fine lines indicate the perimeters of the sub-plots.

Another way of considering a split-plot design with three treatment factors is a split-plot design with two main-plot factors, A and B, and one sub-plot factor, C. In this design the treatment factor combination of A and B is allocated to the main-plots. The factor levels of C are randomly allocated to the sub-plots. Each main-plot is split into as many sub-plots as there are levels of C. The effects of factors A and B are tested at the main-plot level and the effect of factor C is tested at the sub-plot level. The effect of the interaction between A and B is tested at the main-plot level and the effects of the interactions of factor C with the other two factors, including the third-order interaction, are tested at the sub-plot level. For example, if factor A has three levels and factors B and C have two levels there would be six

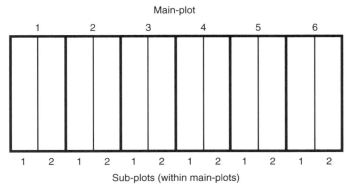

Main-plot

Figure 4.6   Schematic layout for a split-plot design with two treatment factors on the main-plots and one sub-plot treatment factor

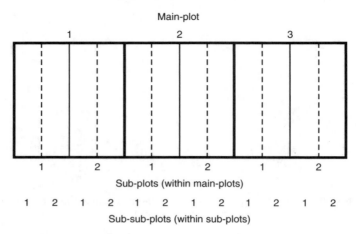

Figure 4.7   Schematic layout for a split-split-plot design

(3×2) main-plots and two sub-plots in each main-plot, as illustrated in Figure 4.6. The thick lines indicate the perimeters of the main-plots and the fine lines indicate the perimeters of the sub-plots.

A further way of arranging the three treatment factors would be to arrange the levels of treatment factor A on the main-plots. The levels of treatment factor B could be arranged on the sub-plots. Each sub-plot could be split into sub-sub-plots and the levels of treatment factor C would be randomized onto the sub-sub-plots. This would be called a SPLIT-SPLIT-PLOT design. For example, if factor A has three levels and factors B and C have two levels there would be three main-plots, two sub-plots in each main-plot and two sub-sub-plots in each sub-plot as illustrated in Figure 4.7. The thick lines indicate the perimeters of the main-plots, the fine lines indicate the perimeters of the sub-plots and the dashed lines indicate the sub-sub-plots.

From these examples it can be seen that the power of the tests of the various effects changes according to the design selected. The selection of the main- and sub-plot factors will be determined by the objectives of the study although practical restrictions may also dictate the choice. There are obvious extensions of these three designs using four or more treatment factors but these can quickly get very complex.

As with all the designs discussed so far, any of the blocking structures may be used. Since these designs can become very large, completely randomized and Latin square designs are rarely suitable.

4.3.4.2   Hypotheses

There is one null hypothesis for each of the main effects and one for each of the interaction effects. All these will be tested simultaneously. Examples of hypotheses are shown below. Main effect null hypotheses will be of the following form:

$$H_0 : C_k = \mu_k - \mu = 0 \qquad \text{for } k = 1 \ldots c$$

All treatment factor C population mean values are equal.

First-order interaction effect null hypotheses will be of the following form:

$H_0: BC_{jk} = \mu_{jk} - \mu = 0$ for $j = 1 \ldots b;\ k = 1 \ldots c$

All BC interaction population mean values are equal.

Second-order interaction effect null hypotheses will be of the form:

$H_0: BCD_{jkl} = \mu_{jkl} - \mu = 0$ for $j = 1 \ldots b;\ k = 1 \ldots c;\ l = 1 \ldots d$

All BCD interaction population mean values are equal.

There is only one possible alternative hypothesis for each null hypothesis. For those illustrated they will be as follows:

$H_1: C_k = \mu_k - \mu \neq 0$ for at least one k

$H_1: BC_{jk} = \mu_{jk} - \mu \neq 0$ for at least one j or k

$H_1: BCD_{jkl} = \mu_{jkl} - \mu \neq 0$ for at least one j or k or l

At least one of the treatment factor C population mean values is not equal to the overall mean.
At least one of the BC interaction population mean values is not equal to the overall mean.
At least one of the BCD interaction population mean values is not equal to the overall mean.

Higher-order interactions will each have their own null hypothesis and alternative hypothesis which will be in similar formats.

4.3.4.3 Randomization

The method of randomization depends on the design of the experiment. Main-plot treatment factors or treatment factor combinations will be randomized as previously described as the appropriate method is determined by the blocking structure. However, within each main-plot, the sub-plot treatment factors or treatment factor combinations are randomly allocated to the sub-plots. This technique can be extended if there are further 'splits'. The complete randomization process is a multi-stage process with as many stages as there are strata in the final analysis of variance table. At each stage the treatment factor levels or treatment combinations applied to that stratum are randomly allocated to the units in that stratum.

Example Consider a three-factor split-split-plot in two randomized blocks with one replicate of all treatment combinations in each block. Factor A has three levels and is the main-plot factor. Factor B is the sub-plot factor with two levels. Factor C has three levels and is the sub-sub-plot factor. There are a total of 18 treatment combinations and 36 experiment units.

For each block the three main-plot treatments are randomly allocated to the three main-plots according to the randomized blocks randomization procedure. Then within each main-plot the two sub-plot treatments are randomly allocated to the sub-plots as per the randomized blocks randomization procedure. The final stage is to use, for each sub-plot, the randomized blocks randomization procedure to randomly allocate the three sub-sub-plot treatments to the sub-sub-plots. The complete randomization is shown below.

Step	Block 1																		
1	Main-plot	1						2						3					
1	Sub-plot	1			2			1			2			1			2		
1	Sub-sub-plot	1	2	3	1	2	3	1	2	3	1	2	3	1	2	3	1	2	3
	Plot number	1	2	3	4	5	6	7	8	9	10	11	12	13	14	15	16	17	18
2 and 3	Random no.	44						81						69					
4	Rank	1						3						2					
5 and 6	Main-plot treatment	a1						a3						a2					
2 and 3	Random no.	88			35			68			60			22			75		
4	Rank	2			1			2			1			1			2		
5 and 6	Sub-plot treatment	b2			b1			b2			b1			b1			b2		
2 and 3	Random no.	15	94	31	50	63	31	2	41	51	97	33	70	86	97	27	22	8	6
4	Rank	1	3	2	2	3	1	1	2	3	3	1	2	2	3	1	3	2	1
5 and 6	Sub-sub-plot treatment	c1	c3	c2	c2	c3	c1	c1	c2	c3	c3	c1	c2	c2	c3	c1	c3	c2	c1

Step	Block 2																		
1	Main-plot	1						2						3					
1	Sub-plot	1			2			1			2			1			2		
1	Sub-sub-plot	1	2	3	1	2	3	1	2	3	1	2	3	1	2	3	1	2	3
	Plot number	19	20	21	22	23	24	25	26	27	28	29	30	31	32	33	34	35	36
2 and 3	Random no.	26						32						25					
4	Rank	2						3						1					
5 and 6	Main-plot treatment	a2						a3						a1					
2 and 3	Random no.	87			77			82			53			99			68		
4	Rank	2			1			2			1			2			1		
5 and 6	Sub-plot treatment	b2			b1			b2			b1			b2			b1		
2 and 3	Random no.	19	66	95	66	45	75	27	25	65	10	84	26	34	41	3	20	25	92
4	Rank	1	2	3	2	1	3	2	1	3	1	3	2	2	3	1	1	2	3
5 and 6	Sub-sub-plot treatment	c1	c2	c3	c2	c1	c3	c2	c1	c3	c1	c3	c2	c2	c3	c1	c1	c2	c3

4.3.4.4 Advantages

Treatment factors that require different amounts of experiment material can be tested in the same experiment. Interaction terms can be estimated. Many main effects can be investigated in an experiment provided no interaction terms are significant.

4.3.4.5 Disadvantages

Experiments can become complex. High-order interactions are difficult to interpret. Main-plot treatment effects are estimated with less precision than sub-plot treatment effects, which in turn are estimated with less precision than sub-sub-plot treatment effects and so on. Equal replication of all treatment factor combinations is essential. Estimation techniques can be used if there are relatively few missing values.

Although these designs are mentioned in many statistical texts as an extension of the two-treatment factor split-plot design, usually only brief details are given. Two texts that provide slightly more information are Montgomery (1991) who presents a good description of these designs with model formulae, and Cochran and Cox (1957) who provide details on a variety of these designs.

4.3.4.6 Example: a three-factor split-plot design

Description of experiment   The treatments to be used in this experiment are potato tubers at two physiological ages, $A1$ and $A2$, nitrogen treatments to be applied at six different rates, $N1$ to $N6$, and two different planting dates, $P1$ and $P2$, for the tubers. Using six different levels of nitrogen application will permit an investigation into the effects of physiological age and planting date on the crop response to increasing rates of nitrogen application that may lead to the identification of an optimum rate of nitrogen for use with these ages and sowing dates. The different physiological ages of the tubers are achieved by storing the tubers under different temperature and light conditions. It is impractical to plant very small plots of potatoes but nitrogen can be applied to small plots. It makes sense, therefore, to put the treatments involving the actual potato tubers on the main-plots of a split-plot design and put the nitrogen treatments on the sub-plots of the design. The factorial combination of the two treatment factors, planting date and physiological age, are randomized on the main-plots of the experiment with each of the four main-plots being split into six sub-plots for the nitrogen treatments. The main effects of interest are the effects of physiological age and its interaction with nitrogen on total tuber yield so it is a little unfortunate that the physiological age treatments cannot be applied to the sub-plots. However, past experience has shown that four replicates should be sufficient in order to detect the required size of yield differences between the two physiological ages and so a larger difference could be detected between the first-order interaction mean values. Each replicate of the treatment combinations is positioned in a separate block of 24 plots. The experiment is a three-factor split-plot design with two main-plot factors and a single sub-plot factor. The treatment factors are planting, age and nitrogen; the treatment factor levels are the two planting dates, $A1$ and $A2$, the two physiological ages, $P1$ and $P2$, and the six rates of nitrogen, $N1$ to $N6$.

Specific objective   To determine the effect on potato yield of altering the physiological age of the tuber and the planting date, and of using increasing rates of nitrogen application.

Hypotheses   The three main effect null hypotheses are as follows:

There is no difference between the two physiological age mean potato yields.
There is no difference between the two planting date mean potato yields.
There are no differences in the six nitrogen rate mean potato yields.

The three first-order interaction effect null hypotheses are as follows:

There are no differences in the four physiological age/planting date interaction mean potato yields.
There are no differences in the 12 physiological age/nitrogen rate interaction mean potato yields.

There are no differences in the 12 planting date/nitrogen rate interaction mean potato yields.

The second-order interaction null hypothesis is as follows:

There are no differences in the 24 physiological age/planting date/nitrogen rate interaction mean potato yields.

The alternative hypotheses are as follows:

The two physiological age mean potato yields are different.
The two planting date mean potato yields are different.
At least one of the six nitrogen rate mean potato yields is different from the others.
At least one of the four physiological age/planting date interaction mean potato yields is different from the others.
At least one of the 12 physiological age/nitrogen rate interaction mean potato yields is different from the others.

Table 4.7   Treatment randomization plan for the potato trial, a three-factor split-plot design with a single main-plot factor

Block 3	Main-plot				Sub-plot			
Units	Plot no.	Random no.	Rank	Treatment	Plot no.	Random no.	Rank	Treatment
49					1	976	6	N6
50					2	177	2	N2
51					3	161	1	N1
52	1	55	1	A1P1	4	742	5	N5
53					5	730	4	N5
54					6	222	3	N3
55					1	975	5	N5
56					2	113	2	N2
57					3	356	4	N4
58					4	43	1	N1
59	2	698	3	A1P2	5	996	6	N6
60					6	178	3	N3
61					1	38	1	N1
62					2	295	3	N3
63					3	642	5	N5
64	3	219	2	A2P1	4	623	4	N4
65					5	698	6	N6
66					6	269	2	N2
67					1	721	4	N4
68					2	167	1	N1
69					3	751	5	N5
70	4	703	4	A2P2	4	696	3	N3
71					5	941	6	N6
72					6	214	2	N2

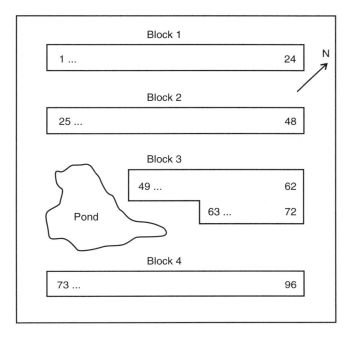

Figure 4.8   Outline site plan for the three-factor split-plot in randomized blocks experiment described in Section 4.3.4.6

At least one of the 12 planting date/nitrogen rate interaction mean potato yields is different from the others.

At least one of the 24 physiological age/planting date/nitrogen rate interaction mean potato yields is different from the others.

Randomization   The randomization is shown in Table 4.7 for block 3 only. Blocks 1, 2 and 4 will be randomized in a similar manner. The main-plot treatments are numbered from 1 to 4 respectively as first physiological age at first planting date, second physiological age at first planting date, first physiological age at second planting date and second physiological age at second planting date.

Site plan   The outline site plan is given in Figure 4.8.

4.3.5   Factorial plus control designs

Keypoints

- Factorial plus control designs use a factorial arrangement of treatments with an additional treatment.
- Any of the blocking structures can be used.
- Missing values need to be estimated to maintain the balance of the design.

4.3.5.1 Design

The FACTORIAL PLUS CONTROL design is a factorial design with an additional single treatment, and is considered in the same way as a factorial experiment. The full list of treatment combinations has two parts. One part consists of a fully balanced factorial design. The other part is a single treatment that is not combined with all the other treatment combinations. This single treatment is often an untreated control treatment but it need not always be so.

The factorial arrangement can be multi-factor. As with the other factorial treatment structures, this one can be applied to any of the three block structures. There can be more than one plot receiving the control treatment in a replicate. Cochran and Cox (1957) and Genstat 5 Committee (1993) both describe a factorial plus control experiment. Cochran and Cox illustrate the analysis ignoring the treatment structure and the Genstat 5 Committee present a more complete analysis that takes account of the factorial treatment structure.

4.3.5.2 Hypotheses

There is one null hypothesis for the control effect, one for each of the main effects and one for each of the interaction effects. All these will be tested simultaneously. Examples of hypotheses are shown below.

The control effect null hypothesis will be as follows:

$$H_0: \text{CONTROL}_i = \mu_i - \mu = 0 \qquad \text{for } i = 1, 2$$

There is no difference between the control treatment mean and the overall mean of all treatments including the control.

The main effect null hypotheses will be of the following form:

$$H_0: C_k = \mu_k - \mu_T = 0 \qquad \text{for } k = 1 \ldots c$$

All treatment factor C population mean values are equal.

The interaction effect null hypotheses will be of the following form:

$$H_0: BC_{jk} = \mu_{jk} - \mu_T = 0 \qquad \text{for } j = 1 \ldots b; \, k = 1 \ldots c$$

All BC interaction population mean values are equal.

There is only one possible alternative hypothesis for each null hypothesis. For those illustrated they will be as follows:

$$H_1: \text{CONTROL}_i = \mu_i - \mu \neq 0 \qquad \text{for at least one } i$$
$$H_1: C_k = \mu_k - \mu_T \neq 0 \qquad \text{for at least one } k$$
$$H_1: BC_{jk} = \mu_{jk} - \mu_T \neq 0 \qquad \text{for at least one } j \text{ or } k$$

The control treatment mean is not equal to the overall mean.
At least one of the treatment factor C population mean values is not equal to the overall treatment mean.
At least one of the BC interaction population mean values is not equal to the overall treatment mean.

4.3.5.3   Randomization

The treatment factor combinations and control treatment(s) are randomized to the experiment units in the same way as for a factorial experiment and in accordance with the block structure (Cochran and Cox, 1957).

4.3.5.4   Advantages

Many main effects can be compared against a control.

4.3.5.5   Disadvantages

As with all factorial designs the experiments can become very large and complex. The additional control treatment can cause problems in interpreting the results. Estimation techniques can be used if there are relatively few missing values.

4.3.5.6   Example: a two-way factorial plus control design

Description of experiment   The experiment is designed to investigate the residue effects of nitrogen fertilizer applications to a potato crop on the yield and uptake of the subsequent cereal crop. Poultry manure is applied at three different rates ($R1$, $R2$ and $R3$) and in the autumn and spring ($T1$ and $T2$) to a potato crop. In addition to the various harvest measurements of the following cereal crop, a variety of soil measurements are to be measured after the potato crop is harvested. These measurements will provide information about the levels of residue left in the soil by the potato crop after the different fertilizer applications. It is also of interest to know what the residue levels would be if there had been no fertilizer applied and so an untreated control treatment (C) is to be included in the experiment for comparison. Thus there are a total of seven treatments in the experiment. From previous experiments it is thought that three replicates should suffice. A total of 21 experiment units are therefore required. The field in which the experiment is to be run had been used in an experiment the previous year although it is not thought that this would have any effect on this experiment. However, it is not possible to fit 21 plots either into the part of the field that had not been used in the previous experiment or into the part that had been used. Consequently, just in case there is an effect due to the previous cropping, this experiment is to be blocked according to the previous cropping. Two of the blocks are on areas of land used in the previous experiment and one is on an area not previously used. The design is therefore a two-way factorial plus control in three randomized blocks; the treatment factors are rate and timing, with the treatment factor levels being the actual rates and times of application.

Specific objective   To assess the effect of mineral fertilizer nitrogen applied to the previous year's potato crop on the nitrogen uptake of the following year's cereal crop.

Hypotheses   There are four null hypotheses as follows:

There is no difference in control mean nitrogen uptake and the overall manure treatment mean nitrogen uptake.
There is no difference in mean nitrogen uptake of the two timings of manure application.
There are no differences in mean nitrogen uptake of the three rates of manure application.
There are no differences in the six interaction mean nitrogen uptake values.

Table 4.8   Treatment randomization plan for the nitrogen fertilizer experiment, a factorial plus control
in randomized blocks design

Block 1

Plot	1	2	3	4	5	6	7
Random no.	72	284	15	879	960	365	546
Rank	5	6	1	3	4	7	2
Treatment	$t2r1$	$t2r2$	C	$t1r2$	$t1r3$	$t2r3$	$t1r1$

Block 2

Plot	8	9	10	11	12	13	14
Random no.	761	599	570	48	677	413	572
Rank	4	2	6	1	3	5	7
Treatment	$t1r3$	$t1r1$	$t2r2$	C	$t1r2$	$t2r1$	$t2r3$

Block 3

Plot	15	16	17	18	19	20	21
Random no.	913	821	788	606	343	806	751
Rank	4	3	7	5	1	2	6
Treatment	$t1r3$	$t1r2$	$t2r3$	$t2r1$	C	$t1r1$	$t2r2$

There are four alternative hypotheses as follows:

There is a difference in control mean nitrogen uptake and the treatment mean nitrogen uptake.
There is a difference in two timings of manure application mean nitrogen uptake values.
At least one of the mean nitrogen uptake values for the three manure application rates is different from the rest.
At least one of the six interaction mean nitrogen uptake values is different from the rest.

Randomization   The treatments are combined to create a single treatment factor which is then randomly allocated to the plots as per the randomized blocks design. Each treatment is allocated to the plot assigned the same random number rank. The final randomization is presented in Table 4.8.

Treatment number 1: the untreated control treatment, C
Treatment number 2: the first timing and lowest application rate, $t1r1$
Treatment number 3: the first timing and middle application rate, $t1r2$
Treatment number 4: the first timing and highest application rate, $t1r3$
Treatment number 5: the second timing and lowest application rate, $t2r1$
Treatment number 6: the second timing and middle application rate, $t2r2$
Treatment number 7: the second timing and highest application rate, $t2r3$

Site plan   The outline site plan is given in Figure 4.9.

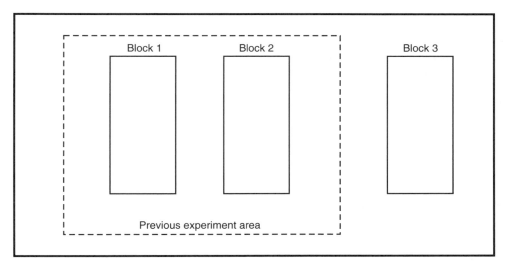

Figure 4.9   Outline site plan for the factorial plus control in randomized blocks experiment described
in Section 4.3.5.6

4.3.6   Summary of multi-treatment factor designs

- *Overall*: Interactions estimated; can use any blocking structure; missing values need to be estimated.
- *Factorial*: estimates all effects with equal precision; all treatments on same size units.
- *Split-plot*: estimates main-plot effects with less precision; treatments can have different size units.
- *Factorial plus control*: compares treatments with a control; more complex interpretation.

4.4   SOME OTHER DESIGNS

In some instances the above designs may not be suitable for the experiment in mind. If this is due to an over-complicated treatment structure running several smaller designs will usually be preferable. If the designs are still not suitable, possibly due to practical reasons, more complex designs are available, some of which are discussed below. Only brief descriptions are provided and further details of these and other designs can be found in other texts.

All the designs mentioned previously can be combined in various ways to produce much more complex designs. This can be undertaken to enable many treatments to be compared in one large complex experiment. If several smaller simpler experiments were conducted which incorporated all the required treatments, statistical analysis and interpretation would be much simpler. Practical problems are probably less likely to occur in a smaller experiment but if they do the remaining experiments will be totally unaffected. If problems occur in the larger experiment, it may have an effect on all the results. Before embarking on a very complex experiment it should confirmed that a valid statistical analysis can be undertaken. There are cases where more complex designs may be very useful and necessary to achieve the desired result. If the designs are carefully thought out and planned before any practical work is started they can be run successfully and be problem free. The actual analysis may be more complex mathematically but the interpretation can be as easy as the more straightforward designs. For example, 2^n and 3^n designs are very useful in the early stages of experimental work when there are many factors to be investigated and it is not known which factors are of importance. Also Latin square designs, including Graeco-Latin squares and partial squares, are invaluable designs when resources (such as number of animals) are very limited, particularly for experiments involving a lot of measurement recording, such as required in 24 hour monitoring. These designs need few experiment units (animals) but gain replication by allowing the units to be their own replicates.

Brief details of some more complex designs other than those already discussed are given below. These cover standard designs that allow incomplete replication of treatments in a block and increase the level of replication in Latin squares.

4.4.1   Repeated Latin square designs

In a single Latin square design the number of replicates per treatment is the same as the number of treatments. It is frequently the case that more replicates than this will be required to achieve the necessary precision. By repeating the basic Latin square the required precision can be obtained by increasing the number of treatment replicates. Each of the squares is constructed as described in Section 4.2.3.

The rows and columns of each square can be used to control variability but the squares themselves can also be used to account for sources of variability. The extent to which they do this will depend on exactly how the squares are constructed. Each square will contain the same treatments, but may contain different experiment units in either the rows or the columns, or different experiment units in both the rows and columns, or may use the same experiment units in all squares. For example, consider an experiment investigating the effects of vitamin supplements to a sheep's usual diet. The experiment could be designed with batches of feed being the column effect and the individual sheep being the row effect. If the same sheep were used in each repetition of the square with the same batches of feed then the repetitions of the square are being run with common rows and common columns. Alternatively different sheep could be used in each repetition but the same batches of feed used. These squares would have different rows and common columns. Squares having different columns and common rows would use the same set of animals in each square but different batches of feed in each square. These different situations will influence the results obtained from analysis. A clear illustration of the analysis of variance for these situations is given in Montgomery (1991), and Box, Hunter and Hunter (1978) give worked examples of each of the different forms of replication.

4.4.2 2^n and 3^n factorial designs

These designs are particularly useful to investigate the effects of a large number of treatment factors. They are most commonly used at the initial stages of an investigation and are a special case of factorial designs. Each treatment factor is only included at two or three levels so the design is limited to detecting whether an effect between the two factor levels is linear or not. The designs are usually performed when it would use too many resources to include the treatment factors at more levels and the objective would be limited to investigating which of the factors had an effect.

The designs can get very large; for example, a single replicate of a 2^5 experiment would need 32 experiment units. It is for this reason that single replicates or even partial replicates of these designs are often performed. In a single replicate experiment it is not possible to obtain an estimate of the true experiment error so there is nothing to test the other effects against. However, when considering factorial designs, high-order interactions can have little practical meaning and if they are assumed to be negligible then these high-order interactions can be combined to form a residual mean square against which the other treatment effects can be tested. This idea can be extended so that only part of a replicate of the design is run. In these designs only a subset of the treatment effects can be estimated as others are confounded. The confounded effects are combined to form a residual term so that the effects which are not confounded can be tested. There are various procedures used to design these experiments so that a useful subset of treatment effects can be tested.

Box, Hunter and Hunter (1978) give a brief description of 2^n designs with a number of worked examples. Montgomery (1991) devotes four chapters to this topic, mentioning a variety of situations from fully replicated designs to designs with single replication and confounding. Cochran and Cox (1957) cover the plans and analysis of a variety of 2^n and 3^n designs.

4.4.3   Balanced incomplete blocks designs

It is not always possible to run a randomized block design for various practical reasons. It could be that a block of sufficient homogeneous units could not be found, or that time or lack of labour does not allow the fully randomized block experiment to be performed. For example, in an experiment to investigate the amount of gas produced by eight different substances the bottles of substances need to be incubated in a water bath, but if the water bath only has the capacity to hold four bottles at a time a complete replicate cannot be incubated together. The problem could be solved by omitting some of the substances from the treatment list, which may mean that the objective of the experiment could not fully be met. The experiment could be run in two stages but this assumes that putting the bottles in the water bath at different times will produce similar results. However, the problem could be solved by using a balanced incomplete blocks design.

In incomplete blocks designs the number of treatments exceeds the number of experiment units in a block. So if the bath in the above example is regarded as a block of four experiment units a complete replicate of all eight treatments can be run in two blocks (or baths). An incomplete blocks design is referred to as balanced if each pair of treatments occurs the same number of times in the experiment, where a pair of treatments is defined as any two treatments occurring together in the same block. Continuing with the same example, there are a total of 28 pairs with eight treatments. In order to have each pair occurring the same number of times in the experiment there must be 28 or a multiple of 28 experiment units in the whole experiment. There are restrictions on the number of treatments, blocks and replicates that can be used in order to achieve balanced designs. In this example seven replicates, involving 14 blocks and a total of 56 experiment units, must be run in order to achieve the required balance. A pair of treatments will occur three times in the experiment. Guidance on the various combinations of treatment numbers, numbers of blocks and the number of replicates permissible and standard designs can be found in texts and tables. Box, Hunter and Hunter (1978) and Cochran and Cox (1957) give a selection of different size designs and both texts provide worked examples. Cochran and Cox provide a comprehensive coverage of balanced incomplete blocks designs. Montgomery (1991) provides a more gentle introduction with only a brief description of the general design, but does concentrate on the calculations involved.

4.4.4 Cross-over designs

Cross-over designs are experiments in which each experiment unit receives each treatment in sequence. They are particularly useful when there is known variation between experiment units, i.e. some units will perform better than others. In these instances it is desirable to ensure that a treatment is tested equally on the better and worse units, which is achieved by allowing each unit to receive each treatment. In the simplest of these designs there are two treatments, A and B. The experiment units would be randomly assigned so that half received treatment A followed by treatment B and the other half received treatment B followed by treatment A. The designs can be used with any number of treatments but the number of replicates must be a multiple of the number of treatments.

In these designs there could be some effect of the previous treatment applied, i.e. the results from one treatment are affected by the treatment which preceded it. This problem can be resolved by allowing a treatment-free period between the application of treatments. However, this may not always be possible; for example, in a dairy study it may be necessary to complete the whole experiment in one lactation, in which case time will not permit a rest period, or in a feeding trial animals would still need to be fed between treatments. In these instances it is still possible to run these experiments as the analysis can be adjusted to estimate these carry-over effects. In some experiments the objective might even be to estimate the carry-over effects; for example, is the milk yield of a cow better if it is fed diet A in early lactation followed by diet B in late lactation or vice versa. Montgomery (1991) gives a brief description of these designs. Cochran and Cox (1957) provide a more detailed clear description with an example of the statistical analysis.

REFERENCES

Box GEP, Hunter WG and Hunter JS (1978) *Statistics for Experimenters: An Introduction to Design, Data Analysis, and Model Building*. John Wiley, New York, Chapters 9 and 10, Appendices 8B, 8C and 8D.

Cochran WG and Cox GN (1957) *Experimental Designs*, second edition. John Wiley, New York, Chapters 3–7, 9–11.

Fisher RA and Yates F (1963) *Statistical Tables for Biological, Agricultural and Medical Research*, sixth edition. Oliver and Boyd, Edinburgh.

Genstat 5 Committee (1993) *Genstat 5 Release 3 Reference Manual*. Oxford University Press, Oxford, Chapter 9.

Marriott FHC (1990) *A Dictionary of Statistical Terms*, fifth edition. Addison-Wesley Longman.

Montgomery DG (1991) *Design and Analysis of Experiments*, third edition. John Wiley, New York, Chapters 3, 5–7, 9–12, 14.

Snedecor GW and Cochran WG (1989) *Statistical Methods*, eighth edition. Iowa State University Press, Ames, Chapter 16.

Steel RGD and Torrie JH (1980) *Principles and Procedures of Statistics: A Biometrical Approach*, second edition. McGraw-Hill, New York, Chapters 7, 9, 12, 15 and 16.

CHAPTER 5

Data Entry and Exploration*

'Statistics is concerned with collecting, analysing and interpreting data in the best way possible' (Chatfield, 1995). All three aspects are equally important. If data are badly collected and recorded they are worse than useless and the best statistical analysis of 'incorrect' data is not worth interpreting. The importance of data quality cannot be stressed too much. Data should be verified to check that the data recorded are those that were collected and validated to ensure that the data make sense. The adage 'garbage in, garbage out' is applicable to data and should never be far from the experimenter's mind.

Graphs are an ideal way of communicating detailed information about the data set. They can often help with the interpretation of information contained in the data but care must always be taken as they may also distort the 'true' picture. Graphs are useful tools for becoming familiar with a data set. The use of carefully chosen tables and descriptive statistics are fundamental to the initial stages of exploratory analysis. It is important to realize which form of data a variable is as this can determine or influence the statistical techniques that can be used to analyse the data. Data should be examined in this way before formal statistical procedures are employed.

5.1 DATA ENTRY

Keypoints

- Make back-up copies of data and keep them secure.
- Hard copy data must be durable and legible.

Data are expensive to collect so once the time and resources have been spent in their collection it is sensible to take all necessary precautions to ensure that they are not 'lost' or distorted. This involves selecting an appropriate long-term storage medium and taking additional copies, known as back-up copies, of important or large data sets. Whatever medium is used the data set should be clearly identified, for example with a unique reference number associating it with a particular trial and the year to which the data refer. Back-up copies of data, regardless of the medium used, should be securely stored in a different location to the original data.

In most cases storage will be in electronic form. All files held on mainframe and minicomputers are usually regularly backed-up by operating staff with no involvement required by the user. Local area networks (LANs) should have a procedure for storing and backing-up data. If PC floppy disks are used for storage it is usually up to the user to take back-up copies. It should also be remembered that floppy disks have a limited shelf-life of

---

* Note: unless otherwise stated, all data are provided by ADAS, with permission.

approximately two years. If it is necessary to store data on the hard drive of a PC then they should be regularly backed-up on to floppy disks or to tape or other medium to prevent data loss through machine failure. Another advantage of storing data in computer format is that it is easily available for data manipulation or analysis at a later date. Data manipulation is much more difficult if the only copy of the data is stored on paper. Paper storage must be legible, durable and easy to photocopy. Ink fades with time, paper gets torn and the ink on fax paper very quickly fades, all faxes (other than plain-paper faxes) should be photocopied if they are required for any length of time.

When entering data into a computer data file remember to save the data regularly. Power failures and computer crashes are not unknown and will result in the loss of all data entered since the last save. Think of how much time you are willing to spend re-entering data if the editing should be lost. If you are only prepared to lose 5 minutes worth of data entry then save your file every 5 minutes. If you are happy risking having to re-type longer periods of data entry then save less frequently. In all events if you have to leave the machine at any point in the data entry stage for whatever time period then save the data before you leave.

5.2 DATA

Keypoints

- Qualitative data have no numerical ordering.
- Quantitative data can be placed in numerical order.

All data can be classified as being qualitative or quantitative data. *Qualitative data* are data 'whose values cannot be put in any numerical order' (Clarke and Cook, 1992). The values for these data can either be numerical or text and are used simply to classify an object, person or characteristic into two or more mutually exclusive groups. Mathematical operations, such as addition or multiplication, cannot be performed on these data. One example of this type of data is the sex of animals. Numerical values can be assigned to the different groups for ease of data handling. For example, the value 1 could be assigned to males and the value 2 assigned to females but there is no natural ordering to the values and the average is nonsensical. The classification of apples, according to colour say, would be a qualitative variable. The variable could hold the values 1, 2, 3, etc., indicating the different colours or it could hold a word describing the colour, such as green, yellow and red.

Quantitative data are data 'whose values can be put in numerical order' (Clarke and Cook, 1992). Birth weight, condition scores, and number of eggs laid per day by hens are all examples of quantitative data. Mathematical operations can be performed on such data though the results may not always be meaningful.

5.2.1 Data types

Keypoints

- Discrete data have a finite set of possible values.
- Continuous data have an infinite set of possible values.

As well as being classified as either qualitative or quantitative, data can be further specified as being of a particular type. There are two main types of data: discrete and continuous.

These types are not coincident with the data classification mentioned previously. Quantitative data can be discrete or continuous but qualitative data can only be discrete.

5.2.1.1 Discrete data

Discrete data are data that have 'a finite number of possible values' from a range of values (Clarke and Cook, 1992). The values do not necessarily have to be integer values. The set of positive whole numbers above zero and the number of lambs born to a ewe are both examples of discrete data.

A special form of discrete data is *binary data*. These can take only two values. For example, a variable indicating whether weeds are present or absent in a plot of land could have the values 1 (for present) or 2 (for absent). The sex of a pre-weaned calf is binary as it can only be male or female. These would be binary variables as there are only two possible values the variable can hold.

5.2.1.2 Continuous data

Continuous data are data that 'may take all values within a given range' of values (Clarke and Cook, 1992). The weight of an animal, the yield of a crop, and the quantity of pesticide in a sample of water are all examples of continuous data. Since it is not possible to record a value to an infinite number of significant figures or decimal places, the values of a continuous variable will actually be recorded to a restricted number of decimal places. Provided there are 10 or more distinct values the variable can still be thought of as continuous (Chatfield, 1995). The data should be recorded to the same level of precision as that at which they were measured, so this will be determined by practical considerations. For example, if scales can record a pig's weight to two decimal places then the weight would be recorded as 80.08 kg, whereas in reality the weight might be 80.07893458579843567 kg if it were possible to measure such a weight. Condition scores recorded to the nearest 0.5 with a range of 0 to 10 inclusive may strictly speaking be discrete data as they are not taking all possible values within the range (they can only take values ending in .0 or .5 within the range) but if there are more than 10 distinct scores they could be considered continuous data. Percentage disease cover data where the percentage area covered by disease is recorded to the nearest 5% is another example of discrete data that could be considered continuous providing there are sufficient distinct values.

5.2.2   Scales of measurement

Keypoints

- There are four scales of measurement: nominal, ordinal, interval and ratio.
- The scale of measurement is determined by the experiment material and objectives.
- Different scales of measurement can involve different statistical techniques.

Data have been described previously as qualitative or quantitative, and falling into two types, discrete and continuous. This, however, is not the end of the story because there are differences in the way in which numerical values can be assigned to the elements in a data variable. There are four scales of measurement: nominal, ordinal, interval and ratio.

The scale of measurement used to record data helps determine the statistical tests that are appropriate for analysing the data. Techniques that are suitable for analysing nominal or ordinal data may not be appropriate for analysing data recorded on an interval scale. The scale of measurement should be determined at the outset and is usually determined by the experiment material, experiment objectives and limitations of the measuring equipment. A thorough and detailed discussion of the different scales of measurement can be found in Siegal and Castellan (1988). These scales of measurement are also covered by Chatfield (1995) but much more briefly.

5.2.2.1 Nominal scale

On a nominal scale of measurement numbers are used to identify mutually exclusive categories. This is the weakest scale of measurement. For example, farm buildings could be classified by type (storage barn, tractor shed, cow shed, etc.) and a number assigned to each type (1, 2 and 3, etc.). There is no ordering to the classification. It makes no sense to perform mathematical operations such as addition or multiplication on the assigned numbers as they are used merely as symbols. This scale of measurement is qualitative and discrete.

5.2.2.2 Ordinal scale or ranks

The ordinal scale of measurement provides a little more detail than the nominal scale. Numbers are used to classify objects and characteristics and there is a fixed order to the categories. In other words, each category will always have the same neighbours. The inherent order of the numbers assigned to each category indicates the 'pecking order' of the objects or characteristics and so must match the natural ordering of the categories. For example, size grades of potatoes can be assigned numerals ordered according to the order of size, with the smallest number being assigned to the smallest size grade and the largest number being assigned to the largest size grade. Scores, such as body condition scores, are also an example of data on this scale of measurement. Mathematical operations such as addition and multiplication are nonsensical as the actual value of the assigned number has no meaning other than to indicate the order of the categories.

5.2.2.3 Interval scale

Data recorded on an interval scale are continuous data which have a natural numerical order 'where there are equal differences between successive integers but where the zero point is arbitrary' (Chatfield, 1995). Unlike the previous scales of measurement, the differences between numbers on this scale have some meaning; the ratio of any two differences is independent of the actual unit of measurement used. Temperature is measured on an interval scale. The 'zero' is arbitrary, either 0 or 32 depending on the units, Celsius or Fahrenheit, and the ratio of any two *intervals* on the Fahrenheit scale is the same as the corresponding ratio on the Celsius scale. This scale of measurement is quantitative. Mathematical operations can be sensibly performed.

5.2.2.4 Ratio scale

The ratio scale is the most complex and informative scale of measurement. Ratio data are continuous data that have a natural numerical order 'where it is possible to compare the relative magnitude of the values as well as the differences' (Chatfield, 1995). A ratio scale has all the characteristics of an interval scale but the zero is absolute and the ratio of any two *values* measured in one set of units is identical to the corresponding ratio in another set of units. For example, a weight of zero is zero whether it is measured in grams, kilograms or tonnes, and a weight in pounds that is double a second weight in pounds is also double if it was measured in grams. Another example of the ratio scale is logarithmic data. Logarithmic values are measured to a base, and the two most commonly used bases are base 10 and base e (natural logs). Both these logarithmic scales ($\log_{10}$ and natural log) satisfy the requirements of equal ratios of differences for corresponding intervals and equal ratios of values, and have an absolute zero. The difference between the ratio and interval scales tends to be of little importance from a practical viewpoint.

5.3   DATA CHECKING

Keypoints

- Data must be verified to check they have been correctly recorded.
- Data must be validated to check for credibility and consistency.
- Extreme values must be checked; they may be errors or outliers.

This section mentions some basic techniques that can be used to check data before undertaking any analysis. Only some of the most straightforward checks are covered. Further detailed coverage can be found in Chatfield (1995), which includes much practical advice.

The importance of data quality cannot be stressed too much. Poor quality data will lead to poor or invalid conclusions being drawn. It is vital that the quality of data is checked prior to any type of statistical analyses. It is also important to take note of any anomalies in the data and ensure these are documented so that others do not waste valuable time rechecking the data. Missing and outlying values should be noted and, if possible, reasons for them found or data corrected. Computers can record data to many decimal places, more than can accurately be measured. Note should be taken of the number of decimal places recorded for each variable and whether it is sensible or not.

The first stage is *data verification*, i.e. checking that the data recorded are those which were collected. Even before looking at the data set, potential problems can be found or sorted at the data entry stage. The person entering the data should be calling attention to any obvious mistakes or omissions on the original data sheets. Where possible, the data should be re-entered to verify the data. It is very easy to transpose two digits or misread hand-written forms.

Data validation is the process used to ensure that the data make sense. For example, birth weights greater than weaning weights would need to be checked as an animal is expected to gain weight after birth. Data should be checked for unexpected missing values and to make sure all variables expected in a data set are present, i.e. that the data set is as complete as expected. It is also important to check that the values in the data set are all physically feasible, e.g. a wheat crop yielding 156 tonnes per hectare would need to be questioned. The easiest way to detect such values is to look at the minimum and maximum values in the data set; if these values are feasible (in magnitude) then all values between them will also be feasible. It is possible to make these checks electronically and specify a range outside which values will be classed as questionable. The range must be carefully chosen otherwise errors will either be missed or too many values will be questioned. It is also necessary to check that values are consistent. This is particularly important when several values are measured over time; for example, if plant heights decreased during a trial or an animal's birth weight was greater than a weight several weeks into a trial it may indicate a problem with the data.

A very simple check of data can be made by printing the data out in neat columns and simply scanning the data. Shifts in the decimal place can be picked up in this manner and missing values are quite obvious. For example, 105 could have been mistakenly entered as 1.05 or 1.5. Suspect values may also be detected by eye from plots of the data. This approach will be impractical with large set of data but in such cases subsets could be examined to get an overall idea of data quality.

Any strange data values should be checked and such values may become apparent at any stage of data checking. The values may occur due to error in data entry, or may even have been incorrectly recorded. An *outlier* is 'a "wild" or extreme observation which does not appear to be consistent with the rest of the data' (Chatfield, 1995). Outliers may be genuine values or they may indicate errors. When an outlier has been detected it should not be ignored or deleted. A reason for its existence must be found. The first step is to go back to the original data recorded and check that the value in the data file is the one that was recorded. If it is not, the value should be replaced with the correct value. If it is not possible to check the values in this way but the value is known to be incorrect (e.g. if such a reading is physically impossible), the value should be changed to a missing value and this documented in such a way that anyone who examines the data is clear what has happened. If the suspect value is correct it must be retained. There are statistical techniques that can deal with outliers.

Many data exploration techniques are useful tools for data checking. These are described in more detail in the next section.

5.4   DATA EXPLORATION

Familiarization with the data is necessary before a formal analysis can be undertaken. Exploratory data analysis is the term used to describe the process of becoming familiar with a data set. This includes finding out whether data are discrete or continuous, what scale of measurement was used, and how precisely the data were recorded. If data are continuous then the distribution of the final digit recorded will be uniform. So, for example, if continuous data have been recorded to two decimal places then the digits 0 to 9 should appear approximately equally in the second decimal place. If there are too many zeros or fives it would indicate that the data have been rounded and hence recorded to more decimal places than appropriate for the true precision of the data. Exploratory data analysis is also used to get an initial idea about the distributions of the data and the types of more formal analysis that may be appropriate.

The way in which data were collected should be clearly understood as this can help determine the data type and hence which statistical techniques may be appropriate or not. It should also be known which variables were measured and which were derived. If a data entry error in a measured variable is corrected then the corresponding entry in any variables derived from the corrected measured variable must also be corrected. An experiment design may not have been randomized or may have been randomized differently from that expected, which can have a serious impact on the statistical analysis. Mean data may have been recorded rather than raw data. Is it possible to get back to the raw data? Have any calculations been performed on the original recorded observations? Are the calculations correct? If an experiment includes data from several sources then care should be taken; the same variable may have been measured with different precision or different measuring techniques may have been used at the different sources. If this has happened all data to be analysed as a single variable should be corrected to the level of least precision if there is no other way of resolving it. This is particularly important in multi-site studies or if data have been analysed by different laboratories.

5.4.1 Descriptive statistics

Descriptive statistics are useful at all stages of statistical analysis, from checking and summarizing the data to initial exploratory analysis through to the final presentation of results and conclusions. Using descriptive statistics involves the calculation of summary statistics, such as mean, minimum and maximum values, standard deviations, numbers of observations and missing values.

These statistics are used to describe the location, dispersion and shape of the populations from which the data are sampled. No one statistic will ever completely describe a set of data. To gain an insight into the information contained in the data the following would be useful:

– *Number of observations*: the larger the sample size, the more representative it is of the population.
– *Minimum and maximum*: help detect outliers and errors.
– *Measures of location*: give an indication of where the observations lie.
– *Measures of dispersion*: give an indication of the variability of the data.
– *Number of missing values*: gives an indication of the completeness of the data.

All the examples in the following two sections use fictitious data sets to illustrate the various calculations involved and other aspects of the statistics. Set 1 is of discrete data, set 2 is of continuous data, set 3 is of continuous data with an extreme value, and set 4 is of continuous data from two groups. Most, if not all, statistical texts will mention descriptive statistics in some form. Good descriptions of the basic measures of location and dispersion can be found in texts such as those by Clarke and Cook (1992), Sokal and Rohlf (1995) and Steel and Torrie (1980) along with details of any calculations and formulae involved.

5.4.1.1 Measures of location

Keypoints

- *Mode*: not all observations are used in the calculation; used with discrete data; it is not influenced by extreme values; can be used on skewed distributions.
- *Median*: uses middle observations in the calculation; usually used with discrete data; it is not influenced by extreme values; better than the arithmetic mean for skewed distributions.
- *Quartiles*: not all observations are used in the calculation; used with discrete and continuous data; they are not influenced by extreme values; can be used on skewed distributions.
- *Arithmetic mean*: all observations are used in the calculation; usually used on continuous data; it is influenced by extreme values; works best with symmetrical distributions.

It has been established that data are variable. It is useful to be able to give a single value that is representative of all individual values for the population being considered. Statistically, such a measure of location is called an average, of which there are several types. Averages indicate the point or location about which the individual observations cluster. Since it is rare to have data for a whole population, samples must be taken. The average of the sample is the best estimate of the average of the population. It is usual to record measures of location to the same number of decimal places as the original data. The following section concentrates on sample statistics as these are the most commonly used but the same methodology is used for population statistics although the symbols used in the formulae differ slightly.

Mode   A measure of location that is not affected by extreme values is the mode. The *mode* of a group of observations is the value of the most frequently occurring observation. Hence, this measure of location is most useful for count data. In some cases there may be more than one mode and these data sets are known as multimodal. The bimodal distribution which has two modes is a special case of the multimodal distribution.

   For continuous data it is unlikely that a modal value will occur since it is unlikely that two observations will be exactly the same. Despite this, the mode can be used on continuous data. Class intervals are selected and the class which contains the highest frequency of data is known as the modal class. This gives a useful indication of the overall shape of the data distribution but is not a particularly useful summary in itself because of the rather subjective nature of selecting class-intervals. It may help highlight situations where discrete data are actually obtained but continuous data were expected. If a mode occurs in data recorded to a number of decimal places it may be an indication that there are only a few distinct values in the data set. The data would then have to be treated as discrete rather than continuous.

Example
Consider the following set of discrete data (set 1):

$$3, 5, 6, 8, 1, 6, 6, 4, 1, 2, 5, 2, 6$$

The frequency of occurrence of the actual values is as follows:

Value	1	2	3	4	5	6	7	8
Frequency	2	2	1	1	2	4	0	1

Hence, the mode is 6.

Consider the following set of continuous data (set 2):

$$81.5, 69.9, 58.9, 74.0, 89.9, 54.8, 72.9, 73.9, 64.7, 64.4, 79.3, 83.1, 89.1, 78.3, 74.4$$

The frequency of observations in the following modal classes is indicated below:

Value	50.00–54.95	55.00–59.95	60.00–64.95	65.00–69.95	70.00–74.95	75.00–79.95	80.00–84.95	85.00–89.95
Frequency	1	1	2	1	4	2	2	2

Hence, the modal class is 70.00–74.95.

Median   This measure of location should be used for discrete data but can be used with continuous data, particularly when skewed. The median value uses only the values of the observations in the middle when the data are sorted in rank order thus the median value is not influenced by any extreme high or low values.

The median is particularly useful in situations where it may be impossible, uneconomic or too time-consuming to take all measurements. For example, it is often not cost-effective in laboratory measurements to try and detect quantities of substances below a given level; these measurements are simply recorded as 'below the level of detection'. Similar circumstances arise if it would take too many resources to count the number of weeds occurring in a plot above a given level, so such values would be recorded as above 100 plants per metre. This statistic is also widely used in behaviour studies where the measurement to be taken is the time before an animal performs a particular behavioural pattern. In such studies it would be too time-consuming, or even impossible, to wait for all animals to perform the required act but once a time has been recorded for 50% of the animals a median value can be calculated.

The *median* is the middle value of the observations when they are arranged in order of magnitude. By its definition the median is the value that has an equal number of observations on either side of it. This is obviously the value of the central observation when there are an odd number of observations. When there are an equal number of observations the median value is half the sum of the two middle observations. It is calculated as follows:

$$M = x_{(n+1)/2} \qquad \text{for } n \text{ is odd}$$

$$M = \frac{x_{n/2} + x_{1+n/2}}{2} \qquad \text{for } n \text{ is even}$$

where x_n is the value of the nth sample observation, M is the median, and n is the number of observations in the sample.

Example
Consider the following three data sets:

set 1: 3, 5, 6, 8, 1, 6, 6, 4, 1, 2, 5, 2, 6

Rank order of data = 1, 1, 2, 2, 3, 4, 5, 5, 6, 6, 6, 6, 8.
Number of observations, $n = 13$.
Median value = the value of the $(13+1)/2$th observation = the value of the 7th observation = 5.

set 2: 81.5, 69.9, 58.9, 74.0, 89.9, 54.8, 72.9, 73.9, 64.7, 64.4, 79.3, 83.1, 89.1, 78.3, 74.4

Rank order of data = 54.8, 58.9, 64.4, 64.7, 69.9, 72.9, 73.9, 74.0, 74.4, 78.3, 79.3, 81.5, 83.1, 89.1, 89.9.
Number of observations, $n = 15$.
Median value = the value of the $(15+1)/2$th observation = the value of the 8th observation = 74.0.

set 3: 11, 20, 18, 20, 90, 22, 28, 32, 46, 45, 27, 31

Rank order of data = 11, 18, 20, 20, 22, 27, 28, 31, 32, 45, 46, 90.
Number of observations, $n = 12$.
Median value = half the value of the sum of the (12/2)th and (12/2+1)th observation = half the value of the sum of the 6th and 7th observations = $(27 + 28)/2 = 28$.

This value is representative of the majority of the sample. It has not been influenced by the extreme value of 90. Since the data are skewed the median is the best measure of location. The median has been quoted to the same number of decimal places as the original data.

Quartiles   Quartiles are a measure of location that can be used on discrete or continuous data. These split a data set into four equal-sized parts. The *first quartile* or *lower quartile* is the value below which one-quarter of the sample observations lie. The *third quartile* or *upper quartile* is the value above which one-quarter of the sample observations lie. It follows that 50% of the data will lie between the lower and upper quartiles. The median value is the second quartile, the value above which, and below which, half the sample observations lie. These quartiles ignore the majority of the actual observed values and are not influenced by extreme values.

To calculate the quartiles the data first need to be sorted into ascending order. The first quartile is the value of the observation in the $(N+1)/4$th position and the third quartile is the value of the observation in the $3(N+1)/4$th position where N is the number of observations in the sample. If these do not result in whole numbers interpolation can be used to estimate the actual quartile.

Example
Consider the ranked data (set 2):

54.8, 58.9, 64.4, 64.7, 69.9, 72.9, 73.9, 74.0, 74.4, 78.3, 79.3, 81.5, 83.1, 89.1, 89.9

Number of observations, $n = 15$.
First quartile: the value of the (15+1)/4th observation = the value of the 4th observation = 64.7.
Third quartile: the value of the 3(15+1)/4th observation = the value of the 12th observation = 81.5.

Consider the ranked data (set 3):

11, 18, 20, 20, 22, 27, 28, 31, 32, 45, 46, 90

Number of observations, $n = 12$.
First quartile: the value of the (12+1)/4th observation = the value of the 3.25th observation. Therefore the value lies one-quarter of the way between the third and fourth ranked observations. The first quartile is the value of the 3rd observation + 0.25*(4th observation−3rd observation) = 20+0.25*(20−20) = 20.
Third quartile: the value of the 3(12+1)/4th observation = the value of the 9.75th observation. Therefore the value lies three-quarters of the way between the ninth and tenth ranked observations. The third quartile is the value of the 9th observation + 0.75*(10th observation−9th observation) = 32 + 0.75*(45−32) = 42.

Arithmetic mean   The most commonly thought of average value is the mean value or arithmetic mean. It is used extensively in statistical analysis and it uses all the actual observations in the data set. It can, however, be unduly influenced by one or more very extreme values. As such it is not always the most suitable average to use for skewed data. It is most appropriate for calculating the location of data that are sampled from symmetrical distributions. If the sample observations were symmetrical about the arithmetic mean value then the median value and the arithmetic mean would be the same.

The calculation of mean values is not appropriate for qualitative data. When data have obvious groupings the arithmetic mean value of all the data makes little sense although it can be suitable for each group separately. For example, the slaughter weights of a breed of female cattle are noticeably lighter than those for the same breed of male cattle. An arithmetic mean could be calculated for all the cattle but the results would make more sense if separate arithmetic means were calculated for male and female cattle.

The *arithmetic mean* is 'the sum of the observations divided by the number of observations' (Clarke and Cook, 1992). It is calculated as follows:

$$\bar{x} = \frac{\sum_{i=1}^{n} x_i}{n}$$

where x_i is the value of the ith individual sample observation; $\bar{x}$ is the sample arithmetic mean; and n is the number of observations in the sample.

Example
Consider the following three sets of data:

set 2: 81.5, 69.9, 58.9, 74.0, 89.9, 54.8, 72.9, 73.9, 64.7, 64.4, 79.3, 83.1, 89.1, 78.3, 74.4

Number of observations, $n = 15$.
Arithmetic mean value = sum of the observations/15 = 1109.1/15 = 73.9.

The arithmetic mean value and the median value are similar for these data indicating a symmetrical distribution.

set 3: 11, 20, 18, 20, 90, 22, 28, 32, 46, 45, 27, 31

Number of observations, $n = 12$.
Mean value = sum of the observations/12 = 390/12 = 33.

This value is quite different from the median value. It has been influenced by the extreme value of 90. Since the data are skewed the median is the best measure of location.

set 4: 14.8, 15.6, 15.2, 14.8, 13.5, 14.2, 15.3, 25.0, 23.3, 26.2, 27.5, 24.9, 24.6, 25.0, 26.1

Number of observations, $n = 15$.
Arithmetic mean value = sum of the observations/15 = 306.0/15 = 20.4.

This value is not representative of the sample which seems to form two groups. If there was a reason to believe the data came from two groups, with the first seven observations forming one group and the next eight forming the second group, it would be better to calculate the arithmetic mean of each group. The arithmetic mean for the first group is 14.8 and for the second group is 25.3.

5.4.1.2   Measures of dispersion

Keypoints

- *Range*: used for identifying outliers; used with discrete and continuous data; it is affected by extreme values; it increases with sample size.
- *Inter-quartile range*: it is the interval containing 50% of the number of observations; used with discrete and continuous data; it is not affected by extreme values; used with skewed distributions.
- *Standard deviation*: all observations are used in the calculation; used with continuous data; decreases with increasing sample size; should be recorded in same units as the data.
- *Standard error*: it takes several forms (mean or difference); all observations are used in the calculation; used with continuous data; should be recorded in same units as the data.
- *Confidence limits and intervals*; it is the interval containing true mean value with a stated certainty; used with continuous data; the interval reduces in size as sample size increases.
- *Coefficient of variation*: a measure of the spread of observations in relative terms; it is a ratio value with no units.

The measures of location described do not provide any information on how well the measure of location represents the sample observations. Measures of dispersion are used to describe the 'shape' of the observations. This is particularly important when trying to assess if the observations from different variables are from the same population when an idea of the extent to which the individual observations within a sample vary is required. The observations may be clustered tightly together, they may spread over a very wide range of values or they may form clusters about different points. Measures of dispersion should be quoted in the same units as the original data.

Range   This is the simplest of the measures of dispersion and is usually associated with discrete data. If the appropriate measure of location is the median then the range can be quoted as the measure of dispersion. This measure of dispersion can be used to compare the variability of data sets of similar size, however since the range tends to increase with increasing sample size it is not directly applicable for samples of different sizes. It is mainly used as a tool for identifying errors and outlying values.

The *range* of a set of observations is 'the difference in value between the largest and the smallest observations in the set' (Clarke and Cook, 1992). One extreme value can greatly inflate the range, and since it only makes use of the maximum and minimum values it will only ever give an idea of the overall dispersion of a data set so tends to be uninformative for large data sets. For example, the range of two data sets could be the same but in one all observations may be tightly clustered together with one extreme value and in the other they may all be equally spaced.

Example
Consider the following two sets of data:

set 2 (in ranked order): 54.8, 58.9, 64.4, 64.7, 69.9, 72.9, 73.9, 74.0, 74.4, 78.3, 79.3, 81.5, 83.1, 89.1, 89.9

Minimum value $= 54.8$
Maximum value $= 89.9$
Range $= 89.9 - 54.8 = 35.1$

set 3 (in ranked order): 11, 18, 20, 20, 22, 27, 28, 31, 32, 45, 46, 90

Minimum value = 11
Maximum value = 90
Range = 90−11 = 79

This range has been influenced by the extreme value of 90; without it the range would have been 35.

Inter-quartile range   An alternative more robust measure of spread than the range is the inter-quartile range. This measure of dispersion is also associated with the median as a measure of location because it also uses the data in ranked order.

This is a useful measure of dispersion when the upper or lower limits of the sample values may be unbounded or undefined. For example, there is no upper limit to the length of time it takes an animal to perform a certain activity, such as head-butting. The inter-quartile range is less affected by extreme values than the range. Consequently it is a more useful measure of variability when the sample contains extreme values or is skewed.

The *inter-quartile range* is 'the difference between the upper and lower quartiles' (Clarke and Cook, 1992). Since quartiles split the ranked data set into four equal parts the inter-quartile range contains the central 50% of the observations.

Example
Consider the following two sets of data:

set 2 (in ranked order): 54.8, 58.9, 64.4, 64.7, 69.9, 72.9, 73.9, 74.0, 74.4, 78.3, 79.3, 81.5, 83.1, 89.1, 89.9

First quartile = 64.7
Third quartile = 81.5
Inter-quartile range = 81.5−64.7 = 16.8

set 3 (in ranked order): 11, 18, 20, 20, 22, 27, 28, 31, 32, 45, 46, 90

First quartile = 20
Third quartile = 42
Inter-quartile range = 42−20 = 22

The inter-quartile range has been less affected than the range by the extreme value.

Standard deviation   This is the measure of variability that most people are familiar with and is, probably, the most widely used measure of variability. The standard deviation is usually associated with the arithmetic mean and is really designed for continuous data and is not unduly affected by extreme values. It assumes that the mean of the sample is equal to the population mean and uses the observation values from the whole data set. It should be noted that the variance is the square of the standard deviation and is therefore not in the same units of measurement as the data. The variance should therefore never be used as a descriptive statistic.

As the number of observations increases, the measure of variability will decrease and the estimate of the population average (the sample mean) will be made with greater precision. Doubling the number of observations does not halve the standard deviation. It should be noted that reducing the standard deviation increases the precision of the population mean estimate, but it does not necessarily follow that the estimate is accurate as accuracy refers to the average value. A small standard deviation in comparison with the size of the mean value implies that the sample observations are clustered tightly about their mean value. A large standard deviation in comparison with the mean value implies that the observations are widely scattered about their mean value. Thus the standard deviation gives an indication of how the observations lie around the mean value.

The *standard deviation* of a set of observations is the positive square root of the mean of the squared deviations from the arithmetic mean value of the observations. The sample standard deviation is calculated as follows:

$$s = \sqrt{\frac{\sum_{i=1}^{n}(x_i - \bar{x})^2}{n - 1}}$$

where s is the sample standard deviation; x_i is the ith individual sample observation; $\bar{x}$ is the sample mean; and n is the number of observations in the sample.

Note: If using a scientific calculator to calculate the standard deviation of a set of numbers, the standard deviation is provided by the button marked σ_{n-1}. Be careful not to confuse this with the σ_n button which gives the population standard deviation.

A computationally easier way to calculate the sample standard deviation is as follows:

$$s = \sqrt{\frac{\sum_{i=1}^{n}x_i^2 - \frac{\left(\sum_{i=1}^{n}x_i\right)^2}{n}}{n - 1}}$$

Example
Consider the following data set (set 2):

81.5, 69.9, 58.9, 74.0, 89.9, 54.8, 72.9, 73.9, 64.7, 64.4, 79.3, 83.1, 89.1, 78.3, 74.4

Number of observations, $n = 15$

$$\sum_{i=1}^{15}x_i^2 = (81.5^2 + 69.9^2 + 58.9^2 + \ldots + 83.1^2 + 89.1^2 + 78.3^2 + 74.4^2) = 83\,466.75$$

$$\sum_{i=1}^{15}x_i = (81.5 + 69.9 + 58.9 + 74.0 + \ldots + 79.3 + 83.1 + 89.1 + 78.3 + 74.4) = 1109.1$$

$$\text{Sample standard deviation} = \sqrt{\frac{83\,466.75 - \frac{1109.1^2}{15}}{15 - 1}} = 10.21$$

Standard error   Standard deviations of various statistics, such as means and differences, are known as standard errors. They use all data observations and are used with continuous data. They can be unduly influenced by extreme values and decrease with increasing sample size for a given population. If the term standard error is used without any further explanation it is implied that the statistic referred to is the standard error of the mean; any other standard error would need further qualification. In practice, the most commonly used standard errors are the standard error of the mean and the standard error of the difference between two means.

More than one random sample can be selected from a population, leading to more than one estimate of the population mean value. Each sample will have its own mean value (the estimate of the population mean) and standard deviation. The different sample means will form their own distribution which is a normal distribution regardless of the original population distribution. This distribution of mean values will, of course, have its own mean value and standard deviation. The standard deviation of the distribution of mean values for samples of a given size, is called the *standard error of the mean*.

The above principles still apply when sampling from more than one population. In this instance there is more than one distribution of mean values; there will be one for each population. Comparisons can be made between two mean values to test whether or not their populations are the same. In other words, is the difference between two mean values zero? Since each of the two populations will have a distribution of means, there will also be a distribution of differences between two populations. The standard deviation of a distribution of differences for samples of a given size, is called the *standard error of the difference*.

The standard error of a single sample mean value is calculated as follows:

$$s_{\bar{x}} = \frac{s}{\sqrt{n}}$$

where $s_{\bar{x}}$ is the standard error of the mean; s is the standard deviation for the sample; and n is the number of observations in the sample.

The standard error of the difference between two sample mean values is calculated as follows:

$$s_{\bar{x}_1 - \bar{x}_2} = \sqrt{\frac{(n_1 - 1) s_1^2 + (n_2 - 1) s_2^2}{n_1 + n_2 - 2} \left(\frac{1}{n_1} + \frac{1}{n_2} \right)}$$

where $s_{\bar{x}_1 - \bar{x}_2}$ is the standard error of the difference between two sample mean values; s_1 is the standard deviation for the first sample; n_1 is the number of observations in the first sample; s_2 is the standard deviation for the second sample; n_2 is the number of observations in the second sample.

Example
Consider the following two data sets:

set 2: 81.5, 69.9, 58.9, 74.0, 89.9, 54.8, 72.9, 73.9, 64.7, 64.4, 79.3, 83.1, 89.1, 78.3, 74.4
Number of observations, $n = 15$
Sample standard deviation $= 10.21$

set 4: 14.8, 15.6, 15.2, 14.8, 13.5, 14.2, 15.3, 25.0, 23.3, 26.2, 27.5, 24.9, 24.6, 25.0, 26.1
Number of observations, $n = 15$
Sample standard deviation $= 5.54$

The standard error of the sample mean for set 2 $= \dfrac{10.21}{\sqrt{15}} = 2.64$

The standard error of the sample mean for set 4 $= \dfrac{5.54}{\sqrt{15}} = 1.43$

The standard error of the difference between the two sample mean values

$$= \sqrt{\frac{(15-1)10.21^2 + (15-1)5.54^2}{15+15-2}\left(\frac{1}{15}+\frac{1}{15}\right)} = 3.00$$

Confidence limits and intervals   The various sample statistics such as mean values and standard deviations are estimates of population parameters (population mean and standard deviation). True population values for these parameters will not be known unless data are collected for the whole population. It is, however, possible to estimate the reliability of the calculated sample statistics used to estimate the population parameters by setting confidence limits on the statistic estimate.

Confidence limits define the upper and lower bounds of an interval which has a given probability of covering the true population mean value. For example, if the lower and upper 95% confidence limits were represented by L_1 and L_u respectively then there is a 0.95 chance that the true population mean will lie between L_1 and L_u. The *confidence interval* is the distance between the upper and lower confidence limits. Therefore, a 95% confidence interval states that there is a 95% chance that the interval contains the true population mean value.

Different degrees of certainty can be applied to the confidence limits or interval. The higher the degree of certainty required, the wider the confidence interval becomes. Thus, for a given data set and sample size the confidence interval will be wider to be, say, 99% certain than to be 95% certain that the mean value lies within the interval. This provides greater confidence that the population mean lies between the confidence limits but there is less certainty about the actual true value of the mean because the limits are further apart.

The width of the confidence interval can be reduced by reducing the standard error of the mean. This could be done by increasing the sample size. Clearly this should be considered at the design stage.

Confidence limits for a normal distribution are symmetrical about the sample mean. They can, however, be asymmetrical about the sample mean for other distributions. Care should therefore be taken if the data have been transformed or if the distribution is known to be asymmetrical.

The formulae for calculating the $100(1-\alpha)\%$ confidence limit about a mean value are as follows:

$$\text{lower } 100(1-\alpha)\% \text{ confidence limit, } L_1 = \bar{x} - t_{n-1,\alpha/2}s_{\bar{x}}$$

$$\text{upper } 100(1-\alpha)\% \text{ confidence limit, } L_u = \bar{x} + t_{n-1,\alpha/2}s_{\bar{x}}$$

where

L_1 is the lower confidence limit,
L_u is the upper confidence limit,
$\bar{x}$ is the sample mean,
$t_{n-1,\alpha/2}$ is the t-statistic with $n-1$ degrees of freedom,
α is the probability level, i.e. 0.05 for 95% probability,

$s_{\bar{x}}$ is the standard error of the mean, and

n is the number of observations in the sample.

Example

Consider the following data set:

set 2: 81.5, 69.9, 58.9, 74.0, 89.9, 54.8, 72.9, 73.9, 64.7, 64.4, 79.3, 83.1, 89.1, 78.3, 74.4

The sample mean value $= 73.9$

The standard error of the sample mean $= 2.64$

The 95% *t*-statistic with 14 degrees of freedom $= 2.145$

The lower 95% confidence limit $= 73.9 - 2.145 * 2.64 = 68.2$

The upper 95% confidence limit $= 73.9 + 2.145 * 2.64 = 79.6$

The 95% confidence interval $= 79.6 - 68.2 = 11.4$

Hence there is a 95% chance that the interval, 68.2 to 79.6, contains the true population mean value, estimated as 73.9.

Coefficient of variation   All measures of dispersion so far described are in the same units as the individual observations. For some purposes it is more useful to measure the spread in relative terms. The *coefficient of variation* is a measure of sample variability in relation to its location. It is the ratio of the standard deviation and the absolute value of the sample mean. It can be expressed as a percentage. This measure can be used to compare directly the variability of samples of different sizes.

A small coefficient of variation implies that the sample observations are clustered tightly about their mean value. A large coefficient of variation implies that the observations are widely scattered about their mean value. An advantage of the coefficient of variation is that it is independent of the units in which the variable is measured provided that the scales of measurement begin at zero. It is calculated as follows:

$$\%CV = \frac{s}{|\bar{x}|} 100$$

where %CV is the percentage coefficient of variation; s is the sample standard deviation; and $\bar{x}$ is the sample mean.

Example

Consider the following data set:

set 2: 81.5, 69.9, 58.9, 74.0, 89.9, 54.8, 72.9, 73.9, 64.7, 64.4, 79.3, 83.1, 89.1, 78.3, 74.4

The sample mean value $= 73.9$

The sample standard deviation $= 10.21$

The %CV $= (10.21/73.9)100 = 13.82$

5.4.2   Graphical presentation

Keypoints

- *Bar-charts*: show the shape of the data distribution; can be used with discrete data; can be used to detect extreme values.

- *Histograms*: show the shape of the data distribution; can be used with continuous data; can be used to detect extreme values.
- *Box-and-whisker plots*: summarize some descriptive statistics of the data; can be used with quantitative data; can be used to detect extreme values.
- *Scatter plots*: identify possible relationships between data-variables; can be used with discrete and continuous data; can detect clusters of data; can be used to detect unusual observations.

'A picture is worth a thousand words.' This may be an old cliché but a glance at newspapers or magazines will confirm that a picture can convey a large quantity of information in a way that can be easily received. This is also true in statistical analysis: it is far easier to look at a graph than to try to understand a page of figures or read a lengthy description. A graph or figure can display information clearly and precisely, and the use of colour, shading, scales and symbols can be used to highlight a particular message in the data. However, graphs can also be presented in such a way that the facts are distorted and the same features that highlight messages can also hide them. The intention should always be to show the information contained in the data in a clear and concise manner. It may sometimes be better to present the information in tabular form.

Graphs can be used to show apparent relationships between variables, or give an idea of the distribution of a data set, and can also be very useful in indicating outliers or other anomalies in the data. They can also help to promote visual comparison between two or more groups of objects, either on the one graph or across a series of graphs. Care must be taken in interpreting the information within a graph as inappropriate use of axes may distort the 'true' picture.

Care should be taken to make a graph clear and easily understood. All axes should be clearly labelled with the name of the variable and the units of measurement used. If a scale does not start at zero this should be clearly indicated. A scale break is usually indicated on the axes by a zigzag (~~) in the axes or by physically breaking the line with parallel sloping lines. It should be noted that many computer packages do not offer the facility of showing a scale break in such a way, and if presenting graphics from such a package care should be taken to indicate if a scale has not been started at zero. It is also important to take considerable care when choosing the scale of the axes. If the scale is too small information can be suppressed but if it is too large information can be distorted by exaggerating small differences. A useful rule of thumb when plotting linear trends is to choose the scales so that the slope of the line is between 30° and 45°.

The way in which a graph is presented and the type of graph presented should be given careful consideration. One form of a graph can give a very different picture to the same data presented in a different form. A graph in which the points are connected will have a different meaning from the same graph with points not connected. Symbols such as asterisks and dots can be used but if more than one symbol is used they should be easily distinguishable. Points can be connected, if appropriate, with dotted or solid lines, curves or straight lines, for example, and again it is important that these can be easily identified from each other.

All of the above points and more are mentioned by Chatfield (1995) in his excellent discussion of the pros and cons of graphics. Some of the most useful types of graphs are considered in more detail in the following section. Clear descriptions of the main characteristics and features of bar-charts, histograms and stem-and-leaf plots are given in the introductory texts by Clarke and Cook (1992) and Clarke (1994). The essential differences between these different graphical presentations are also identified.

Bar-charts   A bar-chart is used to look at the distribution of a variable of discrete data. Bar-charts provide information on the symmetry of the distribution and the range of values, and may also help in identifying extreme values or outliers.

Each bar represents a distinct value and the height of a bar is proportional to the frequency of that value. The bar can be of any width, although it can be misleading if some bars are wider than others so it is usually best to keep all bars the same width. Each bar should be separated from the next bar by a gap as it is this that indicates that each bar is distinct and the data are discrete.

The simplest way to plot a bar-chart of a discrete variable is to plot actual frequency against the range of possible values the variable takes, then the height of a bar represents the actual frequency. If the actual frequencies are used, direct comparisons between bar-charts can only be made if each chart uses variables with the same number of total observations. For example, a frequency of five has a different interpretation if it comes from a total of 10 observations than if it was obtained from 100 observations. Using relative frequencies enables comparisons to be made with bar-charts of variables with differing numbers of total observations. In this case a bar height is the proportion of observations which have that value.

Comparisons between bar-charts are simplified by plotting more than one variable per chart. Each variable should have a similar range of values. The set of bars for each distinct value can be stacked, overlaid or adjacent, depending on the information to be presented.

Figure 5.1 shows a bar-chart of calving difficulty scores (integer values between 0 and 5 inclusive) for a group of cows.

Figure 5.1   Bar-chart of calving difficulty scores

Histograms   A histogram is the continuous data variable equivalent of a bar-chart. It provides information on the symmetry of the distribution and the range of values and may also help in identifying extreme values or outliers.

The range of variable values is split into groups and each group is called a class-interval. The histogram differs from the bar-chart in that a rectangle is drawn above each class-interval. In the simplest case, the *area* of a rectangle, not its height, is the frequency of observations falling in the class-interval and no gap is left between the rectangles. If the class-interval is of unit width then the bar height will equal the area and will therefore represent the actual frequency. Class-intervals are not necessarily of equal width and in these cases the vertical axis should be labelled 'frequency density'.

As with bar-charts, if the actual frequencies are used direct comparisons between histograms can only be made if each histogram uses variables with the same number of total observations. The class-intervals of all histograms being compared must be equal. In practice, provided the class-intervals are all equal, frequency can be plotted on the vertical axis without affecting the overall shape of the plot. This is the case with most statistical packages. Care must obviously be taken with the vertical axis labelling when dealing with histograms to avoid confusion when making comparisons.

Using relative frequencies enables comparisons to be made with histograms of variables with differing numbers of total observations. In this case the area of the rectangle is the proportion of observations which lie in the class-interval.

The class-intervals chosen for a histogram should be mutually exclusive so every data point should fall into one and only one interval. This is achieved by ensuring that the intervals cover the whole range of data and by expressing intervals to one more decimal place or significant figure than was used for the data. Each rectangle is centred over the mid-point of its interval. Either the mid-points or the interval ends are identified below the rectangles. Sometimes the end class-intervals will be open-ended because it is undesirable to include several intervals to cover one or two extreme values. In these cases the end intervals cannot be plotted on the histogram. For example, with the ages of cattle the class intervals could be <5 days, $6-10$ days, $11-20$ days, and 21 days or more.

There are no hard and fast rules for deciding on the number of class-intervals to use but there are a few rules of thumb. One of the most frequently quoted of these is to use 10 class-intervals, and another is to use $\sqrt{n}$ where n is the total number of observations (Clarke, 1994). Alternatively, the data can lend themselves to 'obvious' intervals, e.g. interval widths of 1 may be useful for data ranging between the values 0 to 15. Whichever 'rule' is used it is important not to choose intervals that are too narrow or too broad as this affects the shape of the histogram. If the intervals are too fine then they will contain few observations and the histogram will appear very irregular, whilst if they are too wide only a very coarse representation of the sample distribution will be achieved.

Figure 5.2 shows a histogram of percent mildew on a leaf (integer values between 0 to 100 inclusive).

Box-and-whisker plots   Box-and-whisker plots can be used to display the main features of a quantitative variable. They are particularly useful tools when comparing groups of data of roughly equal size as differences in variability and location are clearly illustrated by the plots.

There are variations on exactly what is plotted but the basic idea is always the same. Most commonly these plots consist of a box, with the length of the box equal to the inter-quartile

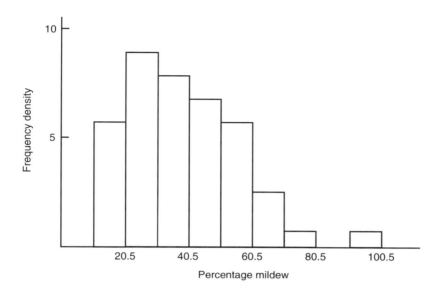

Figure 5.2   Histogram of percentage mildew on a leaf (%)

range and the sides of the box at the lower and upper quartiles so 50% of the observations lie within the box. From both sides of the box whiskers extend to the minimum and maximum values. The position of the median within the box is marked by a line. If the median lies halfway along the length of the box this is indicative of a symmetrical distribution. There is no meaning attached to the height of the box. The actual values of the descriptive statistics used to define the plot are written on the plot or indicated using a scale below.

The whiskers can stop in a variety of different places and it is important to understand which method has been used. Some plots have the whiskers extending to the largest and smallest values within 1.5 inter-quartile ranges of the sides of the box. Individual data points lying outside these whiskers are separately identified on the plot, with values outside three inter-quartile ranges being highlighted, thus providing an indication of extreme values or possible outliers.

Comparisons between variables can easily be made by plotting one or more box-and-whisker plots on the same graph. In this case a scale will be used below to indicate the values of the various descriptive statistics. An example of multiple box-and-whisker plots on the same graph can be found in Sokal and Rohlf (1995) using the form indicating extreme values.

Figure 5.3 shows a typical box-and-whisker plot.

Scatter plots   Scatter plots are plots of two variables, one against the other. Hence they can be used to identify potential relationships between variables although they do have other uses. The form of potential relationships between the variables, such as linear or quadratic, must be known before more formal analysis can be performed, which can only be done by plotting the data. These plots can also be used to detect clusters of observations and to assess how well a statistical model has described the data. Scatter plots may be presented in numerous ways according to the message they are intended to put across.

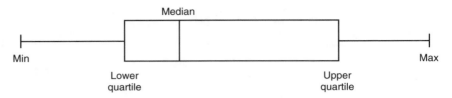

Figure 5.3   The components of a box-and-whisker plot

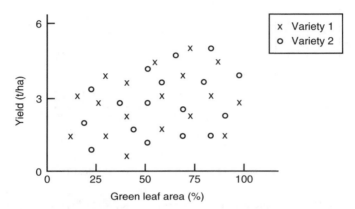

Figure 5.4   An example of a scatter plot of yield (t/ha) against green leaf area (%) (hypothetical data)

A useful technique is to identify groups in the plot by using different plotting symbols. For example, in a livestock trial looking at live weight gain of beef cattle, live weight gain could be plotted against initial weight. Two plotting symbols could be used, one to identify the females and one to identify the male cattle. This may help highlight any differences in live weight gain between male and female cattle.

Another useful technique is to join consecutive points. This is particularly useful when examining interaction effects between factors. Joining points is not always appropriate (e.g. with residual plots), and each case should be examined on its own merits.

Scatter plots may also help to identify any unusual observations, such as a single point away from a main cluster of points or a single point being apart from those lying on a line.

Several lines can be plotted on the one graph. These lines must all use the same scales on the vertical (y) axis and the horizontal (x) axis, and must be fully labelled. In some cases the lines may fall at the extremes of the axes and in these cases the scale may be broken. These scale breaks must *always* be indicated and this is usually by a zigzag (∿) on the axis. Scale breaks are more common on the y-axis and usually only occur on the x-axis at the origin.

Figure 5.4 illustrates a typical scatter plot of two groups.

Stem-and-leaf plots   Stem-and-leaf plots are similar to histograms in that they present a picture of the shape of the data distribution. The advantage they have over histograms is that they retain the actual values of the data observations and so can be used to calculate estimates of descriptive statistics such as the median and quartiles. Usually the integer part of the numbers form the stem of the plot and the decimal part forms the leaves.

A typical stem-and-leaf diagram is shown in Figure 5.5 (data values range between 0.3 and 4.8).

N = 50
Leaf unit = 1.0

```
0  3
0  57
1  44
1  589
2  11123334444
2  556666677888
3  00001233
3  6678
4  024
4  5678
```

Figure 5.5   A stem-and-leaf diagram (hypothetical data)

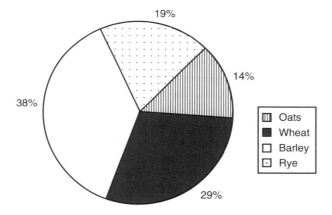

Figure 5.6   A pie chart (hypothetical data)

Pie charts   A pie chart is a circle indicating a total data set that is sliced into segments indicating proportions of groups within the data. A typical pie chart is shown in Figure 5.6. *3-D plots*   These are pseudo three-dimensional plots. They can take a variety of forms, such as three-dimensional histograms or scatter plots. A 3-D bar-chart is shown in Figure 5.7.

Multiple y-axes plots   Multiple y-axes plots are used for plotting two variables using different y-axes scales but the same x-axis. Types of plot can also be mixed, e.g. a histogram and a scatter plot. These plots will have a y-axis on the left of the x-axis and one on the right; both will be labelled. A typical multiple y-axes plot is shown in Figure 5.8.

Error bars   Error bars give an indication of the variability of the data. They can be plotted above histograms or around points on a scatter plot. These bars can be the prediction standard errors if a fitted line is drawn or simply standard errors of the means when mean values are plotted. It would have to be declared with the plot what the bars stood for. Figure 5.9 illustrates a scatter plot of mean values with error bars. The mean values are plotted with error bars showing plus and minus one standard error of the mean.

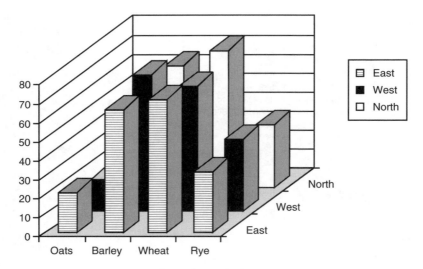

Figure 5.7   Hypothetical example of a 3-D bar-chart

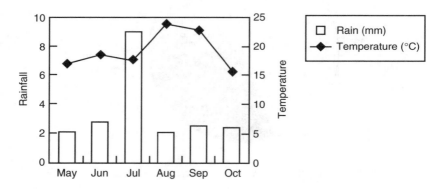

Figure 5.8   Monthly rainfall (mm) and temperature (°C) plotted on a multiple y-axis chart
(hypothetical data)

Others   There are many other ways of presenting data graphically and the above should not
be considered exhaustive.

5.4.3   Tables

Keypoints

- Tables aid interpretation of statistical analysis.
- They are useful for presenting results.
- Tables should be kept simple.

Tables are a useful way of presenting numerical or text data and statistics in a concise
way. They can help to show relationships between variables and also help to interpret

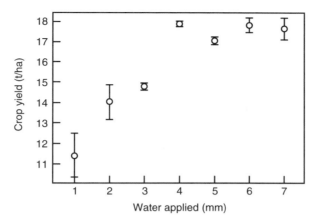

Figure 5.9   Relationship between crop yield (t/ha) and the amount of irrigation water applied (mm), with error bars

results. They can be used to present data summaries, mean values, standard deviations and number of observations, and they can also be used to present raw data. If more than one piece of information is to be presented in a single cell (e.g. the three summary statistics mentioned above could be listed on a line in a single table cell), then the table should be clearly annotated as to the order of the cell contents. It is not always easy to produce a clear table and considerable thought should be given to the following points when designing a table.

The complexity of a table will be determined, to a large extent, by the number of classification factors used. A *classification factor* groups similar units of a variable together. These groupings can then be used to define the number of rows and columns of a table, which in turn identify the cells of the table. The dimensions of a table directly correspond to the number of classification factors used to define the table. Two classification factors are required to define a two-way table, three are required to define a three-way table, and so on. When more than three classification factors are used, columns or rows of a factor will have to be nested within each other to present the information in a two-dimensional format, e.g. on paper or on a computer screen. As more classification factors are used the more complex the table will become and it will be harder to read and understand the table contents. Generally speaking, it becomes too confusing to read and understand a table with more than three or four dimensions. If necessary, one large table should be split into a series of smaller tables to present the information more clearly. Tables should be kept as simple as possible.

Spacing and layout of the table are important. Judicious use of blank rows and columns can help make a table easier to read. The order in which rows and columns are presented can aid interpretation. For example, if different columns hold information for different years then the columns should be ordered chronologically. It is easier to scan columns of a table than look across a row so it helps if tables have more rows than columns, although this is not always possible if information from a number of variables is being presented in the one table. If columns in a table are to be compared they should be positioned close together. Be wary of using computer output directly in a table as it can be very confusing and misleading to present too many decimal places in a table. Measures of location should be quoted to the number of significant figures that was measured. A standard practice is to quote variability measures to one more significant figure than measures of location.

Consider which, if any, summary statistics should be provided with the table. For example, averaging across columns would be nonsensical if the columns represent completely different variables. Whenever tables of averages are quoted the appropriate measures of variability and number of observations used should also be quoted in the same table. For example, if the table is simply summarizing the data then the standard deviation or standard error of the mean and the number of observations should be quoted with each mean value. If the table is being used to make comparisons between mean values then the standard error of the difference between two mean values would be more appropriate.

In reports and papers for publication, especially, tables should be understandable without having to refer to the text. The table title should be clear and self-explanatory, and if there is to be more than one table they should be numbered clearly also. The reader should not be forced to refer to the text to know the contents of the table. Rows and columns should be clearly and uniquely labelled, with the units of measurement if they differ from column to column or row to row. If the units are all the same then the table will look less cluttered if they are stated in the title. Any abbreviations or user terminology used in headings or as cell contents should be annotated in table footnotes. For example, '*' could be used to indicate a missing value or could indicate the level at which a test was significant, and 'na' could indicate that the recorded statistic was not applicable for that particular row and column combination. Usually all cells will be filled but if cells are left blank this should be explained, e.g. it could be another way of indicating missing data.

Both Clarke and Cook (1992) and Chatfield (1995) give useful lists of points to consider when constructing tables. An example of turning a 'bad' table into a 'good' table can also be found in Chatfield.

Example 1 The analysis presented in this example is from the experiment designed in Section 4.3.3.6 and analysed in Section 6.1.5.6. This experiment investigated the effect of seven nitrogen rates, N1 to N7, and four fallow treatments, A to D, on the specific weight of a winter wheat crop. The experiment is a two-factor split-plot design in four randomized blocks with fallow treatments on the main-plots and nitrogen treatments on the sub-plots. Table 5.1 presents a summary of the raw data.

Table 5.1   Summary table of specific weight data (kg/hl) of a winter wheat crop treated with different levels of nitrogen and fallow treatments

Nitrogen treatment	Fallow treatment			
	A	B	C	D
	Mean (SD) N	Mean (SD) N	Mean (SD) N	Mean (SD) N
N1	75.95 (1.292) 4	73.81 (1.517) 4	73.72 (0.999) 4	75.09 (0.309) 4
N2	75.07 (0.374) 4	73.45 (0.578) 4	74.00 (0.852) 4	75.53 (0.917) 4
N3	76.01 (0.938) 4	74.64 (0.406) 4	75.07 (1.152) 4	75.74 (1.133) 4
N4	77.09 (0.515) 4	76.61 (0.465) 4	75.90 (0.749) 4	77.61 (0.111) 4
N5	77.51 (0.298) 4	77.01 (0.805) 4	76.28 (0.926) 4	77.47 (0.512) 4
N6	77.95 (0.294) 4	77.00 (1.062) 4	76.16 (0.601) 4	77.78 (0.504) 4
N7	77.99 (0.459) 4	77.84 (0.772) 4	76.18 (0.726) 4	77.99 (0.529) 4

SD, Standard deviation; N, number of observations.

Table 5.2   Table of mean milk protein values (%) for heifers fed on two diets and housed in different environments

| Diet | Milk protein (%) | | | |
| | Housing | | Mean | SED (df) |
	H1	H2		
D1	3.34	3.23	3.28	0.060 (24)
D2	3.04	2.97	3.01	
Mean	3.19	3.10	3.14	
SED (df)		0.060 (24)		0.085 (24)

SED, Standard error of the difference; df, residual degrees of freedom.

Example 2   The data analysed in this example are from the experiment designed in Section 4.3.1.6 and analysed in Section 6.1.5.4. This experiment investigates the effect on protein of heifers' milk when fed with two diets (D1 and D2) and housed in two different environments (H1 and H2). The experiment uses nine randomized blocks with one complete replicate of the four treatments in each block. Table 5.2 presents a summary of the analysed data in a way that assists treatment comparisons.

REFERENCES

Chatfield C (1995) *Problem Solving: A Statistician's Guide*, second edition. Chapman and Hall, London, Chapters 1 and 6, Appendix 1.
Clarke GM (1994) *Statistics and Experimental Design: An Introduction for Biologists and Biochemists*, third edition. Edward Arnold, London, Chapters 1 and 2.
Clarke GM and Cooke D (1992) *A Basic Course in Statistics*, third edition. Edward Arnold, London, Chapters 1, 2 and 4.
Siegel S and Castellan NJ, Jr (1988) *Nonparametric Statistics for the Behavioral Sciences*, second edition. McGraw-Hill, New York, Chapter 3.
Sokal RR and Rohlf FJ (1995) *Biometry*, third edition. Freeman, New York, Chapters 4 and 7.
Steel RGD and Torrie JH (1980) *Principles and Procedures of Statistics: A Biometrical Approach*, second edition. McGraw-Hill, New York, Chapter 2.

CHAPTER 6

Analytical Techniques*

Any statistical analysis involving the use of hypothesis testing, such as analysis of variance or comparison of means, usually involves calculating a test statistic from the actual data. This is then compared with values published in statistical tables to determine whether the test is statistically significant or not; in other words, whether the null hypothesis is to be accepted or rejected. The degrees of freedom and the significance level determine which value to use from the statistical tables.

The following sections detail the hypotheses and general conclusions that can be inferred from a variety of techniques. Many different techniques exist that cover a wide variety of situations but only the most commonly used are covered here.

6.1 PARAMETRIC TECHNIQUES

Parametric statistical techniques make many assumptions about the nature of the populations from which observations or data were sampled. These techniques are used for estimating parameters and making tests of hypotheses concerning them. They can be used on all forms of quantitative data provided the assumptions being made are valid.

Generally speaking, the assumptions specify the form of the population distribution. For example, it may be assumed that the data were sampled from a normal distribution. Frequently made assumptions are that the errors are normally distributed with zero mean and common variance, and that the observations themselves are independent, normally distributed and randomly sampled. The conclusions made from parametric tests therefore contain the following qualifier: 'if the assumptions made regarding the distribution of the population are correct then it may be concluded that...' (Siegel and Castellan, 1988).

The interpretation of any results from parametric testing is made much easier because of the many assumptions that have been made. The drawback with making the assumptions is that the tests become relevant for only specific types of data. For example, if it is assumed that the data are normally distributed then the data must be measured on at least the interval scale.

Parametric techniques exist for analysing data from a variety of situations. The following sections cover one- and two-sample comparisons, analysis of variance and regression analysis. In many situations these techniques may suffice. Other more complex techniques are mentioned briefly in Chapter 7.

6.1.1 Comparison of one or two means

Keypoints

- *One-sample t-test*: compares a population mean against a specified value; uses continuous data; small data sets must have a normal distribution.

---

* Note: data throughout this chapter are provided with permission from ADAS.

- *Paired t-test*: compares the population mean difference of two related samples against zero; uses continuous data; small data sets must have a normal distribution.
- *Two-sample t-test*: compares the population means of two samples; uses continuous data; small data sets must have a normal distribution; the two populations should have the same variances

Note: t-tables can be presented in a variety of ways. Single-sided tables can present the upper or lower critical regions for single-sided tests only, two-sided tables present both the critical regions for two-sided tests only, and other tables present all this information together. When using single-sided tables to perform a two-sided test the probability at which the table is entered is halved. When using two-sided tables to perform a single-sided test the probability at which the table is entered is doubled.

6.1.1.1 One-sample t-test

The one-sample t-test is one of the most important tests of location whereby the mean value is compared with a specified value, usually zero. It is appropriate for data that are continuous and have been measured on the interval scale. The data must be randomly sampled. This test has several uses; for example, comparing a measured value against an industry standard, testing if a sample could have been taken from a population with a specified mean or testing the significance of parameter estimates in model fitting.

Assumptions The observations must be independent and randomly sampled. The observations must come from a normal distribution.

Limitations This test should not be used if the sample size is small unless the population from which the sample is drawn has a normal distribution. For large samples the test is robust against assumptions of normality. The observations must be measured on the interval scale at least.

Hypotheses

H_0: $\mu = \mu_0$ The null hypothesis, H_0, is that the population mean, μ, is equal to a specified value, μ_0.

There are three possible alternative hypotheses:

H_1: $\mu \neq \mu_0$ The population mean is not equal to the specified value. This is a two-tailed test.

H_1: $\mu > \mu_0$ The population mean is greater than the specified value. This is an upper one-tailed test.

H_1: $\mu < \mu_0$ The population mean is lower than the specified value. This is a lower one-tailed test.

Conclusions If the test was not significant then the null hypothesis would be accepted. The conclusion would be that there is no evidence to say that the population mean is not equal to the specified value provided that the test assumptions are valid. If the test was significant then the null hypothesis would be rejected. In a two-tailed test the conclusion would be that the population mean is not equal to the specified value provided that the test assumptions are valid. In a lower one-tailed test the conclusion would be that the population mean is less than the specified value provided that the test assumptions are valid. In an upper one-tailed test the conclusion would be that the population mean is greater than the specified value provided that the test assumptions are valid.

Formula   The formula for calculating the t-value to be used in the hypothesis test is as follows:

$$t = \frac{\bar{x} - \mu_0}{\sqrt{\dfrac{s^2}{n}}}$$

$$= \frac{\bar{x} - \mu_0}{s_{\bar{x}}}$$

where

t is the calculated t-value,
$\bar{x}$ is the sample mean value,
s^2 is the sample variance,
n is the total number of observations,
μ_0 is the specified value, and
$s_{\bar{x}}$ is the standard error of the mean.

A critical t-value is obtained from t-tables with $n-1$ degrees of freedom which defines the critical region. The test is rejected if the calculated value lies inside the critical region. The critical value selected depends on the alternative hypothesis, i.e. whether it is a one-tailed or two-tailed test.

Example   A batch of silage bales are to be tested against the industry standard of 8% for percentage dry matter loss to ensure storage methods are adequate. It is imperative that dry matter is no more than 8%, otherwise the bales have failed a quality test. The following dry matter values were recorded: 7.00 8.86 7.88 6.94 8.00 8.02 7.88 7.82.

H_0: $\mu = 8$
H_1: $\mu < 8$
The null hypothesis states the per cent dry matter loss mean is equal to 8 and the alternative hypothesis that it is less than 8.

Significance level, $\alpha = 0.025$
Degrees of freedom $= 7$
Critical region is '$t_{7,0.025} \leq -2.365$'
Sample mean value $= 7.38$
Sample standard deviation $= 0.536$
Calculated t-statistic $= -3.27$

The calculated t-statistic lies within the critical region and therefore the test is significant so the null hypothesis is rejected. If the assumptions of the test are valid the conclusion is that, at the 2.5% level of significance, the mean per cent dry matter loss of the population of bales is less than 8%.

6.1.1.2   Paired t-test

This is a useful test for determining if the means of two related samples are estimates from the same population. Related samples can arise in one of two ways. Each unit can be used as its own control so each unit results in two measurements; for example, a plot could receive a single treatment application and be sampled before and after the treatment application. The second way of achieving related samples is to match two samples so that they are as similar as possible with respect to any characteristic that may affect the variable being tested. For example, twin lambs would form one related sample and each could receive one of two treatments, or two cows that are expected to calve on the same day could form a related sample. Given that it is not always possible to know or to measure which characteristics may affect the variable being tested, it is better to form related samples using the first approach whenever possible.

Assumptions   The differences between the values for the two matched samples are independently drawn from a normal distribution. The observations must be selected randomly.

Limitations   This test should not be used if the sample size is small unless the population differences follow a normal distribution. For large samples the test is robust against assumptions of normality. The observations must be measured on the interval scale at least.

Hypotheses

H_0: $\mu_d = 0$ The null hypothesis, H_0, is that the population mean difference for the two related samples is zero. The population mean difference is the mean of the differences between the pairs of observations, μ_d.

There are three possible alternative hypotheses:

H_1: $\mu_d \neq 0$ The two population means are not equal. This is a two-tailed test.
H_1: $\mu_d > 0$ The first population mean is greater than the second population mean. This is an upper one-tailed test.
H_1: $\mu_d < 0$ The first population mean is lower than the second population mean. This is a lower one-tailed test.

Conclusions   If the test was not significant then the null hypothesis would be accepted. The conclusion would be that there is no evidence to say that the population means are not equal provided that the test assumptions are valid. If the test was significant then the null hypothesis would be rejected. In a two-tailed test the conclusion would be that the population means are not equal provided that the test assumptions are valid. In a one-tailed test the conclusion would be that the two population means differ in the direction tested provided that the test assumptions are valid.

Formula   The formula for calculating the t-value to be used in the hypothesis test is as follows:

$$t = \frac{\bar{d}}{\sqrt{\dfrac{s_d^2}{n}}}$$

$$= \frac{\bar{d}}{s_{\bar{d}}}$$

where

t is the calculated t-value,
$\bar{d}$ is the sample mean value of the differences between the paired observations,

s_d^2 is the sample variance of the calculated differences,
n is the total number of differences,
$s_{\bar{d}}$ is the standard error of the mean of the differences.

A critical t-value is obtained from t-tables with $n-1$ degrees of freedom which defines the critical region. The test is rejected if the calculated value lies inside the critical region. The critical value selected depends on the alternative hypothesis, i.e. whether it is a one-tailed or a two-tailed test.

Example   In order to compare two observers using a quadrat for assessing percentage ground cover of a particular species of weed, the quadrat was placed on the ground and each observer made an assessment from the same quadrat. This was repeated with 10 quadrat placements. The data collected are detailed in Table 6.1.

H_0: $\mu_d = 0$
H_1: $\mu_d \neq 0$
The null hypothesis states that there is no difference in percent ground cover as recorded by the two observers and the alternative hypothesis is that there is a difference.

Significance level, $\alpha = 0.05$
Degrees of freedom $= 9$
Critical region is '$t_{9,0.025} \leq -2.262$ and $t_{9,0.025} \geq 2.262$'
Mean difference in the samples, $\mu_d = 1.3$
Standard deviation of the difference $= 1.77$
Calculated t-statistic $= 2.33$

The calculated t-statistic lies within the critical region and therefore the test is significant at the 5% level so the null hypothesis is rejected. If the assumptions of the test are valid the conclusion is that the two observers are producing different results for percentage ground cover for a particular weed species.

6.1.1.3   Two-sample t-test

The two-sample t-test is a very robust and powerful test that is used to test for differences between two population mean values from unrelated samples. It is appropriate for data that are continuous and have been measured on the interval or ratio scale. The data for each sample must be randomly sampled. This test has several uses, e.g. comparing the mean yield for two varieties of wheat or comparing the mean weight gain of cattle on two different

Table 6.1   Percentage ground cover of a particular weed species as recorded by two observers for 10 quadrat placements (hypothetical data)

	Sample									
	1	2	3	4	5	6	7	8	9	10
Observer A	33	76	41	72	19	37	51	79	87	16
Observer B	34	79	41	70	21	40	51	80	88	20
Difference	1	3	0	−2	2	3	0	1	1	4

feeding regimes. It is not necessary for the two samples to be of equal size but they should come from populations with equal variance.

Assumptions The observations in each of the two samples must be independent and randomly sampled. The variances of each of the two populations being tested should be equal. The observations in each sample should come from normal distributions.

Limitations This test should not be used if the two population variances are different. If the sample size is small the two samples must be drawn from a normal distribution. For large samples the test is robust against assumptions of normality. The observations must be measured on the interval scale at least.

Hypotheses

H_0: $\mu_1 = \mu_2$ The null hypothesis, H_0, is that the two population means, μ_1 and μ_2, are equal.

There are three possible alternative hypotheses:

H_1: $\mu_1 \neq \mu_2$ The two population means are not equal. This is a two-tailed test.
H_1: $\mu_1 > \mu_2$ The first population mean is greater than the second population mean. This is an upper one-tailed test.
H_1: $\mu_1 < \mu_2$ The first population mean is lower than the second population mean. This is a lower one-tailed test.

Conclusions If the test was not significant then the null hypothesis would be accepted. The conclusion would be that there is no evidence to say that the two population mean values are not equal provided that the test assumptions are valid. If the test was significant then the null hypothesis would be rejected. In a two-tailed test the conclusion would be that the two population means are not equal provided that the test assumptions are valid. In a lower one-tailed test the conclusion would be that the first population mean is less than the second provided that the test assumptions are valid. In an upper one-tailed test the conclusion is that the second population mean value is greater than the first provided that the test assumptions are valid.

Formula The formula for calculating the t-value to be used in the hypothesis test is as follows:

$$t = \frac{\bar{x}_1 - \bar{x}_2}{\sqrt{\frac{(n_1 - 1)s_1^2 + (n_2 - 1)s_2^2}{n_1 + n_2 - 2}}\sqrt{\left(\frac{1}{n_1} + \frac{1}{n_2}\right)}} = \frac{\bar{x}_1 - \bar{x}_2}{s_{\bar{x}_1 - \bar{x}_2}}$$

where

t is the calculated t-value,
$\bar{x}_1$ is the sample mean value of the first sample,
s_1^2 is the sample variance of the observations in the first sample,
n_1 is the number of observations in the first sample,
$\bar{x}_2$ is the sample mean value of the second sample,

s_2^2 is the sample variance of the observations in the second sample,
n_2 is the number of observations in the second sample
$s_{\bar{x}_1 - \bar{x}_2}$ is the standard error of the difference of the mean values

A critical t-value is obtained from t-tables with $n_1 + n_2 - 2$ degrees of freedom which defines the critical region. The test is rejected if the calculated value lies inside the critical region. The critical value selected depends on the alternative hypothesis, i.e. whether it is a one-tailed or a two-tailed test.

Example   In an experiment looking at wireworm populations in grassland fields, one of the factors thought potentially to influence the presence or absence of wireworm populations was the soil pH value. From a total of 62 grassland fields, 38 were found to be infested and 24 were not infested. The soil pH value was assessed in each field: the average pH value for the infested fields was 6.3 with a standard deviation of 0.69, and the average value for the uninfested fields was 6.1 with a standard deviation of 0.61.

H_0: $\mu_i - \mu_u = 0$
H_1: $\mu_i - \mu_u \neq 0$
where μ_i is the population mean for the infested fields and μ_u is the population mean for the uninfested fields.
The null hypothesis states that there is no difference in the average soil pH value for infested and uninfested grassland fields and the alternative hypothesis is that there is a difference.

Significance level, $\alpha = 0.05$
Degrees of freedom $= 60$
Critical region is '$t_{18,0.025} \leq -2.000$ and $t_{18,0.025} \geq 2.000$'
Uninfested fields sample mean $= 6.3$
Uninfested fields sample standard deviation $= 0.69$
Infested fields sample mean $= 6.1$
Infested fields sample standard deviation $= 0.61$
Calculated t-statistic $= 0.941$

The calculated t-statistic lies outside the critical region and therefore the test is not significant at the 5% level so the null hypothesis is accepted. If the assumptions of the test are valid the conclusion is that there is no evidence to believe that the soil pH value in fields infested by wireworms and fields not infested by wireworms are different.
    Note: A formal test should have been carried out to test the assumption of equal population variances before performing the t-test. This is performed using an F-test which is described in Section 6.1.2.

6.1.2   Comparison of two variances

Keypoints

- *F-test*: compares two population variances; uses continuous data; the data must have normal distributions.

6.1.2.1 F-tests

An F-test is used to test the difference between two population variances and forms the basis of the tests in an analysis of variance. It is appropriate for data that are continuous and have been measured on the interval scale or ratio. The data for each sample must be randomly sampled and it is not necessary for the samples to be of equal size. This test is used to test the assumption of equal variance made when using a two-sample t-test.

Assumptions The observations must be independent and randomly sampled. The observations from the two samples must be normally distributed.

Limitations The observations must be measured on at least an interval scale. The test is not very robust against the violation of normality.

Hypotheses

$H_0: \sigma_1^2 = \sigma_2^2$ The null hypothesis, H_0 is that the two population variances, σ_1^2 and σ_2^2, are equal.

There are three possible alternative hypotheses:

$H_1: \sigma_1^2 \neq \sigma_2^2$ The two population variances are not equal. This is a two-tailed test.
$H_1: \sigma_1^2 > \sigma_2^2$ The first population variance is greater than the second population variance. This is an upper one-tailed test.
$H_1: \sigma_1^2 < \sigma_2^2$ The first population variance is lower than the second population variance. This is a lower one-tailed test.

Conclusions If the test was not significant then the null hypothesis would be accepted. The conclusion would be that there is no evidence to say that the population variances are not equal provided that the test assumptions are valid. If the test was significant then the null hypothesis would be rejected. In a two-tailed test the conclusion would be that the population variances are not equal provided that the test assumptions are valid. In a lower one-tailed test the conclusion would be that the first population variance is less than the second provided that the test assumptions are valid. In an upper one-tailed test the conclusion would be that the second population variance is larger than the first provided the test assumptions are valid.

Formula The formula for calculating the F-value to be used in the hypothesis test is as follows:

$$F = \frac{s_1^2}{s_2^2}$$

where

F is the calculated F-value,
s_1^2 is the sample variance calculated from the first sample,
s_2^2 is the sample variance calculated from the second sample,

The calculated F-value is tested against a critical F-value obtained from F-tables with (n_1-1) and (n_2-1) degrees of freedom. The exact value obtained depends on the alternative hypothesis, i.e. whether it is a one-tailed or two-tailed test. In practice, because F-tables usually only present the upper-tailed test critical values, it helps make the use of the tables simpler if the largest variance is put on the numerator and the smallest on the denominator. The alternative hypotheses used then are either the upper one-tailed or the two-tailed. If a one-tailed test is used the tables are entered at the stated probability level. If a two-tailed test is used the tables are entered at half the required probability.

Example   In the example used for the two-sample t-test concerning soil pH value in fields infested and not infested by wireworms, the variances of the two samples should have been tested to check that they are equal. This is done in the following example.

H_0: $\sigma_i^2 = \sigma_u^2$

H_1: $\sigma_i^2 \neq \sigma_u^2$

where σ_i^2 is the population variance of the infested fields and σ_u^2 is the population variance of the uninfested fields.

The null hypothesis is stating that there is no difference in the two population variances of soil pH for infested and uninfested fields and the alternative hypothesis is that there is a difference.

Significance level, $\alpha = 0.05$
Critical region is '$F_{9,9,0.025} \geq 2.0513$'
Uninfested fields degrees of freedom $= 23$
Uninfested fields sample standard deviation $= 0.69$
Infested fields degrees of freedom $= 37$
Infested fields sample standard deviation $= 0.61$
Calculated F-statistic $= 1.279$

The calculated F-statistic lies outside the critical region and therefore the test is not significant at the 5% level of significance so the null hypothesis is accepted. If the assumptions of the test are valid the conclusion is that there is no evidence to believe that the two population variances of the soil pH values of fields infested or not infested with wireworms are different. Therefore this assumption for the two-sample t-test was valid.

6.1.3   Analysis of variance: comparison of more than two means

Keypoints

- Observations are assumed to be mutually independent.
- Treatment variability is assumed to be constant.
- Errors are assumed to be independent and normally distributed.
- Several null hypotheses are tested simultaneously.

Each statistical test mentioned so far has had one null hypothesis statement. In the analysis of variance several null hypotheses can be tested simultaneously. The number of null hypotheses tested by a design is dictated by the design. For example, a completely randomized design with only one treatment factor will test the null hypothesis of 'no difference between the treatment population means' whereas a two-treatment factor completely randomized design can test the two null hypotheses of 'no difference between the treatment population means' and a further one of 'no difference between the interaction means'. All null hypotheses will test for 'no difference between population means'. There can only be one alternative hypothesis for each null hypothesis – the one that states that 'a difference exists between two or more treatment population means' – so all alternative hypotheses for mean values in the analysis of variance are two-tailed.

Essentially, the analysis of variance technique depends upon the partitioning of the degrees of freedom and the sum of squared deviations into a component called 'error' or 'residual' and a component or components which may be termed the treatment effect. The mean squares in an analysis of variance are simply the sum of squares divided by their respective degrees of freedom. The mean square for the 'error' component is frequently referred to as the 'background error variance' or 'error variance' or 'residual variance' and is a measure of the variation in response between identically treated units estimated across all treatments due to experiment error (see Section 3.4) and so the background error variance can be estimated from each treatment. If the null hypothesis of no treatment effect is true, the treatment effect mean square will equal the error mean square. If there is a real treatment effect then the variability between the treatment observations will be inflated and the treatment mean square will be greater than the error mean square. Thus the treatment sum of squares is influenced by both the treatment effect and the error term. The mean squares can be tested for similarity using an F-test as they are variance estimates. In the analysis of variance table the ratio of the treatment mean square to the error mean square is usually called the variance ratio and this value is compared against a critical value taken from F-tables. If the F-test is statistically significant it indicates there is a real difference in the treatment mean values because the treatment mean square is significantly greater than the error mean square. When performing the F-test in the analysis of variance table the treatment mean square is always on the numerator and the error mean square is always on the denominator and a one-tailed F-test is used. The analysis is looking for evidence of a treatment effect which would be indicated by the treatment mean square being larger than the error mean square so the F-test is a one-tailed alternative test. However, the actual alternative hypothesis on the mean values is two-tailed because the larger treatment mean square can be caused by differences in treatment mean values in any direction. If the variance ratio is less than one because the treatment mean square is less than the error mean square the result is always considered to be non-significant. Steel and Torrie (1980) present a good discussion of the hypothesis testing in the analysis of variance with examples.

An analysis of variance is in effect fitting a model to the data. These models are specified in each of the appropriate following sections in this text. In each of these there is a response variable, y, which is modelled in terms of an overall population mean, treatment effects and block effects. The model will result in a fitted value for each of the observed values. The differences between the observed and the fitted values are known as the residuals which are represented in the model by the random experimental error term, e. The specific analysis of variance tables resulting from the analysis of the different statistical designs are detailed under each design heading.

The treatments to be used in an experiment can be selected in different ways, which will influence the inferences that can be drawn from any hypothesis testing. If the experimenter selects a specific set of treatments (unstructured or structured) the model is termed a 'fixed effects' model. Any inferences drawn from the resultant analysis will apply only to that set of treatments and if the experiment were to be repeated the same set of treatments would be used. If the experimenter randomly selects treatments to represent a wider population of treatments then the model is termed a 'random effects' model. Any inferences drawn from such a model apply to the wider treatment population from which the treatments actually used were drawn and if the experiment were to be repeated a different selection of treatments from the same wider population of treatments would be used. It is also possible to have combined models with both fixed and random effects. Whilst fixed effects models are concerned with the differences between treatments, random effects models are concerned with the treatment variability. The type of model influences not only the sort of conclusions that can be made from the analysis but also the actual form of the decision rule, although the sum of squares calculations for an effect will be the same irrespective of it being 'fixed' or 'random'. This text considers only fixed effects models so formulae and conclusions presented only apply to these models. Further detail on 'fixed' and 'random' effects models can be found in Montgomery (1991) and Steel and Torrie (1980). Both authors present examples of the decision rules for both fixed and random effects versions of the same models for the designs mentioned in this text.

6.1.3.1 *General assumptions and detecting violations*

There are several assumptions which underlie the technique of analysis of variance. If these assumptions cannot be met then the analysis and conclusions drawn from it may not be valid. There are four assumptions: the mutual independence of observations, constant treatment variability, normally distributed experiment errors and independence of experiment errors. There are various techniques for detecting violations of the analysis of variance assumptions. Basic methods for checking the above assumptions are detailed here and, where appropriate, potential remedies are suggested but some of the assumptions can only be met by following the correct experiment procedure. More detailed discussion of the topic can be found in Sokal and Rohlf (1995) and Montgomery (1991) who also suggest a few more formal methods for checking the model assumptions. Cochran and Cox (1957) mention briefly the assumptions and provide further more detailed references.

The assumption of mutually independent observations requires random sampling and that the results from one experiment unit cannot influence results from neighbouring units. In some cases this may mean that experiment units need to be physically separated in space or time; for example in an arable trial where the treatment is to be sprayed, buffer plots between experiment units may be necessary to ensure that the spray applied to one experiment unit will not drift onto the next experiment unit. The assumption is also met by ensuring that the treatments are randomized to the experiment units using the appropriate randomization procedure. If this assumption is violated it cannot be corrected for in the analysis; however, the assumption can be made if correct experiment procedure has been followed.

The assumption of constant treatment variability is concerned with ensuring that experiment units respond to the various treatments in a consistent manner. This may not always be true with the data collected. It could be that the inherent variability between observations increases as the observations themselves increase in magnitude. Other causes of non-constant variance may be erratic responses to a treatment or skewed data distributions. The simplest way to check this is to produce a plot of the model residuals against the fitted values. The points should be randomly scattered on the plot and there

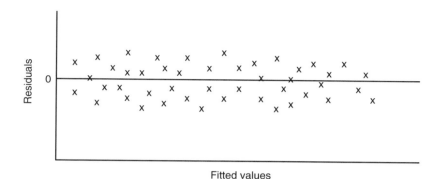

Figure 6.1   A residual plot that looks acceptable, i.e. a random scatter of points lying in a horizontal band around the zero residual. Reproduced by permission from Weisberg (1985)

should be no evidence of *any* patterns. The points on the plot should form a horizontal band around the zero residual as shown in Figure 6.1. It is possible that the points could be grouped but in such cases within each group the points should be randomly scattered and the dispersion of the points around the zero residual should be constant for each group. For example, if the points on the plot are funnel-shaped, as shown in Figure 6.2, then the assumption of constant treatment variance is violated. Violation of this assumption may be resolved by transforming the data and re-analysing. There are formal statistical tests that test for homogeneity of variance but they tend to be sensitive to departures from normality.

The assumption of independent errors is concerned with the distribution of the model residuals. In field trials adjacent plots of land tend to give similar yields and if all replicate plots of one treatment were contiguous the errors could not be assumed independent. Following the correct randomization of treatments for an experiment design will ensure that the errors are independent. Non-independence of errors cannot be corrected for after the experiment has been run. This assumption can be checked by arranging the residuals in a logical order independent of magnitude, such as plot sequence or in ascending order of fitted value magnitude; the residual values would then be expected to occur in a random order. For example, large sequences of positive or negative values would be unlikely and would indicate non-independence of residuals. The sign test, covered in more detail in Section 6.2.2, can be used on the sign of the residuals after they have been logically ordered, and a non-significant result would indicating the residuals are independent.

The assumption of normally distributed errors is also concerned with the distribution of the model residuals. A histogram of the residuals will give a visual indication of whether this assumption is met. If the errors are normally distributed then most of the residuals will lie

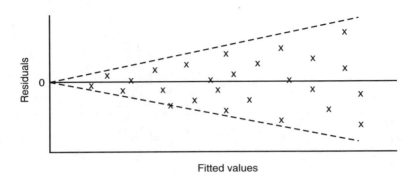

Figure 6.2   A residual plot with a pattern indicating non-constant treatment variance. Reproduced by permission from Weisberg (1985)

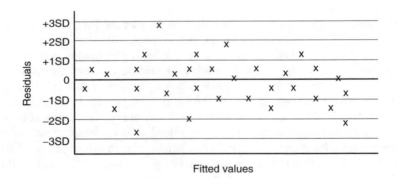

SD = standard deviation

Figure 6.3   A residual plot with standard deviation (SD) limits drawn

between plus or minus one standard deviation, the majority will be between plus or minus two standard deviations and few, if any, residuals should be outside plus or minus three standard deviations. A typical plot of residuals versus fitted values illustrating this is shown in Figure 6.3. Normality of residuals is a feature of the data rather than the design, and violation may be due to an asymmetrical data distribution. In these cases lack of normality of residuals may be corrected for by transforming the data. There are a number of statistical tests that can be used to test for normality of data though these will not be covered in any detail in this text.

6.1.4 Complications associated with analysis of variance

A few common complications that may arise when analysing data from designed experiments are discussed below. These cover the problems of missing values, outliers and transformations. These affect *all* designed experiments when the statistical analysis is of the analysis of variance form. The way in which each analysis is affected by the complications may depend on the actual design used.

There are no hard and fast rules about how to cope with these complications. Often experience is the only way of obtaining a sensible solution to the problem, if a solution exists. Consideration is given to the various methodologies involved and some practical advice is offered.

6.1.4.1 Missing values

Keypoints

- Only a few missing values should be estimated for any one variable.
- Missing value estimates should not bias the results.
- Missing values are best estimated as the corresponding fitted value.
- Missing values need to be estimated for all designs except single-treatment completely randomized designs to maintain the design balance.

Sometimes measurements cannot be made on particular experiment units, or the results are lost or are unusable. The only design for which missing values are not a problem is a single-treatment factor completely randomized design. In this design treatments can be unequally replicated so the analysis can proceed with the reduced number of observations. With all other designs missing values are a problem because every treatment no longer occurs equally. This lack of balance or non-orthogonality has to be taken into account in the subsequent statistical analysis. The analysis of incomplete data due to missing values could be accomplished using generalized linear models. This is a more complex technique than analysis of variance and beyond the scope of this text. The following discussion on handling missing values is from the viewpoint of being able to use an analysis of variance to analyse the data.

Missing data can occur in an experiment for various reasons. Sometimes they occur intentionally and it should then be possible to analyse a balanced subset of data without estimating any missing values. For example, treatments may be applied at different times in the season with sampling dates occurring at regular intervals throughout the season. On the earlier sampling dates there will be missing observations from all those plots receiving treatments that have not yet been applied but the data from those plots receiving treatments by that date can be analysed. These missing values are treated differently from those missing unintentionally. The standard way to deal with unintended missing values in most designed experiments to be analysed by the analysis of variance technique is to estimate a value for each affected experiment unit. Inclusion of these missing value estimates results in a fully balanced data set which to all intents and purposes can be analysed as though the values where genuine observations. If, however, the missing values occur unintentionally, possibly due to faulty measuring equipment, or human error, or an animal on the trial dying, or some other unforeseen circumstance, then they have to be handled differently.

When performing an analysis of variance a linear model is actually being fitted to the data. Fitted values and residuals can be calculated from the model for each experiment unit. The calculation for the fitted values depends on the experiment design being used but in all cases the fitted value of an experiment unit provides an unbiased estimate for that unit. Thus, a missing value in a design will be best estimated by the fitted value for that experiment unit. As residual values are the difference between the actual value and the fitted value, the residual for an experiment unit with a missing value is zero. Consequently the residual sum of squares in the analysis of variance table is not affected by these unbiased estimates of missing values.

The technique for estimating missing values is such that the interpretation of the analysis of variance table is unaffected by the missing value estimates. However, some alterations to the calculations involved in the analysis take place. The total and residual degrees of freedom in the analysis of variance table will be reduced by the number of missing values present. The degrees of freedom for each treatment factor will be as if the data set had been complete, unless observations in all experiment units for a factor level are missing. The mean values of any treatments with a missing value also need to be adjusted. The adjusted treatment mean values are calculated as the arithmetic mean of the actual observations and the missing value estimates.

The simplest instance of missing value estimation is when there is only a single missing value in the data variable. This can be estimated from the formula appropriate to the design of the experiment. Formulae for a variety of designs are given in Cochran and Cox (1957) in their discussion on missing observations. The estimation of missing values becomes more complicated when there is more than one in a variable. Basically the estimation procedure becomes an iterative process and the whole procedure for estimating multiple missing values in a variable is described by Steel and Torrie (1980). Only a few missing values per variable should be estimated for a given design. The precise number of missing values allowed may often depend upon the circumstances involved. If there are many missing values the form of the analysis should be reconsidered. For example, if the majority of observations for a treatment are missing it may be wise to remove the treatment from the analysis altogether and certainly if all the observations for a treatment are missing the whole treatment must be omitted from the analysis. This may well alter the design of the experiment requiring a completely different model fitting to the data. Missing values are estimated from the actual observed data, so if many missing values are to be estimated in comparison to the amount of actual data present the many artificially created observations will exaggerate confidence in the data. If all observations from one complete replicate are missing they should not be estimated and the analysis should proceed with the reduced level of replication.

6.1.4.2   Outliers

Keypoints

- Outliers should not be automatically discarded.
- The handling of outliers should be fully documented.
- Extreme values should be classified and their cause found, if possible.

Extreme values can occur for a number of reasons, e.g. mis-recording of data observations, outlying values, or errors occurring in transcription. Such values can be found at the data recording stage, the data entry and validation stages or may not materialize until during the statistical analysis. Most extreme values will generally become apparent throughout the stages mentioned above, with most being found at the data validation stage.

These values should *not* be discarded. A reason for the existence of the extreme value should be sought. If the extreme value is known to be an erroneous value, the correct value should be substituted where possible. If the correct value is not available the extreme value should be replaced with a missing value and this should be reported. If the extreme value is believed to be a genuine observation but a valid reason for its occurrence is identified, which is unrelated to the treatment, then the extreme value may be removed. For example, if an animal fell ill during the trial or a piece of machinery malfunctioned for the observation, the extreme value and the reason for its exclusion should be fully documented. However, it may also be prudent to follow the advice given below when no explanation for the extreme value can be found. If an outlier is detected and it is a genuine observation but no explanation can be found, the value should not be discarded. A practical way round this problem is suggested by Chatfield (1995); analyse the data with and without the outlying value. If the conclusions drawn from these two analyses are very similar then the outlying value can be retained and the results from the analysis of the complete data set can be reported. However, if the conclusions greatly differ then results from both analyses should be reported.

6.1.4.3   Transforming data

Keypoints

- Transformations may be used to correct violations of the analysis of variance assumptions.
- Standard errors must not be back-transformed.
- Transformed means, transformed standard errors and the number of observations should be reported.

If they are violated the two assumptions of constant treatment variance and normality of errors may be corrected for by transforming the data. In this section a few of the more frequently used transformations are discussed. Patterns found in plots of residuals against fitted values are illustrated; dotted lines have been used to highlight the pattern being shown and do not appear in the usual plot. An indication is given, in each instance, of the transformation that may correct the problem and achieve a data distribution that satisfies the analysis of variance assumptions. Further details on why specific transformations may work for certain types of data or data distribution are given in Sokal and Rohlf (1995) who discuss each of the following transformations in turn. They also mention a larger family of transformations of which the selection below is a subset.

Often skewed distributions will lead to non-constant variance. Skewness can be identified from a histogram of the untransformed data and an examination of the position of the mean value in relation to the maximum and minimum values. Skewness alone should not be used as the justification for transforming. The decision to transform data should only be taken after examining the plot of residuals versus fitted values produced from the analysis of the untransformed data. If the data do not need transforming the histogram will be approximately normal (bell-shaped), the mean will be roughly halfway between the

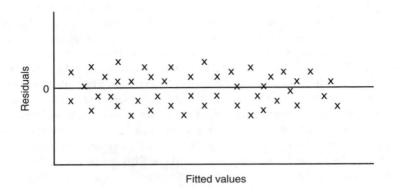

Figure 6.4   A residual plot that looks acceptable, i.e. a random scatter of points lying in a horizontal band around the zero residual. Reproduced by permission from Weisberg (1985)

minimum and the maximum, and the points on the residual plot will form a horizontal band about the zero residual as shown in Figure 6.4.

Transforming data involves applying a mathematical function to the observations. There are a number of standard functions that are often used, such as log, sin and square. The choice of transformation depends on the pattern displayed by the residual plot and the form of the data. After data have been transformed the analysis of the transformed data is repeated in exactly the same way as if they had not been transformed and the assumptions concerning the residuals, now from the transformed data analysis, are again checked. Transformations should be selected on the basis of the one that produces a data distribution that best satisfies the analysis of variance assumptions. They should not be selected on the basis of giving the 'best' or 'desired' results in terms of which treatment effects are found to be statistically significant. The most common patterns found in residual plots are discussed later.

Choosing transformations   If the data are positively skewed and the residual plot exhibits an outward 'funnel' shape as shown in Figure 6.5 then the square root, logarithmic, reciprocal square root or reciprocal transformations can be tried in the order given (Montgomery, 1991). If the data variable contains zero values then applying all these transformations to the variable will result in the generation of missing values. This is obviously a loss of valuable data. The problem is overcome by adding a constant value to *all* data values in the variable. This simply shifts the location of the data without affecting its distribution and therefore does not affect the conclusions drawn from the analysis. It is usual to add 0.5 if the square root or reciprocal square root transformations are used. It is usual to add 1.0 if the logarithmic or reciprocal transformations are used. The same principle applies when the variable contains negative values. The size of the constant to be added will depend

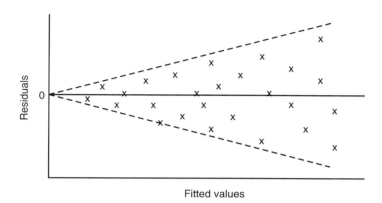

Figure 6.5   A residual plot with a pattern indicating non-constant treatment variance. Square root, logarithmic, reciprocal square root and reciprocal transformations may transform the data into a variable with constant treatment variance. Reproduced by permission from Weisberg (1985)

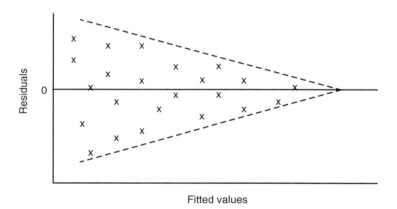

Figure 6.6   A residual plot with a pattern indicating non-constant treatment variance. Square, cube and anti-logarithmic transformations may transform the data into a variable with constant treatment variance. Reproduced by permission from Weisberg (1985)

on the magnitude of the largest absolute negative value and needs to make all variable observations positive.

If the data are negatively skewed or the residual plot exhibits an inward 'funnel' shape as illustrated in Figure 6.6 then the square, cube, etc., can be tried in order. With these

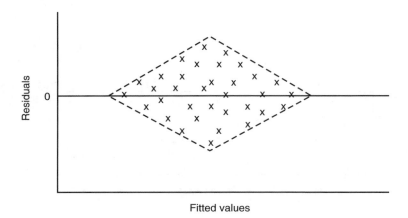

Figure 6.7   A residual plot with a pattern indicating non-constant treatment variance. The angular transformation may transform the data into a variable with constant treatment variance. Reproduced by permission from Weisberg (1985)

transformations there is no need to add constants since applying these functions to zero or negative data will give a valid result rather than a missing value.

If the residual plot exhibits a diamond shape as shown in Figure 6.7 then the arcsine (inverse sine; $\sin^{-1}$ on the calculator) or angular transformation may be useful. The data values must range between 0 and 1 inclusive.

In some cases an 'obvious' transformation to use may be determined by the form of the data, particularly when the theoretical distribution of the observations is known. The arcsine transformation can be used with proportion or percentage data from a binomial distribution but percentages need to be converted to proportions first. The square root transformation is frequently suitable for count data because they frequently follow a Poisson distribution. However, data should never be transformed without first referencing the histogram and residual plot of the untransformed analysis.

If the residual plot has a striped appearance as shown in Figure 6.8 then discrete data are indicated. This residual pattern can occur even for data that were expected to be continuous. In such cases there will be few distinct values, with each value being represented by a single stripe on the plot; in Figure 6.8 there would be just three distinct values in the 12 observations. This cannot be rectified by any transformation, and a different method of analysis must be considered as analysis of variance is not valid.

Table 6.2 indicates the actual calculation to be performed on the raw data when using one of the above transformations and how the back-transformation is achieved. A constant term, c, has been included in the formulae but is only required when there are negative or zero observations in the raw data variable, X. The transformed variable is Y and the back-transformed variable is X^T.

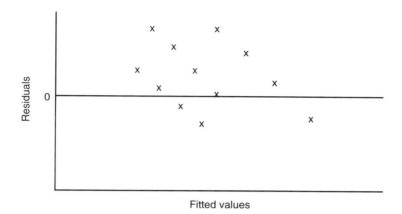

Figure 6.8   A residual plot pattern suggesting the data are discrete

Table 6.2   Table of transformations and the corresponding back-transformations

Transformation name	Transformation calculation	Back-transformation calculation
Square root	$Y = \sqrt{X + c}$	$X^T = Y^2 - c$
Logarithmic	$Y = \ln(X + c)$	$X^T = e^Y - c$
Reciprocal square root	$Y = 1/\sqrt{X + c}$	$X^T = (1/Y^2) - c$
Reciprocal	$Y = 1/(X + c)$	$X^T = (1/Y) - c$
Arcsine	$Y = \sin^{-1}(\sqrt{X})$	$X^T = (\sin(Y))^2$
Square	$Y = X^*X$	$X^T = \sqrt{Y}$
Cube	$Y = X^*X^*X$	$X^T = Y^{1/3}$

Reporting transformed data   If data are transformed as part of the analysis the resulting statistics are only valid for the transformed data and only the transformed data and statistics on the transformed data need be reported. If required for clarity, back-transformed means (not means of the untransformed data) may additionally be shown. The transform of the treatment mean must not be reported as it is not the same as the treatment mean of the transformed data. It is important when reporting data that have been transformed before using an analysis of variance to quote the actual transformation used and whether any constants were added, and to quote the significance of the treatment effects from the analysis on the transformed data.

If the back-transformation of data is required this is done by applying the reverse of the transformation function to the transformed data. For example, if data have been transformed by adding 1 and taking the natural logarithm then it is back-transformed by taking the natural anti-logarithm and then subtracting 1.

If the variability of the back-transformed means needs to be shown confidence limits should be calculated for each of the transformed means and these limits should then be back-transformed. Back-transformed confidence limits for non-linear transformations such as those mentioned in this text will not be symmetrical about the back-transformed mean. The standard error must not be back-transformed.

Example   The histogram shown in Figure 6.9 is of untransformed data and shows the data to have a skewed distribution. A histogram of the residuals and a plot of the residuals versus fitted values are shown in Figure 6.10. Although the histogram looks to be fairly normally distributed there is definite evidence of an outward funnel shape in the residual plot. This indicates that the assumption of normality of errors is valid but the assumption of constant treatment variance is violated. It is concluded that the data need transforming.

Since the plot of residuals against fitted values shows an outward funnel shape, the square root transformation is tried. The histogram of the transformed data is shown in Figure 6.11 and no longer shows such a skewed distribution. The histogram of residuals in Figure 6.12 still looks to be normally distributed but the residual plot is still showing an outward funnel shape. The indication is that the assumption of constant treatment variance is violated. A stronger transformation is tried.

The more powerful logarithmic transformation is tried next and a histogram of these transformed data is shown in Figure 6.13. The shape is not particularly bell-shaped but this alone is never used as evidence for transforming data, therefore it is still worth looking at the graphs of the residuals in Figure 6.14. The histogram of the residuals indicates the standard bell-shape of a normal distribution. The funnelling effect in the scatter plot of the residuals against the fitted values is far less noticeable but there is still some evidence of this effect. It is decided to try an even more powerful transformation.

The next in the sequence of transformations is the reciprocal square root transformation. The histogram of the transformed data in Figure 6.15 is reasonable. The histogram of the

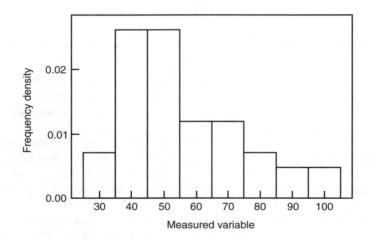

Figure 6.9   Histogram of raw data variate

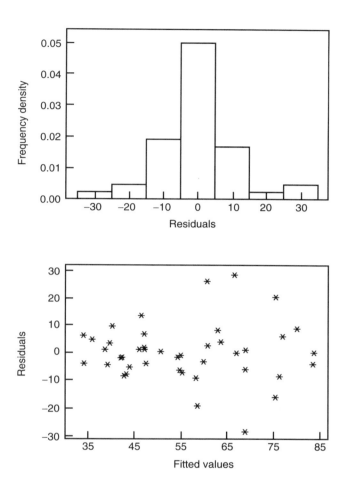

Figure 6.10   A histogram of the model residuals and a residual plot obtained from fitting a model to the raw data

residuals in Figure 6.16 still illustrates an acceptable normal distribution. The plot of residuals versus fitted values in the same figure shows a random scatter around the zero residual although there is evidence of a potential outlier with a large positive residual of over 0.03. It is thought that this transformation is acceptable. Whilst it is not recommended that transformations are performed on the strength of a single observation, it is worthwhile having a look at the most powerful transformation in the sequence.

The reciprocal transformation is applied and a reasonably acceptable histogram of the transformed data is illustrated in Figure 6.17, although it does not appear to be as good an

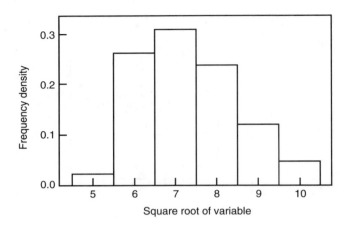

Figure 6.11   Square root transformed data. The raw data were given in Figure 6.9

approximation to a normal distribution as the one in Figure 6.15. The histogram of the residuals in Figure 6.18 is still acceptable although there is evidence that it has started to skew in the opposite direction so it is possible that this transformation has over-transformed the data. The scatter plot of the residuals against the fitted values looks no better than the one shown in Figure 6.16. It is decided to accept the previous analysis. The original data need transforming using the reciprocal square root transformation.

6.1.5   Basic experiment designs

In this section details of the analysis of variance used for the basic experiment designs detailed in Sections 4.2 to 4.4 are presented along with the general conclusions that can be inferred from each design. Discussion and testing methodology has been restricted to a simple consideration of whether or not an overall treatment effect exists, leading to the minimum conclusions that can be formed from an analysis of variance. In practice the experimenter is unlikely to be satisfied with such vague conclusions and will wish to investigate treatment effects in a more detailed manner. Such investigations are introduced in Sections 6.1.6 and 6.1.7 with further detail being provided by referenced texts. Numerical examples are presented and indications are given where further more detailed investigations of treatment effects are warranted. In these examples, although no formal statistical testing has been presented on the model residuals, each set of residuals was subjected to a sign test and a test for normality. Details of these tests have been omitted in the interests of brevity but in all cases the test results were the desired results. For the single-treatment factor designs the general forms of the analysis of variance tables are detailed in Montgomery (1991) where the formulae for the calculations are also summarized. Cochran and Cox (1957) illustrate the use of the formulae for these designs.

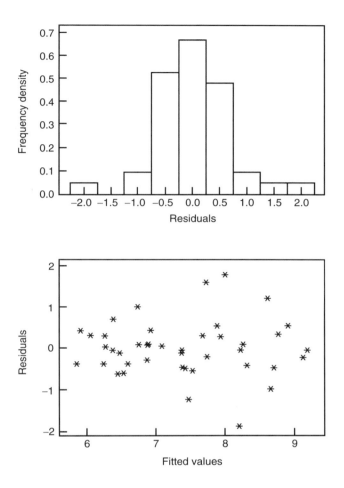

Figure 6.12   A histogram of the model residuals and a residual plot obtained from fitting a model to the square root transformed data

6.1.5.1   *Completely randomized designs*

This is the design described in detail in Section 4.2. This is the simplest of all experiment designs for comparing the effect of t unstructured treatments on some response variable. The statistical model for this design is as follows:

$$y_{ij} = \mu + T_i + e_{ij} \qquad \text{for } i = 1 \ldots t; \; j = 1 \ldots r_i$$

where μ represents the overall population mean response, T_i represents the mean effect of treatment i, r_i represents the number of replicates of treatment i, e_{ij} represents the random experimental error and y_{ij} is the observation of the jth unit receiving the ith treatment.

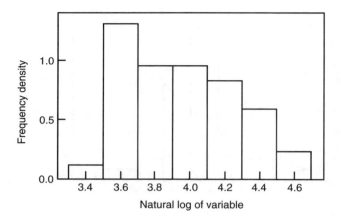

Figure 6.13   Logarithmic transformed data. The raw data were given in Figure 6.9

Hypotheses   It is the treatment effects that are tested by this analysis. There is only one null hypothesis that is tested in a completely randomized design.

H_0: $T_i = \mu_i - \mu = 0$ for $i = 1 \ldots t$
The null hypothesis is that all the treatment population mean values are equal.

H_1: $T_i = \mu_i - \mu \neq 0$ for at least one i
The alternative hypothesis is that at least one of the treatment population mean values is not equal to the overall mean.

T_i is the treatment effect, μ is a constant representing the overall population mean, μ_i is the ith treatment population mean and t is the number of treatments.

Decision rule   The analysis of variance table in Table 6.3 illustrates the effects that are picked out from the analysis together with the residual term and appropriate F-ratios testing for treatment effects. A more detailed coverage of the calculations involved in producing the analysis of variance table is given in Appendix II.

The experimental error variance is the residual mean square as estimated by s^2. The treatment sum of squares is a measure of the variability of the treatment sample means and

Table 6.3   Construction of an analysis of variance table for a completely randomized design

Source	df	ss	ms	vr
Treatment (T_i)	df_T	ss_T	$ms_T = ss_T/df_T$	$F_T = ms_T/s^2$
Residual (e_{ij})	df_E	$ss_E = ss_{TOT} - ss_T$	$s^2 = ss_E/df_E$	
Total	df_{TOT}	ss_{TOT}		

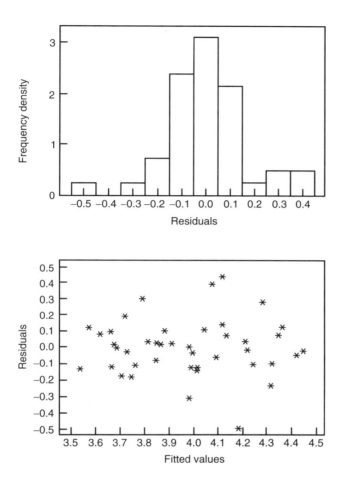

Figure 6.14   A histogram of the model residuals and a residual plot obtained from fitting a model to the logarithmic transformed data

the experiment error. Under the null hypothesis the numerator and denominator of the F-value, or variance ratio, are expected to be equal. If a treatment effect does exist then the treatment mean square will have a higher expected value than the residual mean square so the F-value will produce a significant result when compared against those in tables.

General interpretation   If the null hypothesis is to be accepted then the calculated F-statistic will be less than the appropriate tabulated F-value. In this instance it can be said that there is no statistical evidence to reject the null hypothesis of no difference between the treatment population mean values at the appropriate significance level. It can be concluded

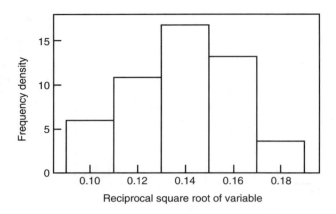

Figure 6.15   Reciprocal square root transformed data. The raw data were given in Figure 6.9

that, at the stated significance level, there is no statistical evidence of any difference in the treatment population mean values provided all the analysis of variance assumptions are met. Although often perceived as a disappointing result, the conclusion of no treatment difference is valid. The experimenter should be aware, however, that an overall non-significant result may be hiding a significant individual treatment effect. Such a scenario may be more likely when treatment factors have many levels.

If the null hypothesis is to be rejected then the calculated F-value will be greater than the appropriate tabulated F-value. In this instance it can be said that there is statistical evidence to reject the null hypothesis of no difference between the treatment population mean values. It can be concluded that, at the stated level of significance, at least one of the treatment population mean values is different from the overall population mean value. The analysis gives no indication of which treatments, or how many, differ from the overall mean. Further analysis will be needed to gain more information from the data with the actual analysis required dependent on the objectives of the experiment. If specific comparisons between treatments or groups of treatments were planned then orthogonal contrasts could be used; if the objective was to compare each treatment with a control treatment then a Dunnett's test would be useful; if no further specific comparisons had been planned but the results suggested a few then multiple range tests may help. These additional tests are described in Sections 6.1.6.1 and 6.1.6.2.

Example   The analysis in this example is from the experiment designed in Section 4.2.1.6. This experiment studied the effect of four stocking densities of hens (A, B, C and D) on egg weight. The results analysed are the total egg weight in the 21st week of the trial.

Hypotheses

H_0: $T_i = \mu_i - \mu = 0$ for $i = 1 \ldots 4$

H_1: $T_i = \mu_i - \mu \neq 0$ for at least one i

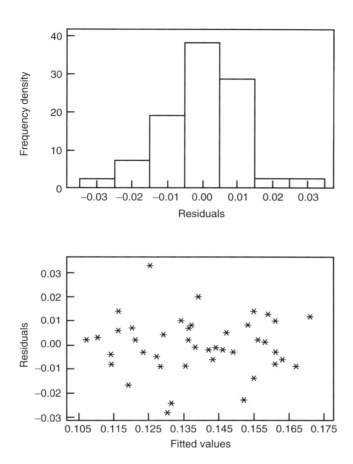

Figure 6.16   A histogram of the model residuals and a residual plot obtained from fitting a model to the reciprocal square root transformed data

The null hypothesis is that there are no differences in the four stocking density mean total egg weights. The alternative hypothesis is that there are one or more differences between the mean values.

Graphs   A histogram of the total egg weights is given in Figure 6.19. The data appear to be reasonably normally distributed; there is no obvious skewness or outlying values.

The model assumptions of constant treatment variance and normally distributed errors with zero mean need to be checked. A histogram of the model residuals is given in Figure 6.20. This gives no indication of any outlying values and illustrates a normal distribution. The residuals and fitted values are plotted in Figure 6.21. There are no obvious indications

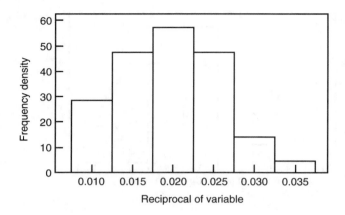

Figure 6.17   Reciprocal transformed data. The raw data were given in Figure 6.9

of non-constant treatment variance nor of any other systematic pattern in the residuals. Both the histogram and the residual plot seem acceptable, implying that the assumptions are satisfied.

Tables   Tables 6.4 and 6.5 give the analysis of variance table and basic summary statistics calculated from the total egg weight data.

Conclusions   The F-test is not significant at the 0.05 level, $p = 0.954$, so the null hypothesis is accepted. It is concluded that, at the 5% level of significance, there is no evidence of a difference between the four population mean total egg weight values. Any further testing is not appropriate in this case since no differences between the treatments were found.

Table 6.4   Analysis of variance table for the total egg weight data

Source	df	ss	ms	vr	$F_{probability}$
Stocking density (T_i)	3	2.013	0.671	0.10	0.954
Residual (e_{ij})	6	38.423	6.404		
Total	9	40.436			

Table 6.5   Table of mean values and number of observations, n, for the total egg weight data

Treatment	A	B	C	D
Total egg weight (g)	47.1	47.4	46.1	46.9
n	3	2	2	3

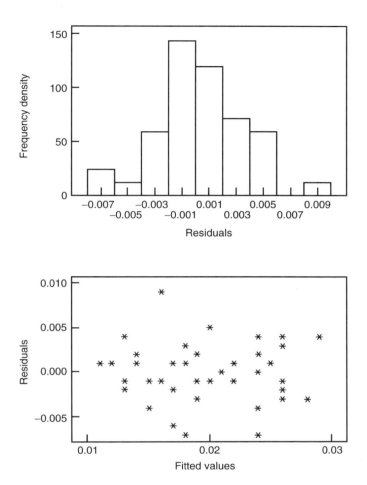

Figure 6.18   A histogram of the model residuals and a residual plot obtained from fitting a model to the reciprocal transformed data

6.1.5.2   Randomized blocks designs

This is the design described in Section 4.3. The purpose of the design is to compare the effects of t unstructured treatments when they are arranged in b randomized blocks on some response variable, y. Each treatment is to appear exactly once in each block so that each block will contain one complete replicate of the entire set of treatments. The appropriate statistical model for this design is as follows:

$$y_{hi} = \mu + T_i + Bl_h + e_{hi} \qquad \text{for } i = 1 \ldots t; h = 1 \ldots b$$

where μ represents the overall population mean response, T_i represents the mean effects of

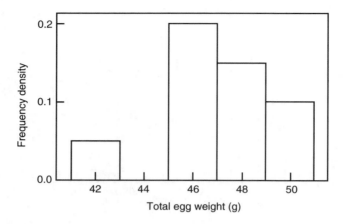

Figure 6.19   Total egg weight (g)

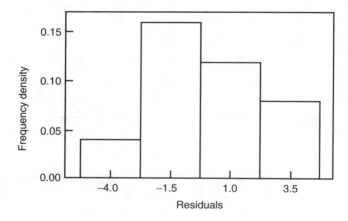

Figure 6.20   Residuals obtained from fitting a completely randomized model to the total egg weight
data

treatment i, Bl_h represents the effect of block h, e_{hi} represents the random experimental error
and y_{hi} is the observation from the unit in block h receiving the ith treatment.

Hypotheses   Both treatment effects and block effects are accounted for in this analysis
although there may be no null hypothesis relating to the block effects. The objective of the
experiment is to look for treatment effects so this is the only hypothesis that it is necessary to
test in a randomized blocks design.

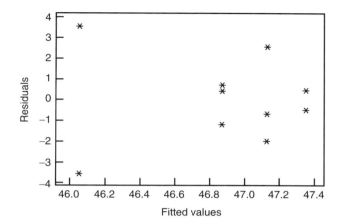

Figure 6.21   Residuals and fitted values obtained from fitting a completely randomized model to the total egg weight data

H_0: $T_i = \mu_i - \mu = 0$ for $i = 1 \ldots t$
The null hypothesis is that all the treatment population mean values are equal.

H_1: $T_i = \mu_i - \mu \neq 0$ for at least one i
The alternative hypothesis is that at least one of the treatment population mean values is not equal to the overall mean.

Where T_i is the treatment effect, μ is a constant representing the overall population mean, μ_i is the ith treatment population mean and t is the number of treatments.

Decision rule   Table 6.6 illustrates the effects that are picked out from the analysis together with the residual term and appropriate F-ratios testing for treatments effects. A more detailed coverage of the calculations involved in producing the analysis of variance table is given in Appendix II.

The experimental error variance is the residual mean square as estimated by s^2. The treatment sum of squares is a measure of the variability of the treatment sample means and the experiment error. Under the null hypothesis the numerator and denominator of the F-value or variance ratio are expected to be equal. If a treatment effect does exist then the

Table 6.6   Construction of an analysis of variance table for a randomized blocks design

Source	df	SS	ms	vr
Block (Bl_h)	df_{Bl}	SS_{Bl}	$ms_{Bl} = ss_{Bl}/df_{Bl}$	
Treatment (T_i)	df_T	SS_T	$ms_T = ss_T/df_T$	$F_T = ms_T/s^2$
Residual (e_{hi})	df_E	$SS_E = ss_{TOT} - ss_T - ss_{Bl}$	$s^2 = ss_E/df_E$	
Total	df_{TOT}	SS_{TOT}		

treatment mean square will have a higher expected value than the error mean square so the F-value will produce a significant result when compared against those in tables.

Although the block effect has been accounted for in the analysis by separating out its sum of squares from the residual sum of squares, a variance ratio has not been calculated. Knowledge of block differences does not affect the interpretation of the treatment effect. If not accounted for in the analysis the block effect will be included in the residual term. This will artificially inflate the experimental error. As treatment effects are tested against the experimental error this will make it less likely that effects will be detected if they exist.

General interpretation   If the null hypothesis is to be accepted then the calculated F-statistic will be less than the appropriate tabulated F-value. In this instance it can be said that there is no statistical evidence to reject the null hypothesis of no difference between the treatment population mean values at the appropriate significance level. It can be concluded that, at the stated significance level, there is no statistical evidence of any difference in the treatment population mean values provided all the analysis of variance assumptions are met. Although often perceived as a disappointing result, the conclusion of no treatment difference is valid. The experimenter should be aware, however, that an overall non-significant result may be hiding a significant individual treatment effect. Such a scenario may be more likely when treatment factors have many levels.

If the null hypothesis is to be rejected then the calculated F-value will be greater than the appropriate tabulated F-value. In this instance it can be said that there is statistical evidence to reject the null hypothesis of no difference between the treatment population mean values. It can be concluded that, at the stated level of significance, at least one of the treatment population mean values is different from the overall population mean value. The analysis gives no indication of which treatments, or how many, differ from the overall mean. Further analysis will be needed to gain more information from the data, with the actual analysis required dependent on the objectives of the experiment. If specific comparisons between treatments or groups of treatments were planned then orthogonal contrasts could be used; if the objective was to compare each treatment with a control treatment then a Dunnett's test would be useful; if no further specific comparisons had been planned but the results suggested a few then multiple range tests may help. These additional tests are described in Sections 6.1.6.1 and 6.1.6.2.

Example   The analysis in this example is from the experiment designed in Section 4.2.2.6. This experiment studied the effect of an untreated control, A, and nine fungicide treatments, B to J inclusive, on the grain size of variety W winter wheat. Four randomized blocks are used, with each treatment occurring once in each block.

Hypotheses

H_0: $T_i = \mu_i - \mu = 0$ for $i = 1 \ldots 10$
H_1: $T_i = \mu_i - \mu \neq 0$ for at least one i

The null hypothesis is that there are no differences in the 10 population mean thousand grain weights of the harvested ears of wheat. The alternative hypothesis is that there is at least one mean value different from the rest.

Graphs   A histogram of the thousand grain weights is shown in Figure 6.22. The data appear to be reasonably normally distributed with no obvious outlying values.

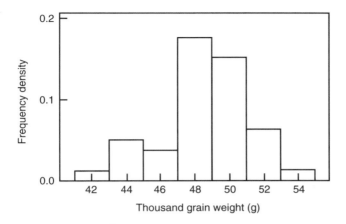

Figure 6.22   Thousand grain weight of a wheat crop (g)

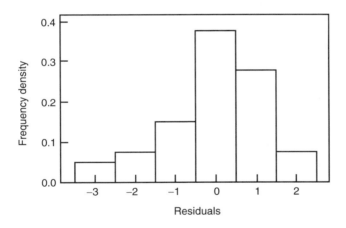

Figure 6.23   Residuals obtained from fitting a randomized blocks model to the thousand grain weight data

The model assumptions of constant treatment variance and normally distributed residuals with zero mean need to be checked. A histogram of the model residuals is drawn in Figure 6.23. There is no definite evidence of a non-normal distribution nor of any outliers. The residuals and fitted values are plotted in Figure 6.24. The points form a random scatter in a horizontal band about the zero residual and there is no obvious systematic pattern although two distinct groups of points are apparent. These seem to be a group for the control treatment and a group for the other treatments. Both graphs seem acceptable, suggesting the two assumptions are satisfied.

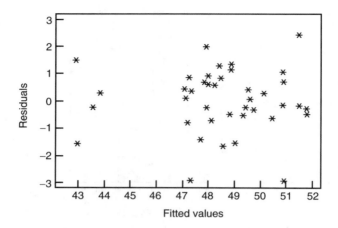

Figure 6.24   Residuals and fitted values obtained from fitting a randomized blocks model to the thousand grain weight data

Table 6.7   Analysis of variance table for the thousand grain weight data

Source	df	ss	ms	vr	$F_{probability}$
Block (Bl_h)	3	6.093	2.031		
Fungicide (T_i)	9	190.504	21.167	11.11	< 0.001
Residual (e_{hi})	27	51.422	1.905		
Total	39	248.019			

Table 6.8   Table of mean values and number of observations, n, for the thousand grain weight data

	Treatment									
	A	B	C	D	E	F	G	H	I	J
Thousand grain weight (g)	43.30	47.47	51.27	47.60	51.27	48.35	48.82	47.70	49.20	49.90
n	4	4	4	4	4	4	4	4	4	4

Tables   Tables 6.7 and 6.8 tabulate the analysis of variance information and basic summary statistics calculated from the thousand grain weight data.

Conclusions   The F-test is significant, $p < 0.001$, so the null hypothesis is rejected. It is concluded that, at the $< 0.1\%$ level of significance, there is at least one difference between the 10 population mean thousand grain weight values. The thousand grain weight for the untreated control is lower than those for all fungicide treatments. Fungicides C and E give the highest thousand grain weight and the control, treatment A, gave the lowest value.

This conclusion may be useful but it is limited. Further information could be gleaned from the data by expanding the analysis to include specific treatment comparisons using tests such as orthogonal contrasts or Dunnett's test. The test chosen will depend on the objectives of the study and whether any particular comparisons were planned prior to the experiment. These tests are covered, with others, in Section 6.1.6 and examples using the above data are provided in Sections 6.1.6.4 and 6.1.6.6.

6.1.5.3 Latin square designs

Latin square designs are described in detail in Section 4.2.3. The Latin square design uses two blocking factors simultaneously to discriminate between the units. This design compares the effects of t unstructured treatments arranged in a Latin square on some response variable, y. The appropriate statistical model for this design is as follows:

$$y_{ijk} = \mu + R_i + C_j + T_k + e_{ijk} \qquad \text{for } i = 1 \ldots t; \ j = 1 \ldots t; \ k = 1 \ldots t$$

where μ represents the overall population mean response, R_i represents the mean effect of row i, C_j represents the mean effect of column j, T_k represents the mean effect of treatment k, e_{ijk} represents the random experimental error and y_{ijk} is the observation from the experiment unit in the ith row and jth column receiving the kth treatment.

Hypotheses   Treatment effects, row effects and column effects are accounted for in this analysis although there may only be a null hypothesis relating to the treatment effects. The objective of the experiment is to look for treatment effects so this is the only hypothesis that it is necessary to test in a Latin square design.

H_0: $T_k = \mu_k - \mu = 0$ for $k = 1 \ldots t$
The null hypothesis is that all the treatment population mean values are equal.

H_1: $T_k = \mu_k - \mu \neq 0$ for at least one k
The alternative hypothesis is that at least one of the treatment population mean values is not equal to the overall mean.

Where T_k is the treatment effect, μ is a constant representing the overall population mean, μ_k is the kth treatment population mean and t is the number of treatments.

Decision rule   Table 6.9 illustrates the effects that are picked out from the analysis together with the residual term and appropriate F-ratios testing for treatment effects. A more detailed coverage of the calculations involved in producing the analysis of variance table is given in Appendix II.

The experimental error variance is the residual mean square as estimated by s^2. The treatment sum of squares is a measure of the variability of the treatment sample means and the experiment error. Under the null hypothesis the numerator and denominator of the F-value or variance ratio are expected to be equal. If a treatment effect does exist then the treatment mean square will have a higher expected value than the residual mean square so the F-value will produce a significant result when compared with those in tables.

Although the row and column effects have been accounted for in the analysis by separating out their sums of squares from the residual sum of squares, variance ratios have not been calculated. Knowledge of such differences does not affect the interpretation of the

Table 6.9   Construction of an analysis of variance table for a Latin square design

Source	df	ss	ms	vr
Row (R_i)	df_R	ss_R	$ms_R = ss_R/df_R$	
Column (C_j)	df_C	ss_C	$ms_C = ss_C/df_C$	
Treatment (T_k)	df_T	ss_T	$ms_T = ss_T/df_T$	$F_T = ms_T/s^2$
Residual (e_{ijk})	df_E	$ss_E = ss_{TOT} - ss_T - ss_C - ss_T$	$s^2 = ss_E/df_E$	
Total	df_{TOT}	ss_{TOT}		

treatment effect. If not accounted for in the analysis the row and column effects will be included in the residual term which will artificially inflate the experimental error. As treatment effects are tested against the experimental error this will make it less likely that effects will be detected if they exist.

General interpretation   If the null hypothesis is to be accepted then the calculated F-statistic will be less than the appropriate tabulated F-value. In this instance it can be said that there is no statistical evidence to reject the null hypothesis of no difference between the treatment population mean values at the appropriate significance level. It can be concluded that, at the stated significance level, there is no statistical evidence of any difference in the treatment population mean values provided all the analysis of variance assumptions are met. Although often perceived as a disappointing result, the conclusion of no treatment difference is valid. The experimenter should be aware, however, that an overall non-significant result may be hiding a significant individual treatment effect. A scenario that may be more likely when treatment factors have many levels.

  If the null hypothesis is to be rejected then the calculated F-value will be greater than the appropriate tabulated F-value. In this instance it can be said that there is statistical evidence to reject the null hypothesis of no difference between the treatment population mean values. It can be concluded that, at the stated level of significance, at least one of the treatment population mean values is different from the overall population mean value. The analysis gives no indication of which treatments, or how many, differ from the overall mean. Further analysis will be needed to gain more information from the data with the actual analysis required dependent on the objectives of the experiment. If specific comparisons between treatments or groups of treatments were planned then orthogonal contrasts could be used; if the objective was to compare each treatment with a control treatment then a Dunnett's test would be useful; if no further specific comparisons had been planned but the results suggested a few then multiple range tests may help. These additional tests are described in Sections 6.1.6.1 and 6.1.6.2.

Example   The analysis in this example is from the experiment designed in Section 4.2.3.6. This experiment studied the effects, on DMD, of supplementing grass hay feed with various proportions of forage. Sheep and period are the simultaneous blocking factors. The four diets are coded A, B, C and D.

Hypotheses

  H_0: $T_k = \mu_k - \mu = 0$ for $k = 1 \ldots 4$
  H_1: $T_k = \mu_k - \mu \neq 0$ for at least one k

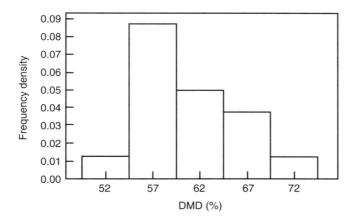

Figure 6.25   DMD (%)

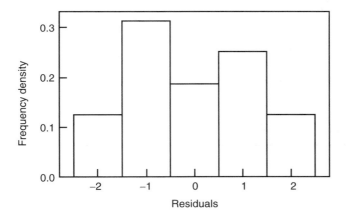

Figure 6.26   Residuals obtained from fitting a Latin square model to the DMD data

The null hypothesis is that there is no difference in the mean dry matter for the four diets. The alternative hypothesis is that there is a difference.

Graphs   A histogram of the DMD values recorded for all sheep is shown in Figure 6.25. The data appear to be reasonably normally distributed with no obvious outlying values.

The analysis assumptions of constant treatment variance and normally distributed errors with zero mean need to be checked. A histogram of the model residuals is plotted in Figure 6.26. There is no indication of a non-normal distribution nor of any outliers. The residuals

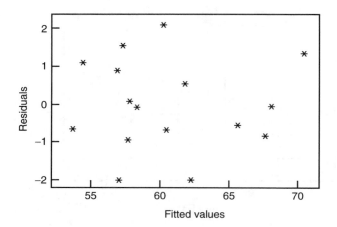

Figure 6.27   Residuals and fitted values obtained from fitting a Latin square blocks model to the DMD data

and fitted values are plotted in Figure 6.27. It can be seen that the points form a random scatter in a horizontal band about the zero residual. Both graphs indicate that the assumptions are satisfied.

Tables   Tables 6.10 and 6.11 are the analysis of variance table and basic summary statistics calculated from the DMD data.

Conclusions   The F-test is significant at the 0.001 level, so the null hypothesis is rejected. It is concluded that, at the <0.1% level of significance, at least one population mean DMD

Table 6.10   Analysis of variance table for the DMD data

Source	df	ss	ms	vr	$F_{probability}$
Row (R_i)	3	12.850	4.283		
Column (C_j)	3	12.673	4.224		
Diet (T_k)	3	349.548	116.516	32.12	<0.001
Residual (e_{ijk})	6	21.767	3.628		
Total	15	396.839			

Table 6.11   Table of mean values and number of observations, *n*, for the DMD data

	Treatment			
	A	B	C	D
DMD (%)	55.55	57.64	61.11	67.87
n	4	4	4	4

value is different from the rest. The analysis does not indicate which of the mean values is different, however it can be seen that diet A has the lowest mean DMD and diet D has the highest mean DMD. To investigate the significant treatment effect further, the techniques mentioned in Section 6.1.6 can be used to analyse the differences between treatment mean values.

6.1.5.4 Two-way factorial design in randomized blocks

This is the design that was described in Section 4.3.1.6. The purpose of the design is to compare the effects of two treatments and their interactions on some response variable, y. Although other blocking structures may be used, the randomized blocks blocking structure is the most commonly used and is therefore illustrated here. As has been discussed above, the blocking structure used does not affect the interpretation of the treatment effects and therefore the treatment null hypotheses. This design has two treatment factors, treatment A with a levels and treatment B with b levels, therefore the number of treatments, t, in this experiment is ab. The appropriate statistical model for the design is as follows:

$$y_{hij} = \mu + Bl_h + A_i + B_j + (AB)_{ij} + e_{hij} \qquad \text{for } h = 1 \ldots r; \ i = 1 \ldots a; \ j = 1 \ldots b$$

where μ represents the overall treatment mean response, Bl_h represents the hth block effect, A_i represents the mean effect of treatment A at level i, B_j represents the mean effect of treatment B at level j, $(AB)_{ij}$ represents the mean effect of the interaction between treatment A at level i and treatment B at level j, e_{hij} represents the random experimental error and y_{hij} is the observation from the unit in the hth block receiving the ith level of treatment A and the jth level of treatment B.

Hypotheses There are three treatment effects and a block effect which are picked out by this analysis although there need not be any null hypotheses relating to the block effect. The objective of the experiment is to look for all the treatment effects so these are the only hypotheses that are tested in a two-way factorial randomized blocks design.

H_0: $A_i = \mu_i - \mu = 0$ for $i = 1 \ldots a$
H_0: $B_j = \mu_i - \mu = 0$ for $j = 1 \ldots b$
H_0: $AB_{ij} = \mu_{ij} - \mu = 0$ for $i = 1 \ldots a, j = 1 \ldots b$
The null hypotheses state as follows:
All treatment factor A population mean values are equal.
All treatment factor B population mean values are equal.
All AB interaction population mean values are equal.

H_1: $A_i = \mu_i - \mu \neq 0$ for at least one i
H_1: $B_j = \mu_j - \mu \neq 0$ for at least one j
H_1: $AB_{ij} = \mu_{ij} - \mu \neq 0$ for at least one i or j
The alternative hypotheses state as follows:
At least one of the treatment factor A population mean values is not equal to the overall mean.
At least one of the treatment factor B population mean values is not equal to the overall mean.
At least one of the interaction population mean values is not equal to the overall mean.

A_i is the factor A treatment effect, B_j is the factor B treatment effect, AB_{ij} is the interaction effect, μ is a constant representing the overall population mean, μ_i is the ith factor A treatment population mean, μ_j is the jth factor B treatment population mean, and a and b are the number of treatments in factors A and B respectively.

Decision rule   Table 6.12 illustrates the effects that are picked out from the analysis together with the residual term and appropriate calculations. A more detailed coverage of the calculations involved in producing the analysis of variance table is given in Appendix II.

The experimental error variance is the residual mean square as estimated by s^2. The factor A sum of squares is a measure of the variability of the factor A sample means and the experiment error. The factor B sum of squares is a measure of the variability of the factor B sample means and the experiment error. The AB interaction sum of squares is a measure of the variability of the AB interaction sample means and the experiment error. Under the null hypotheses the numerators and denominators of the F-values or variance ratios are expected to be equal. If a treatment effect does exist then the treatment mean square will have a higher expected value than the error mean square so the F-value will produce a significant result when compared against those in tables.

Although the block effect has been accounted for in the analysis by separating out its sum of squares from the residual sum of squares, a variance ratio has not been calculated. Knowledge of block differences does not affect the interpretation of the treatment effects. If not accounted for in the analysis the block effect will be included in the residual term. This will artificially inflate the experimental error. As treatment effects are tested against the experimental error this will make it less likely that effects will be detected if they exist.

General interpretation   If any of the null hypotheses are to be accepted then the calculated F-value will be less than the appropriate tabulated F-value. If the F-value for the AB interaction is not statistically significant it can be said that there is no statistical evidence to reject the null hypothesis of no difference between the AB interaction population mean values at the appropriate significance level. If the F-value for factor A is not statistically significant it can be said that there is no statistical evidence to reject the null hypothesis of no difference between the factor A treatment population mean values at the appropriate significance level. Similarly for factor B. Although often perceived as a disappointing result, the conclusion of no treatment difference is valid. The experimenter should be aware, however, that an overall non-significant result may be hiding a significant individual

Table 6.12   Construction of an analysis of variance table for a two-way factorial in randomized blocks design

Source	df	ss	ms	vr
Block (Bl_h)	df_{Bl}	ss_{Bl}	$ms_{Bl} = ss_{Bl}/df_{Bl}$	
Factor A (A_i)	df_A	ss_A	$ms_A = ss_A/df_A$	$F_A = ms_A/s^2$
Factor B (B_j)	df_B	ss_B	$ms_B = ss_B/df_B$	$F_B = ms_B/s^2$
AB interaction (AB_{ij})	df_{AB}	ss_{AB}	$ms_{AB} = ss_{AB}/df_{AB}$	$F_{AB} = ms_{AB}/s^2$
Residual (e_{hij})	df_E	$ss_E = ss_{TOT} - ss_{Bl}$ $- ss_{AB} - ss_A - ss_B$	$s^2 = ss_E/df_E$	
Total	df_{TOT}	ss_{TOT}		

treatment effect. A scenario that may be more likely when treatment factors have many levels.

If any of the null hypotheses are to be rejected then the calculated F-value will be greater than the appropriate tabulated F-value. If there is no significant *AB* interaction the main effects of *A* and *B* can be sensibly interpreted and this will be done in a similar manner to the single-treatment factor in a randomized blocks design. If there is a significant *AB* interaction then it may not be possible to interpret the main effects of *A* and *B*. If any of the treatment effects are found to be significant the analysis gives no indication of which treatments, or how many, differ from the overall mean. Further analysis will be needed to gain more information from the data, with the actual analysis required dependent on the objectives of the experiment. Additional tests for examining individual treatment differences are described in Section 6.1.6.1 and references for examining significant interactions in more detail are provided in the following subsection.

Interpretation of interaction effects   Interpretation of these designs starts with the interaction term and can then move on to the main effects depending on the conclusions drawn from the interaction. If the *AB* interaction effect is statistically significant it can be said that there is statistical evidence to reject the null hypothesis of no difference between the *AB* interaction population mean values. It can be concluded that at the stated level of significance at least one of the *AB* interaction population mean values is different from the overall population mean value. So the response at the different levels of treatment *A* is not the same for all levels of treatment *B*, and vice versa. If there is a significant interaction between *A* and *B* it may make little or no sense to interpret the main effects of *A* and *B*. This is because the main effects of *A* look at the different levels of *A* averaged over all levels of *B*, but if it has already been proven that the results for different levels of *A* depend on the levels of *B* then it makes little sense to average the effects over the levels of *B*. The same will apply to the main effects of *B*.

An interaction plot is a simple way to try and interpret the effects of any two-way interaction. An interaction plot is produced by plotting the interaction, *AB*, means against the levels of one of the factors and doing this for each level of the other factor. The levels of the factor along the x-axis will usually be equispaced and there will be as many lines on the plot as there are levels of the other factor. The plot is easier to interpret if the factor with most levels is put on the x-axis as this will result in more points on fewer lines. If the null hypothesis concerning the interaction is accepted then it will be concluded that there is no significant interaction effect present, which will result in an interaction plot where the lines are approximately parallel, as illustrated in Figure 6.28. If it is concluded that an interaction is present then the lines in the plot will not be parallel. Two such examples are illustrated in Figure 6.29. In the first plot although the lines are not parallel they do not cross. In these circumstances it may be possible to draw some conclusion about the main effects of *A* and *B*, for example there is a trend for values to increase as the levels of *A* increase (across the x-axis). It can also be concluded that level 1 of factor *B* is always above level 2 of factor *B*. In the second plot the two lines do actually cross so no conclusions can be drawn about the main effects.

The analysis and interpretation of these designs is illustrated by example in Steel and Torrie (1980) who also illustrate through example how to examine significant interactions in more depth to identify the factor levels that cause the significant interaction. Montgomery (1991) presents a clear analysis of variance layout with decision rule and a straightforward explanation of interactions.

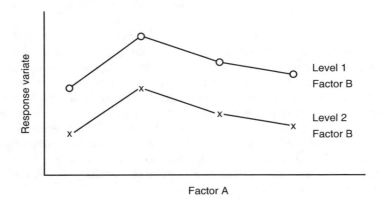

Figure 6.28   Illustration of a non-significant interaction between two treatment factors. After Steel and Torrie (1980)

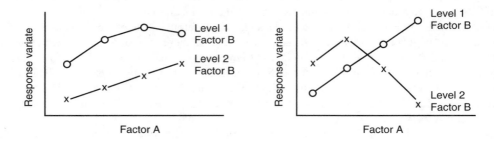

Figure 6.29   Illustration of a significant interaction between two treatment factors. After Steel and Torrie (1980)

Example   The data analysed in this example are from the experiment designed in Section 4.3.1.6. This experiment investigates the effect on protein of heifers' milk of feeding with two diets (D1 and D2) and of housing in two different environments (H1 and H2). The experiment uses nine randomized blocks with one complete replicate of the four treatments in each block. The animals are milked throughout their lactation period and the actual observations used are the mean milk protein values where the protein is expressed as a percentage of the total milk yield.

Hypotheses

Diet main effect

$$H_0: \ D_i = \mu_i - \mu = 0 \quad \text{for } i = 1, 2$$
$$H_1: \ D_i = \mu_i - \mu \neq 0 \quad \text{for at least one } i$$

The null hypothesis is that there is no difference in the mean protein of milk for the two diets. The alternative is that there is a difference.

Housing main effect

H_0: $H_j = \mu_j - \mu = 0$ for $j = 1, 2$
H_1: $H_j = \mu_j - \mu \neq 0$ for at least one j
The null hypothesis is that there is no difference in the two housing environment mean milk protein values. The alternative is that there is a difference.

First-order interaction

H_0: $DH_{ij} = \mu_{ij} - \mu = 0$ for $i = 1 \ldots 2; j = 1 \ldots 2$
H_1: $DH_{ij} = \mu_{ij} - \mu \neq 0$ for at least one i or j
The null hypothesis is that there is no difference in the interaction mean milk protein values. The alternative is that there is a difference

Graphs A histogram of the mean milk protein values is shown in Figure 6.30. The data appear to be reasonably normally distributed with no obvious outlying values.

The analysis assumptions of constant treatment variance and normally distributed errors with zero mean need to be checked. A histogram of the model residuals is plotted in Figure 6.31. There is no indication of a non-normal distribution nor of any outliers. The residuals and fitted values are plotted in Figure 6.32. It can be seen that the points form a random scatter in a horizontal band about the zero residual. Both graphs indicate that the assumptions are satisfied.

Tables The analysis of variance table and basic summary statistics calculated from the mean milk protein data are given in Tables 6.13 and 6.14.

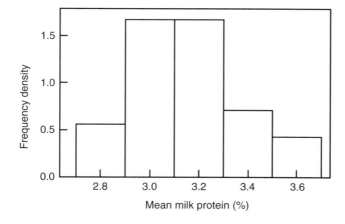

Figure 6.30   Mean milk protein (%)

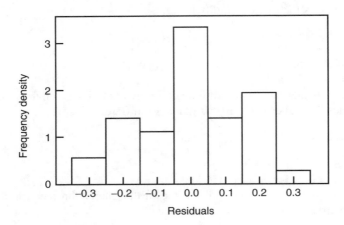

Figure 6.31   Residuals obtained from fitting a two-way factorial in randomized blocks model to the mean milk protein data

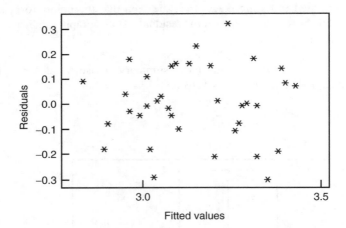

Figure 6.32   Residuals and fitted values obtained from fitting a two-way factorial in randomized blocks model to the mean milk protein data

Conclusions The interaction term is not statistically significant at the 5% level of significance so the null hypothesis of no interaction is accepted. The interaction plot is illustrated in Figure 6.33. The non-significant interaction is indicating that the two lines can be considered parallel and that the difference between them is the same at the two diet points. As there is no interaction between housing environment and diet the main effects of diet and housing can be examined in turn. The effect of housing is not statistically

Table 6.13   Analysis of variance table for the mean milk protein data

Source	df	ss	ms	vr	$F_{probability}$
Block (Bl_h)	8	0.17671	0.02209		
Diet (D_i)	1	0.68890	0.68890	21.40	<0.001
Housing (H_i)	1	0.06760	0.06760	2.10	0.160
DH interaction (DH_{ij})	1	0.00538	0.00538	0.17	0.686
Residual (e_{hij})	24	0.77247	0.03219		
Total	35	1.71106			

Table 6.14   Tables of mean milk protein (%) values and number of observations, n, for the mean milk protein data

Main effects

	Diet		Housing	
Treatment	D1	D2	H1	H2
Mean	3.28	3.01	3.19	3.10
n	10	10	10	10

First-order interaction

$n = 9$	Housing	
Diet	H1	H2
D1	3.34	3.23
D2	3.04	2.97

significant at the 5% level of significance. It is concluded that there is no difference in mean milk protein for the two housing treatments. The effect of diet is statistically significant, $p < 0.001$. It is concluded that the two diet treatments result in different mean milk protein values, with diet treatment D1 giving the higher value. Further analysis is not appropriate in this case since there were only two levels of each treatment and in each case a difference was found.

6.1.5.5   Multi-way factorial designs in randomized blocks

The previous discussion of the two-way factorial design extends to multi-factorial designs arranged in randomized blocks. These designs were described in Section 4.3.2. Although other blocking structures may be used, the randomized blocks blocking structure is the most commonly used and is therefore illustrated here. The blocking structure used does not affect the interpretation of the treatment effects and therefore the null hypotheses. The appropriate statistical model is an extension of that given for the two-way factorial design in randomized blocks in Section 6.1.5.4. The model will differ only in the number of main effects and interaction effects that are to be estimated.

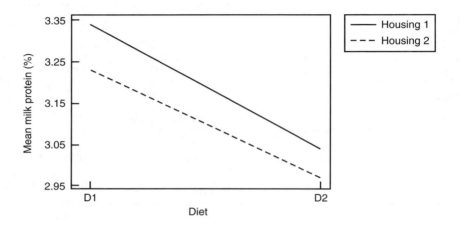

Figure 6.33   Mean milk protein interaction plot for the diet and housing treatment factors

Hypotheses   There are as many treatment effects as there are treatment factors and interactions. These, along with the block effect, are identified by the analysis although there need not be a null hypothesis for the block effect. The objective of the experiment is to look for treatment effects so these are the only effects tested in a multi-factor factorial design. Examples of hypotheses are shown below.

The main effect null hypotheses will be of the form:

H_0: $C_k = \mu_k - \mu = 0$ for $k = 1 \ldots c$
All treatment factor C population mean values are equal.

The first-order interaction effect null hypotheses will be of the following form:

H_0: $BC_{jk} = \mu_{jk} - \mu = 0$ for $j = 1 \ldots b$; $k = 1 \ldots c$
All BC interaction population mean values are equal.

The second-order interaction effect null hypotheses will be of the following form:

H_0: $BCD_{jkl} = \mu_{jkl} - \mu = 0$ for $j = 1 \ldots b$; $k = 1 \ldots c$; $l = 1 \cdots d$
All BCD interaction population mean values are equal.

There is only one possible alternative hypothesis for each null hypothesis. For those illustrated they will be as follows:

H_1: $C_k = \mu_k - \mu \neq 0$ for at least one k
H_1: $BC_{jk} = \mu_{jk} - \mu \neq 0$ for at least one j or k
H_1: $BCD_{jkl} = \mu_{jkl} - \mu \neq 0$ for at least one j or k or l
At least one of the treatment factor C population mean values is not equal to the overall mean.
At least one of the BC interaction population mean values is not equal to the overall mean.
At least one of the BCD interaction population mean values is not equal to the overall mean.

Higher-order interactions will each have their own null hypothesis and alternative hypothesis which will be in similar formats.

Decision rule The analysis of variance is constructed in the same way as for the two-way factorial and is therefore of the same form. The one residual term is used to test all treatment effects. The block effect is considered in the same manner as for the two-way factorial design in randomized blocks.

General interpretation Interpretation of treatment effects is similar to that for the two-way factorial. Non-significant results will be interpreted in the same way. However, if interaction effects are statistically significant then interpretation of any main effects or lower-order interactions associated with the significant interaction may or may not be possible. Therefore interpretation of multi-way factorial designs always starts with the highest-order interaction term. Interpretation then proceeds to the next order interaction terms and so on until the main effects are interpreted. At each stage the presence of a significant interaction may prevent the interpretation of any effects contributing towards that interaction.

The analysis and interpretation of these designs is detailed in both Montgomery (1991) and Steel and Torrie (1980), with complete worked examples. The latter authors also describe and illustrate techniques for examining significant interactions in more detail than can be achieved from the basic analysis of variance table described here.

Example The data analysis considered in this example is from the experiment designed in Section 4.3.2.6. The experiment investigated the effects of two sowing dates (D1 and D2) of five varieties of oats (A to E) with two nitrogen rates (N1 and N2) on the oven dry matter content of the oat grain. The experiment is a three-way factorial in three randomized blocks.

Hypotheses

Variety main effect

H_0: $V_i = \mu_i - \mu = 0$ for $i = 1 \ldots 5$
H_1: $V_i = \mu_i - \mu \neq 0$ for at least one i
The null hypothesis is that there are no differences in the five variety mean oven dry matter values. The alternative is that at least one of the mean values is different from the rest.

Sowing date main effect

H_0: $D_j = \mu_j - \mu = 0$ for $j = 1, 2$
H_1: $D_j = \mu_j - \mu \neq 0$ for at least one j
The null hypothesis is that there is no difference in the two sowing date mean oven dry matter values. The alternative is that there is a difference.

Nitrogen rate main effect

H_0: $N_k = \mu_k - \mu = 0$ for $k = 1, 2$
H_1: $N_k = \mu_k - \mu \neq 0$ for at least one k
The null hypothesis is that there is no difference in the two nitrogen rate mean oven dry matter values. The alternative is that there is a difference.

First-order interactions

H_0: $VD_{ij} = \mu_{ij} - \mu = 0$ for $i = 1 \ldots 5$; $j = 1, 2$
H_1: $VD_{ij} = \mu_{ij} - \mu \neq 0$ for at least one i or j

H_0: $VN_{ik} = \mu_{ik} - \mu = 0$ for $i = 1 \ldots 5$; $k = 1, 2$
H_1: $VN_{ik} = \mu_{ik} - \mu \neq 0$ for at least one i or k

H_0: $DN_{jk} = \mu_{jk} - \mu = 0$ for $j = 1, 2$; $k = 1, 2$
H_1: $DN_{jk} = \mu_{jk} - \mu \neq 0$ for at least one j or k

The null hypotheses are that there are no differences in the first-order interaction mean oven dry matter values. The alternatives are that there are that at least one mean value is different from the rest.

Second-order interaction

H_0: $VDN_{ijk} = \mu_{ijk} - \mu = 0$ for $i = 1 \ldots 5$; $j = 1, 2$; $k = 1, 2$
H_1: $VDN_{ijk} = \mu_{ijk} - \mu \neq 0$ for at least one i or j or k

The null hypothesis is that there are no differences in the second-order interaction mean oven dry matter values. The alternative is that at least one of the interaction mean values is different from the rest.

Graphs A histogram of the oven dry matter values is shown in Figure 6.34. The data appear to be normally distributed with no outlying values.

The model assumptions of constant treatment variance and normally distributed mean values with zero mean need to be checked. A histogram of the residuals is given in Figure 6.35 and the residuals and fitted values are plotted in Figure 6.36. From these there is no evidence that the residuals are not normally distributed about the zero mean. The points on the plot of residuals are randomly scattered in a horizontal band. There are no indications of any unusual observations although the high point in the top right-hand corner of the residual

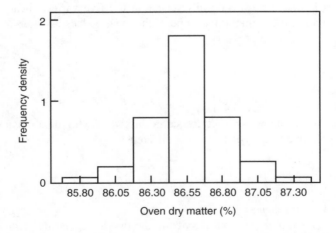

Figure 6.34   Oven dry matter content of oat grain (%)

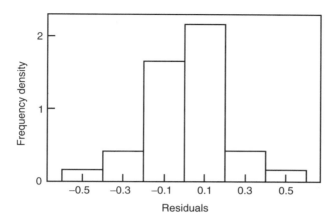

Figure 6.35   Residuals obtained from fitting a three-way factorial in randomized blocks model to the oven dry matter data

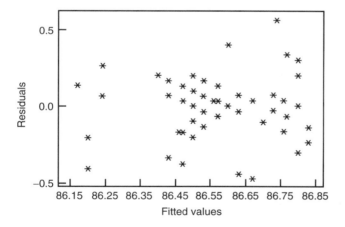

Figure 6.36   Residuals and fitted values obtained from fitting a three-way factorial in randomized blocks model to the oven dry matter data

plot may be worthy of examination. However, it is not so extreme that these two model assumptions can not be accepted.

Tables   The analysis of variance table and basic summary statistics calculated from the oven dry matter data are presented in Tables 6.15 and 6.16.

Conclusions   None of the interaction terms are statistically significant at the 20% level. There is a statistically significant ($p = 0.017$) variety effect. Variety E has the highest oven

Table 6.15   Analysis of variance table for the oven dry matter data

Source	df	ss	ms	vr	$F_{\text{probability}}$
Block (Bl_h)	2	0.04900	0.02450		
Variety (V_i)	4	0.96933	0.21708	3.46	0.017
Sowing date (D_j)	1	0.00067	0.00067	0.01	0.918
Nitrogen (N_k)	1	0.05400	0.05400	0.86	0.359
VD interaction (VD_{ij})	4	0.05100	0.01275	0.20	0.935
VN interaction (VN_{ik})	4	0.27433	0.06858	1.09	0.374
DN interaction (DN_{jk})	1	0.05400	0.05400	0.86	0.359
VDN interaction (VDN_{ijk})	4	0.35433	0.08858	1.41	0.249
Residual (e_{hijk})	38	2.38433	0.06275		
Total	59	4.08998			

dry matter and variety D has the lowest, although this is not much lower than variety C. There are no effects on oven dry matter due to the different sowing dates or to the different rates of nitrogen applied to the oat crop.

The analysis has not identified which varieties are causing the significant treatment effect. The use of contrasts or multiple range tests, covered in Sections 6.1.6.4 and 6.1.6.5, may help determine the statistical significance of specific variety comparisons or give a general impression of which varieties are responsible for the significant variety effect.

6.1.5.6   Two-treatment factor split-plot designs in randomized blocks

This design is described in Section 4.3.3.6. The purpose of the design is to compare the effects of two treatments and their interactions on some response variable, y. Although other blocking structures may be used, the randomized blocks blocking structure is the most commonly used and is therefore illustrated here. As has been discussed above, the blocking structure used does not affect the interpretation of the treatment effects and therefore the treatment null hypotheses. This design has two treatment factors, treatment A with a levels which are applied to the main-plots and treatment B with b levels applied to sub-plots. The appropriate statistical model for the design is as follows:

$$y_{hij} = \mu + Bl_h + A_i + em_{hi} + B_j + (AB)_{ij} + e_{hij} \qquad \text{for } h = 1 \ldots r; \ i = 1 \ldots a; \ j = 1 \ldots b$$

where μ represents the overall treatment mean response, Bl_h represents the effect of the hth block, A_i represents the mean effect of treatment A at level i, B_j represents the mean effect of treatment B at level j, $(AB)_{ij}$ represents the mean effect of the interaction between treatment A at level i and treatment B at level j, em_{hi} represents the main-plot error and e_{hij} the sub-plot error.

Hypotheses   There are three treatment effects and a block effect which are picked out by this analysis although there need not be any null hypothesis relating to the block effect. The objective of the experiment is to look for all the treatment effects so these are the only hypotheses that are tested in a two-factor split-plot in randomized blocks design.

$$H_0: \ A_i = \mu_i - \mu = 0 \qquad \text{for } i = 1; \ \ldots \ a$$

Table 6.16   Tables of oven dry matter (%) mean values and number of observations, n, for the oven dry matter data

Main effects

	Oat variety					Sowing date		Nitrogen rate	
	A	B	C	D	E	D1	D2	N1	N2
Mean	86.5	86.6	86.5	86.4	86.7	86.5	86.6	86.6	86.5
n	12	12	12	12	12	30	30	30	30

First-order interactions

$n = 6$	Sowing date	
Variety	D1	D2
A	86.5	86.5
B	86.6	86.6
C	86.5	86.5
D	86.4	86.5
E	86.8	86.7

$n = 6$	Nitrogen rate	
Variety	D1	D2
A	86.5	86.5
B	86.6	86.7
C	86.6	86.4
D	86.5	86.3
E	86.8	86.7

$n = 15$	Sowing date	
Nitrogen	D1	D2
N1	86.6	86.6
N2	86.5	86.6

Second-order interaction

$n = 3$	Sowing date			
	S1 Nitrogen rate		S2 Nitrogen rate	
Variety	N1	N2	N1	N2
A	86.5	86.5	86.5	86.5
B	86.5	86.8	86.6	86.7
C	86.7	86.2	86.4	86.5
D	86.5	86.2	86.5	86.5
E	86.8	86.8	86.8	86.6

H_0: $B_j = \mu_j - \mu = 0$ for $j = 1 \ldots b$
H_0: $AB_{ij} = \mu_{ij} - \mu = 0$ for $i = 1 \ldots a$; $j = 1 \ldots b$
The null hypotheses state the following:
All treatment factor A population mean values are equal.

All treatment factor B population mean values are equal.
All AB interaction population mean values are equal.

H_1: $A_i = \mu_i - \mu \neq 0$ for at least one i
H_1: $B_j = \mu_j - \mu \neq 0$ for at least one j
H_1: $AB_{ij} = \mu_{ij} - \mu \neq 0$ for at least one i or j
The alternative hypotheses state the following:
At least one of the treatment factor A population mean values is not equal to the overall mean.
At least one of the treatment factor B population mean values is not equal to the overall mean.
At least one of the interaction population mean values is not equal to the overall mean.

A_i is the factor A treatment effect, B_j is the factor B treatment effect, AB_{ij} is the interaction effect, μ is a constant representing the overall population mean, μ_i is the ith factor A treatment population mean, μ_j is the jth factor B treatment population mean, and a and b are the number of treatments in factors A and B respectively.

Decision rule   The effects that are picked out from the analysis together with the residual term and appropriate calculations are illustrated in Table 6.17. A more detailed coverage of the calculations involved in producing the analysis of variance table is given in Appendix II.

The factor A sum of squares is a measure of the variability of the factor A sample means and the experiment error. The factor B sum of squares is a measure of the variability of the factor B sample means and the experiment error. The AB interaction sum of squares is a measure of the variability of the AB interaction sample means and the experiment error. Under the null hypotheses the numerators and denominators of the F-values or variance ratios are expected to be equal. If a treatment effect does exist then the treatment mean square will have a higher expected value than the residual mean square so the F-value will produce a significant result when compared against those in tables.

Although the block effect has been accounted for in the analysis by separating out its sum of squares from the residual sum of squares, a variance ratio has not been calculated. Knowledge of block differences does not affect the interpretation of the treatment effects. If not accounted for in the analysis, the block effect will be included in the main-plot residual term which affects only those effects tested against the main-plot residual.

Table 6.17   Construction of an analysis of variance table for a two factor split-plot design in randomized blocks

Source	df	ss	ms	vr
Block (Bl_h)	df_{Bl}	ss_{Bl}	$ms_{Bl} = ss_{Bl}/df_{Bl}$	
Factor A (A_i)	df_A	ss_A	$ms_A = ss_A/df_A$	$F_A = ms_A/ms_{EM}$
Main-plot residual (em_{hi})	df_{EM}	$ss_{EM} = ss_{MTOT} - ss_A - ss_{Bl}$	$ms_{EM} = ss_{EM}/df_{EM}$	
Main-plot total	df_{MTOT}	ss_{MTOT}		
Factor B (B_j)	df_B	ss_B	$ms_B = ss_B/df_B$	$F_B = ms_B/s^2$
AB interaction (AB_{ij})	df_{AB}	ss_{AB}	$ms_{AB} = ss_{AB}/df_{AB}$	$F_{AB} = ms_{AB}/s^2$
Sub-plot residual (e_{hij})	df_E	$ss_E = ss_{TOT} - ss_B - ss_{AB} - ss_{MTOT}$	$s^2 = ss_E/df_E$	
Total	df_{TOT}	ss_{TOT}		

General interpretation   The interpretation of all treatment effects in the two-factor split-plot design in randomized blocks is identical to that for the two-way factorial design in randomized blocks.

The analysis and interpretation of these designs is detailed in both Cochran and Cox (1957) and Steel and Torrie (1980).

Example   The analysis presented in this example is from the experiment designed in Section 4.3.3.6. This experiment investigated the effect of seven nitrogen rates, N1 to N7, and four fallow treatments, A to D, on specific weight of a winter wheat crop. The experiment is a split-plot design in four randomized blocks with fallow treatments on the main-plots and nitrogen treatments on the sub-plots. There are 28 plots in each block.

Hypotheses

Fallow main effect

H_0: $F_i = \mu_i - \mu = 0$ for $i = 1 \ldots 4$
H_1: $F_i = \mu_i - \mu \neq 0$ for at least one i
The null hypothesis is that there are no differences in the four fallow treatment mean specific weight values. The alternative is that there is at least one difference.

Nitrogen main effect

H_0: $N_j = \mu_j - \mu = 0$ for $j = 1 \ldots 7$
H_1: $N_j = \mu_j - \mu \neq 0$ for at least one j
The null hypothesis is that there are no differences in the seven nitrogen rate mean specific weight values. The alternative is that there is at least one difference.

First-order interaction

H_0: $FN_{ij} = \mu_{ij} - \mu = 0$ for $i = 1 \ldots 4;\ j = 1 \ldots 7$
H_1: $FN_{ij} = \mu_{ij} - \mu \neq 0$ for at least one i or j
The null hypothesis is that there are no differences in the interaction mean specific weight values. The alternative is that there is at least one difference.

Graphs   A histogram of the specific weights is shown in Figure 6.37. The data appear to be reasonably normally distributed with no obvious outlying values although there is a suggestion of a skewed distribution.

The assumptions of constant treatment variance and normally distributed errors with zero mean need to be checked. A histogram of the residuals and a plot of the residuals and fitted values are illustrated in Figures 6.38 and 6.39 respectively. The histogram shows the typical bell-shape associated with a normal distribution. The points on the scatter plot form a random scatter in a horizontal band about the zero residual. There is a suggestion that a few points are slightly extreme, e.g. the point in the bottom left-hand corner and the high point in the middle of the plot. Both points have residuals numerically greater than 1.5. These points may be queried and it would be prudent to go back to the original raw data and check that they are genuine observations. They are not so extreme, however, that the residual plot would be considered unacceptable and it is unlikely that two such values

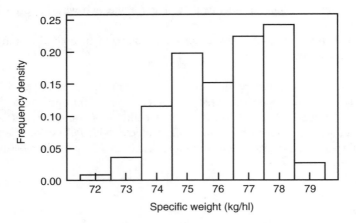

Figure 6.37   Specific weight for a winter wheat (kg/hl)

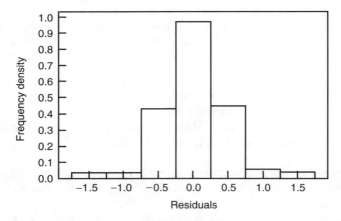

Figure 6.38   Residuals obtained from fitting a two-factor split-plot in randomized blocks model to the specific weight data

would change any conclusions drawn. Both graphs indicate that the two assumptions are satisfied by the data.

Tables   The analysis of variance table and basic summary statistics calculated from the specific weight data are given in Tables 6.18 and 6.19.

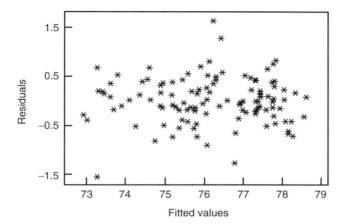

Figure 6.39   Residuals and fitted values obtained from fitting a two-factor split-plot in randomized blocks model to the specific weight data

Table 6.18   Analysis of variance table for the specific weight data

Source	df	ss	ms	vr	$F_{probability}$
Block (Bl_h)	3	14.2054	4.7351		
Fallow (F_i)	3	44.4883	14.8294	10.64	0.003
Main-plot residual (em_{hi})	9	12.5411	1.3935		
Main-plot total	15	71.2348			
Nitrogen (N_j)	6	157.1057	26.1843	76.64	< 0.001
AB interaction (FN_{ij})	18	12.6253	0.7014	2.05	0.017
Sub-plot residual (e_{hij})	72	24.5992	0.3417		
Total	111	265.5650			

Conclusions   The F-test for the interaction term is significant at the 0.05 level, $p = 0.017$, therefore this null hypothesis is rejected. The F-tests for the two main effects are also significant at the 5% level of significance but because their interaction is significant this effect has to be examined first. The interaction plot is given in Figure 6.40

The interaction plot shows a clear trend of increasing specific weight as the nitrogen level increases following an initial reduction at the second nitrogen rate. Fallow treatments A and D produce consistently higher specific weights than the other two treatments at all nitrogen levels. Treatment C tends to produce the lowest specific weights and certainly provides the lowest specific weights at the middle to high nitrogen rates. Treatment B is slightly different from the other treatments in that it covers a much greater range of specific weights, moving from the lowest values at the low nitrogen rates to almost matching the highest values at the high nitrogen rates.

Table 6.19   Tables of specific weight (kg/hl) mean values and numbers of observations, n, for the specific weight data

Main effects

	Nitrogen							Fallow			
	N1	N2	N3	N4	N5	N6	N7	A	B	C	D
Mean	74.64	74.51	75.37	76.80	77.07	77.22	77.50	76.80	75.77	75.33	76.74
n	16	16	16	16	16	16	16	28	28	28	28

First-order interaction

$n=4$	Nitrogen						
Fallow	N1	N2	N3	N4	N5	N6	N7
A	75.95	75.07	76.02	77.09	77.51	77.95	77.99
B	73.81	73.45	74.64	76.62	77.01	77.00	77.84
C	73.73	74.00	75.07	75.90	76.28	76.16	76.19
D	75.09	75.53	75.74	77.61	77.48	77.78	78.00

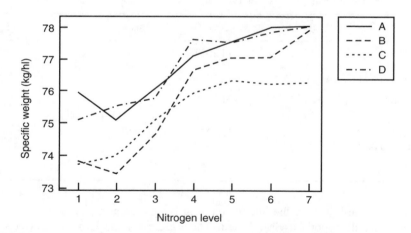

Figure 6.40   Specific weight interaction plot for the nitrogen and fallow treatment factors

Without further investigations and analysis there is little more that can be inferred from the analysis. Modelling of the observed trends must be considered, particularly given one of the experiment aims. Techniques for investigating trends, such as regression analysis (Section 6.1.7) and polynomial contrasts (Sections 6.1.6.1 and 6.1.6.5), are discussed in later sections. There are signs that the specific weight response to nitrogen is reaching or has reached a plateau so it is possible that optimum values of nitrogen for achieving maximum crop specific weight could be derived. Parallel curve analysis could provide additional information, such as indications of whether the optima differ between fallow treatments and details of the form of the nitrogen response curves for each fallow treatment, from these data. The technique is covered briefly in Section 6.1.7.5 and the interested reader is referred to Weisberg (1985) for a more detailed discussion of this technique. An examination of the interaction by looking at the fallow treatment effect for each of the nitrogen levels may help identify the cause of the significant interaction. The method for doing this is covered and illustrated by example in Steel and Torrie (1980).

If the main aim of the experiment had been to examine the response surface caused by the different fallow and nitrogen treatments, a different experiment design may have been more appropriate. The interested reader is referred to Draper and Smith (1981) and Box, Hunter and Hunter (1978) for further details of response surface analysis.

6.1.5.7 Multi-treatment factor split-plot designs in randomized blocks

The above discussion of the two-factor split-plot design extends to multi-factor split-plot designs arranged in randomized blocks. These designs are described in Section 4.3.4. Although other blocking structures may be used, the randomized blocks blocking structure is the most commonly used. The blocking structure used does not affect the interpretation of the treatment effects and therefore the treatment null hypotheses.

The treatment factors in these designs are divided into main-plot treatment factors and sub-plot treatment factors and, if applicable, sub-sub-plot treatment factors and more. It does not matter how these treatment factors are divided, the interpretation of their effects on a response variable is the same. It is the power of the test that changes when factors are placed on sub-plots rather than main-plots or sub-sub-plots rather than sub-plots, etc.

The appropriate statistical model is an extension of that given for two-factor split-plot designs in Section 6.1.5.6. The exact way in which the model is extended depends on the treatment structure.

Hypotheses   There are as many treatment effects as there are treatment factors and interactions. These, along with the block effect, are accounted for in the analysis although there need not be any null hypothesis for the block effect. The objective of the experiment is to look for treatment effects so these are the only effects tested in a multi-factor split-plot design. Examples of hypotheses are shown below.
The main effect null hypotheses will be of the following form:

H_0: $C_k = \mu_k - \mu = 0$ for $k = 1 \ldots c$
All treatment factor C population mean values are equal.

The first-order interaction effect null hypotheses will be of the following form:

H_0: $BC_{jk} = \mu_{jk} - \mu = 0$ for $j = 1 \ldots b$; $k = 1 \ldots c$
All BC interaction population mean values are equal.

The second-order interaction effect null hypotheses will be of the form:

H_0: $BCD_{jkl} = \mu_{jkl} - \mu = 0$ for $j = 1 \ldots b$; $k = 1 \ldots c$; $l = 1 \ldots d$
All BCD interaction population mean values are equal.

There is only one possible alternative hypothesis for each null hypothesis. For those illustrated they will be as follows:

H_1: $C_k = \mu_k - \mu \neq 0$ for at least one k
H_1: $BC_{jk} = \mu_{jk} - \mu \neq 0$ for at least one j or k
H_1: $BCD_{jkl} = \mu_{jkl} - \mu \neq 0$ for at least one j or k or l

At least one of the treatment factor C population mean values is not equal to the overall mean.
At least one of the BC interaction population mean values is not equal to the overall mean.
At least one of the BCD interaction population mean values is not equal to the overall mean.

Higher-order interactions will each have their own null hypothesis and alternative hypothesis which will be in similar formats.

Decision rule   The analysis of variance is constructed in the same way as for the two-factor split-plot and is therefore of the same form. However, if sub-plots have been further divided into sub-sub-plots or more there will be a residual term in the analysis of variance table for each division of the main-plot. The treatment effects estimated in each main-plot division are tested against the residual term associated with that division. For example, in a design with three treatment factors laid out as a split-split-plot design the treatment applied to the split-split-plots will have its effect and the effects of its interaction with all other treatments tested against the split-split-plot residual. The block effect is considered in the same manner as for the two-factor split-plot design in randomized blocks (Genstat 5 Committee, 1993).

The three different forms of three-factor split-plot designs are indicated in Tables 6.20–6.22. Each table indicates the different strata (divisions) that are formed and which treatment effects are estimated and tested in each strata.

General interpretation   Interpretation of treatment effects is similar to that for the two-factor split-plot. Non-significant results will be interpreted in the same way. However, if interaction effects are statistically significant then interpretation of any main effects or lower-order interactions associated with the significant interaction may or may not be possible. Therefore interpretation of multi-way split-plot designs always starts with the highest-order interaction term. Interpretation then proceeds to the next order interaction terms and so on until the main effects are interpreted. At each stage the presence of a significant interaction may prevent the interpretation of any effects contributing towards that interaction.

Examples of the interpretation of these designs can be found in Cochran and Cox (1957). Montgomery (1991) also provides a worked example complete with interpretation.

Example   The analysis considered in this example is from the experiment designed in Section 4.3.4.6. The experiment investigated the effects on potato yield of two physiological ages (A1 and A2) and two planting dates (P1 and P2) of seed potatoes treated with six different equispaced rates of nitrogen (N1 to N6). The experiment uses four randomized blocks with the four planting date/physiological age–treatment combinations randomized to the four main-plots in each block and the six nitrogen rates randomized to the six sub-plots in each main-plot. There are 24 plots in each block.

Table 6.20   Construction of an analysis of variance table for a three-way split-plot design with one treatment factor (A) on the main-plots and two treatment factors (B and C) on the sub-plots

Source	df	ss	ms	vr
Block (Bl_h)	df_{Bl}	ss_{Bl}	$ms_{Bl} = ss_{Bl}/df_{Bl}$	
Factor A (A_i)	df_A	ss_A	$ms_A = ss_A/df_A$	$F_A = ms_A/ms_{EM}$
Main-plot residual (em_{hi})	df_{EM}	$ss_{EM} = ss_{MTOT} - ss_A - ss_{Bl}$	$ms_{EM} = ss_{EM}/df_{EM}$	
Main-plot total	df_{MTOT}	ss_{MTOT}		
Factor B (B_j)	df_B	ss_B	$ms_B = ss_B/df_B$	$F_B = ms_B/s^2$
Factor C (C_k)	df_C	ss_C	$ms_C = ss_C/df_C$	$F_C = ms_C/s^2$
AB interaction (AB_{ij})	df_{AB}	ss_{AB}	$ms_{AB} = ss_{AB}/df_{AB}$	$F_{AB} = ms_{AB}/s^2$
AC interaction (AC_{ik})	df_{AC}	ss_{AC}	$ms_{AC} = ss_{AC}/df_{AC}$	$F_{AC} = ms_{AC}/s^2$
BC interaction (BC_{jk})	df_{BC}	ss_{BC}	$ms_{BC} = ss_{BC}/df_{BC}$	$F_{BC} = ms_{BC}/s^2$
ABC interaction (ABC_{ijk})	df_{ABC}	ss_{ABC}	$ms_{ABC} = ss_{ABC}/df_{ABC}$	$F_{ABC} = ms_{ABC}/s^2$
Sub-plot residual (e_{hijk})	df_E	$ss_E = ss_{TOT} - ss_B - ss_{AB} - ss_{MTOT}$	$s^2 = ss_E/df_E$	
Total	df_{TOT}	ss_{TOT}		

Table 6.21   Construction of an analysis of variance table for a three-way split-plot design with two treatment factors (A and B) on the main-plots and one treatment factor (C) on the sub-plots

Source	df	ss	ms	vr
Block (Bl_h)	df_{Bl}	ss_{Bl}	$ms_{Bl} = ss_{Bl}/df_{Bl}$	
Factor A (A_i)	df_A	ss_A	$ms_A = ss_A/df_A$	$F_A = ms_A/ms_{EM}$
Factor B (B_j)	df_B	ss_B	$ms_B = ss_B/df_B$	$F_B = ms_B/ms_{EM}$
AB interaction (AB_{ij})	df_{AB}	ss_{AB}	$ms_{AB} = ss_{AB}/df_{AB}$	$F_{AB} = ms_{AB}/ms_{EM}$
Main-plot residual (em_{hij})	df_{EM}	$ss_{EM} = ss_{MTOT} - ss_A - ss_{Bl}$	$ms_{EM} = ss_{EM}/df_{EM}$	
Main-plot total	df_{MTOT}	ss_{MTOT}		
Factor C (C_k)	df_C	ss_C	$ms_C = ss_C/df_C$	$F_C = ms_C/s^2$
AC interaction (AC_{ik})	df_{AC}	ss_{AC}	$ms_{AC} = ss_{AC}/df_{AC}$	$F_{AC} = ms_{AC}/s^2$
BC interaction (BC_{jk})	df_{BC}	ss_{BC}	$ms_{BC} = ss_{BC}/df_{BC}$	$F_{BC} = ms_{BC}/s^2$
ABC interaction (ABC_{ijk})	df_{ABC}	ss_{ABC}	$ms_{ABC} = ss_{ABC}/df_{ABC}$	$F_{ABC} = ms_{ABC}/s^2$
Sub-plot residual (e_{hijk})	df_E	$ss_E = ss_{TOT} - ss_B - ss_{AB} - ss_{MTOT}$	$s^2 = ss_E/df_E$	
Total	df_{TOT}	ss_{TOT}		

Hypotheses

Physiological age main effect

H_0: $A_i = \mu_i - \mu = 0$ for $i = 1, 2$
H_1: $A_i = \mu_i - \mu \neq 0$ for at least one i
The null hypothesis is that there is no difference between the two physiological age mean potato yields. The alternative hypothesis is that there is a difference.

Planting date main effect

H_0: $P_j = \mu_j - \mu = 0$ for $j = 1, 2$
H_1: $P_j = \mu_j - \mu \neq 0$ for at least one j

Table 6.22　Construction of an analysis of variance table for a three-way split-plot design with one treatment factor (A) on the main-plots, one treatment factor (B) on the sub-plots and one treatment factor (C) on the sub-sub-plots

Source	df	SS	ms	vr
Block (Bl_h)	df_{Bl}	SS_{Bl}	$ms_{Bl} = SS_{Bl}/df_{Bl}$	
Factor A (A_i)	df_A	SS_A	$ms_A = SS_A/df_A$	$F_A = ms_A/ms_{EM}$
Main-plot residual (em_{hi})	df_{EM}	$SS_{EM} = SS_{MTOT} - SS_A - SS_{Bl}$	$ms_{EM} = SS_{EM}/df_{EM}$	
Main-plot total	df_{MTOT}	SS_{MTOT}		
Factor B (B_j)	df_B	SS_B	$ms_B = SS_B/df_B$	$F_B = ms_B/ms_{ES}$
AB interaction (AB_{ij})	df_{AB}	SS_{AB}	$ms_{AB} = SS_{AB}/df_{AB}$	$F_{AB} = ms_{AB}/ms_{ES}$
Sub-plot residual (es_{ij})	df_{ES}	$SS_{ES} = SS_{STOT} - SS_B - SS_{AB} - SS_{MTOT}$	$ms_{ES} = SS_{ES}/df_{ES}$	
Sub-plot total	df_{STOT}	SS_{STOT}		
Factor C (C_k)	df_C	SS_C	$ms_C = SS_C/df_C$	$F_C = ms_C/s^2$
AC interaction (AC_{ik})	df_{AC}	SS_{AC}	$ms_{AC} = SS_{AC}/df_{AC}$	$F_{AC} = ms_{AC}/s^2$
BC interaction (BC_{jk})	df_{BC}	SS_{BC}	$ms_{BC} = SS_{BC}/df_{BC}$	$F_{BC} = ms_{BC}/s^2$
ABC interaction (ABC_{ijk})	df_{ABC}	SS_{ABC}	$ms_{ABC} = SS_{ABC}/df_{ABC}$	$F_{ABC} = ms_{ABC}/s^2$
Sub-sub-plot residual (e_{hijk})	df_E	$SS_E = SS_{TOT} - SS_C - SS_{AC} - SS_{BC} - SS_{ABC} - SS_{STOT}$	$s^2 = SS_E/df_E$	
Total	df_{TOT}	SS_{TOT}		

The null hypothesis is that there is no difference between the two planting date mean potato yields. The alternative is that there is a difference.

Nitrogen main effect

H_0: $N_k = \mu_k - \mu = 0$　　for $k = 1 \ldots 6$
H_1: $N_k = \mu_k - \mu \neq 0$　　for at least one k
The null hypothesis is that there are no differences in the six nitrogen rate mean potato yields. The alternative is that there is at least one mean value that is different from the rest.

First-order interactions

H_0: $AP_{ij} = \mu_{ij} - \mu = 0$　　for $i = 1, 2; j = 1, 2$
H_1: $AP_{ij} = \mu_{ij} - \mu \neq 0$　　for at least one i or j

H_0: $AN_{ik} = \mu_{ik} - \mu = 0$　　for $i = 1, 2; k = 1 \ldots 6$
H_1: $AN_{ik} = \mu_{ik} - \mu \neq 0$　　for at least one i or k

H_0: $PN_{jk} = \mu_{jk} - \mu = 0$　　for $j = 1, 2; k = 1 \ldots 6$
H_1: $PN_{jk} = \mu_{jk} - \mu \neq 0$　　for at least one j or k
The null hypotheses relate to the first-order interactions and state that there are no differences in the first-order interaction mean potato yields. The alternatives are that there is at least one mean value different from the rest.

Second-order interaction

H_0: $APN_{ijk} = \mu_{ijk} - \mu = 0$　　for $i = 1, 2; j = 1, 2; k = 1 \ldots 6$
H_1: $APN_{ijk} = \mu_{ijk} - \mu \neq 0$　　for at least one i or j or k

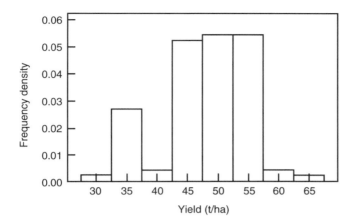

Figure 6.41   Potato yield (t/ha)

The null hypothesis is that there are no differences in the second-order interaction mean potato yields. The alternative is that there is at least one mean value different from the rest.

Graphs   A histogram of potato yield is shown in Figure 6.41. The data appear to be reasonably normally distributed with no obvious outlying values.

The model assumptions of constant treatment variance and normally distributed errors with zero mean need to be checked. A histogram of the model residuals is plotted in Figure 6.42 and indicates they are normally distributed. The residuals and fitted values are plotted in Figure 6.43. It can be seen that the points form a random scatter in a horizontal band around the zero residual. Both graphs of residuals indicate that these two model assumptions are satisfied.

Tables   Detailed in Tables 6.23 and 6.24 are the analysis of variance table and basic summary statistics for the potato yield data.

Conclusions   The F-test for the second-order interaction is not statistically significant at the 0.05 level, therefore this null hypothesis is accepted. It makes sense to next interpret the first-order interaction effects. The only one of these that is statistically significant is the planting date/nitrogen rate interaction. This is the only first-order interaction null hypothesis that is rejected. It is concluded that the differences in mean yield between the six nitrogen rates differ across the two planting dates. A plot of the interaction is given in Figure 6.44. The two lines are diverging at the higher nitrogen levels illustrating the significant interaction effect. However as the curves on the interaction plot are both very similar in shape and only cross at the lowest level, it is possible to interpret these main effects with the interaction effect. Yield at planting date 1 is higher for all nitrogen treatments except N1. The effect of age is not significant at the 0.05 level.

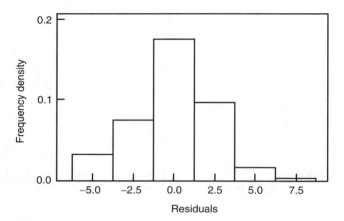

Figure 6.42   Residuals obtained from fitting a three-factor split-split-plot in randomized blocks model
to the potato yield data

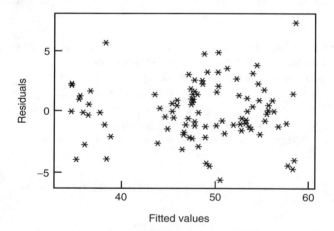

Figure 6.43   Residuals and fitted values obtained from fitting a three-factor split-split-plot in
randomized blocks model to the potato yield data

Further investigations of the interaction effect are warranted, particularly with regard to
identifying the form of the yield response to nitrogen. The statistically significant planting
date/nitrogen interaction is a clear indication that the yield response curves differ for the two
planting dates. The use of orthogonal polynomials and regression analysis would provide
additional useful information and these techniques are covered in Sections 6.1.6.1, 6.1.6.5
and 6.1.6.7.

Table 6.23   Analysis of variance table for the potato yield data

Source	df	ss	ms	vr	$F_{probability}$
Block (Bl_h)	3	69.974	23.325		
Age (A_i)	1	18.123	18.123	1.49	0.254
Planting (P_j)	1	620.730	620.730	50.94	<0.001
AP interaction (AP_{ij})	1	3.592	3.592	0.29	0.600
Main-plot residual (em_{hij})	9	109.667	12.185		
Main-plot total	15	822.088			
Nitrogen (N_k)	5	2994.509	598.902	68.72	<0.001
AN interaction (AN_{ik})	5	26.372	5.274	0.61	0.696
PN interaction (PN_{jk})	5	197.020	39.404	4.52	0.001
APN interaction (APN_{ijk})	5	19.681	3.936	0.45	0.81
Sub-plot residual (e_{hijk})	60	522.914	8.715		
Total	95	4582.583			

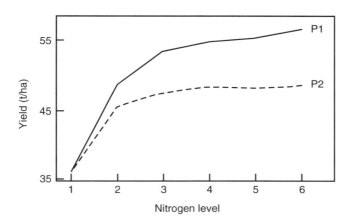

Figure 6.44   Potato yield interaction plot for the nitrogen and planting date treatment factors. P1, planting date 1; P2, planting date 2

6.1.5.8   Factorial plus control designs in randomized blocks

The discussion of the factorial design can be extended to the factorial plus control design. The factorial part of the treatment structure remains unchanged. The differences are in the way in which the control treatment is handled. This treatment may be an untreated control treatment or a standard treatment. Although other blocking structures may be used, the randomized blocks blocking structure is the most commonly used and is therefore illustrated here. The blocking structure used does not affect the interpretation of the treatment effects and therefore the treatment null hypotheses.

Table 6.24   Tables of potato yield (t/ha) mean values and numbers of observations, *n*, for the potato yield data

Main effects

	Nitrogen						Physiological age		Planting date	
	N1	N2	N3	N4	N5	N6	A1	A2	P1	P2
Mean	36.31	47.11	50.36	51.46	51.61	52.41	47.78	48.64	50.75	45.67
n	16	16	16	16	16	16	48	48	48	48

First-order interactions

n = 8	Nitrogen					
Planting date	N1	N2	N3	N4	N5	N6
P1	36.21	48.58	53.33	54.68	55.24	56.48
P2	36.42	45.64	47.39	48.24	47.97	48.34

n = 8	Nitrogen					
Physiological age	N1	N2	N3	N4	N5	N6
A1	36.03	46.93	50.83	50.47	50.51	51.88
A2	36.59	47.28	49.90	52.45	52.70	52.94

n = 24	Planting date	
Physiological age	P1	P2
A1	50.12	45.43
A2	51.38	45.91

Second-order interaction

n = 4		Nitrogen					
Physiological age	Planting date	N1	N2	N3	N4	N5	N6
A1	P1	35.91	48.70	52.77	53.37	53.78	56.22
A1	P2	36.16	45.16	48.88	47.58	47.24	47.53
A2	P1	36.51	48.46	53.90	55.99	56.70	56.74
A2	P2	36.67	46.11	45.91	48.91	48.70	49.15

Hypotheses The treatment combinations have two parts: one for the fully balanced factorial arrangement of treatments and the other for the control treatment. The treatment effects and block effects are identified by the analysis although there need not be any null hypothesis relating to the block effect. The objective of the experiment is to look for treatment effects so these are the only hypotheses that are tested in a factorial plus control design.

H_0: $CONTROL_i = \mu_i - \mu = 0$ for $i = 1, 2$
There is no difference between the control treatment mean and the mean of all the other treatments.

The main effect null hypotheses will be of the following form:

H_0: $C_k = \mu_k - \mu = 0$ for $k = 1 \ldots c$
All treatment factor C population mean values are equal.

The interaction effect null hypotheses will be of the following form:

H_0: $BC_{jk} = \mu_{jk} - \mu = 0$ for $j = 1 \ldots b$; $k = 1 \ldots c$
All BC interaction population mean values are equal.

For the illustrated null hypotheses the alternative hypotheses are as follows:

H_1: $CONTROL_i = \mu_i - \mu \neq 0$ for at least one i
H_1: $C_k = \mu_k - \mu \neq 0$ for at least one k
H_1: $BC_{jk} = \mu_{jk} - \mu \neq 0$ for at least one j or k
The control treatment mean is not equal to the overall mean.
At least one of the treatment factor C population mean values is not equal to the overall mean.
At least one of the BC interaction population mean values is not equal to the overall mean.

Decision rule The analysis of variance table is constructed slightly differently to that for a straightforward factorial design. The effects of the control treatment have to be accounted for in the analysis. Once this effect has been determined the effects of the factorial treatments still have to be accounted for. This gives a nested treatment structure where the effects of the factorial treatments cannot be assessed until the control treatment has been separated out from all effects. Once this has been done the factorial treatment effects are estimated in the same way as for a straightforward factorial design. The one residual term is used to test all treatment and control effects. The block effect is considered in the same manner as for the factorial design in randomized blocks (Genstat 5 Committee, 1993).

General interpretation A non-significant control effect indicates there is no evidence of a difference between the control mean and the mean of all the other treatments. A significant control effect indicates there is a statistical difference between the control mean and the mean of all the other treatments. Interpretation of factorial treatment effects is similar to that for the two-way or multi-way factorial. Non-significant results will be interpreted in the same way. However, if interaction effects are statistically significant then interpretation of any main effects or lower-order interactions associated with the significant interaction may or may not be possible. Therefore interpretation of the factorial part of the treatment effects always starts with the highest-order interaction term. Interpretation then proceeds to the next order interaction terms and so on until the main effects are interpreted. At each stage the presence of a significant interaction may prevent the interpretation of any effects contributing towards that interaction.

Example The experiment analysed in this example is the one that was described in Section 4.3.5.6. This experiment studied the effect on nitrogen uptake of a cereal crop when different rates, R1 to R3, of poultry manure were applied at different times, T1 and T2, to the previous potato crop. There is an untreated control, C. The experiment used three randomized blocks.

Hypotheses

Treatment main effect

H_0: $C_i = \mu_i - \mu = 0$ for $i = 1, 2$
H_1: $C_i = \mu_i - \mu \neq 0$ for $i = 1, 2$
There is no difference in control mean nitrogen uptake and the poultry manure treatment mean nitrogen uptake. The alternative is that there is a difference.

Timing main effect

H_0: $T_j = \mu_j - \mu = 0$ for $j = 1, 2$
H_1: $T_j = \mu_j - \mu \neq 0$ for $j = 1, 2$
There is no difference in the two timing mean nitrogen uptake values. The alternative is that there is a difference.

Manure rate main effect

H_0: $R_k = \mu_k - \mu = 0$ for $k = 1 \ldots 3$
H_1: $R_k = \mu_k - \mu \neq 0$ for $k = 1 \ldots 3$
There are no differences in the three manure rate mean nitrogen uptake values. The alternative is that at least one of the mean values is different from the rest.

First-order interaction

H_0: $TR_{jk} = \mu_{jk} - \mu = 0$ for $j = 1, 2; k = 1 \ldots 3$
H_1: $TR_{jk} = \mu_{jk} - \mu \neq 0$ for $j = 1, 2; k = 1 \ldots 3$
There are no differences in the interaction mean nitrogen uptake values. The alternative is that at least one of the six mean values is different from the rest.

Graphs A histogram of the nitrogen uptake data is shown in Figure 6.45. The data appear to be reasonably normally distributed though there are a few slightly high values. However, the histogram looks acceptable.

The model assumptions of normally distributed errors with zero mean and constant treatment variance need to be checked. A histogram of the model residuals is plotted in Figure 6.46 and the residuals and fitted values are plotted in Figure 6.47. Both graphs indicate that the two assumptions are satisfied. The histogram looks like a normal distribution and the scatter plot points form a random scatter in a horizontal band about the zero residual.

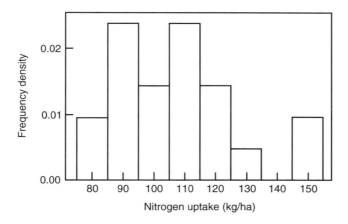

Figure 6.45   Nitrogen uptake of a cereal crop (kg/ha)

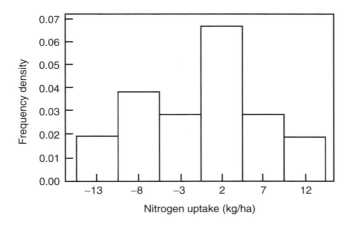

Figure 6.46   Residuals obtained from fitting a two-way factorial plus control in randomized blocks
model to the nitrogen uptake data

Tables   Tables 6.25 and 6.26 detail the analysis of variance table and basic summary
statistics for the nitrogen uptake data.

Conclusions   The control effect is significant at the 0.009 level so the null hypothesis is
rejected. It is concluded that the mean nitrogen uptake for the untreated control treatment
differs from the mean nitrogen uptake of all the poultry manure treatments. Applying a
poultry manure treatment increases the nitrogen uptake of the cereal crop. The interaction

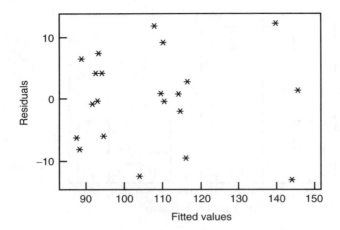

Figure 6.47   Residuals and fitted values obtained from fitting a two-way factorial plus control in randomized blocks model to the nitrogen uptake data

Table 6.25   Analysis of variance table for the nitrogen uptake data

Source	df	ss	ms	vr	$F_{probability}$
Block (Bl_h)	3	128.11	64.06		
Control (C_i)	1	886.90	886.90	9.84	0.009
Timing (T_j)	1	2324.94	2324.94	25.79	< 0.001
Rate (R_k)	2	3046.50	1523.25	16.90	< 0.001
TR interaction (TR_{jk})	2	158.38	79.19	0.88	0.440
Residual (e_{hijk})	12	1081.58	90.13		
Total	20	7626.41			

term is not significant at the 0.05 level so the null hypothesis of no interaction is accepted. It is concluded that there is no interaction between timing and rate of manure application. The main effect of manure timing is significant, $p < 0.001$. It is concluded that the mean nitrogen uptake at the first timing is significantly higher than the mean nitrogen uptake at the second timing. The main effect of rate is significant at the < 0.001 level so it is concluded that the different manure rates affect the mean nitrogen uptake values of the cereal crop in different ways.

With just three manure rates it is not very practical to conduct a modelling exercise to determine the form of the nitrogen uptake response to manure rate. However, the use of polynomial contrasts would be helpful in determining whether the general response was linear or curvilinear over the range of manure rates used. This technique is described briefly in Section 6.1.6.1 with an example in Section 6.1.6.5. A fuller coverage, with a detailed example, can be found in Steel and Torrie (1980).

Table 6.26   Tables of nitrogen uptake mean values (kg/ha) and number of observations, n, for the nitrogen uptake data

Main effects

	Control	Treatment	Manure timing		Manure rate		
			T1	T2	R1	R2	R3
Mean	91.5	110.0	121.4	98.7	99.5	102.2	128.4
n	3	18	9	9	6	6	6

First-order interaction

$n = 3$	Manure rate		
Manure timing	R1	R2	R3
T1	107.1	113.8	143.2
T2	91.9	90.6	113.5

6.1.5.9   Some other designs

Other designs have been mentioned in Section 4.4. The intention here is not to discuss the interpretation of each design in detail. Regardless of the design used, the basic principles of interpreting treatment effects are the same as for the simpler designs mentioned in detail above. The treatment null hypothesis will always test for no difference between treatment means. Care must always be taken when interpreting interaction effects. The principles of interpretation outlined in the factorial models will apply for any interaction effects in these more complex models.

   In many of these designs the statistical analysis can get complicated and may produce very large and complex analysis of variance tables. As with any analysis of variance table, the meaning of all the treatment effects must be understood before any interpretation is attempted. If the designs have been carefully thought out and planned before practical work is started, the objectives clearly stated and the experiment run successfully then although the actual analysis is more mathematically complex the interpretation can be as easy as for more straightforward designs.

6.1.6   ANALYSIS OF MEANS

Keypoints

- An overall treatment effect is required if any further testing of treatment means is required.
- Contrasts are planned independent comparisons.
- The number of planned comparisons is limited by the number of degrees of freedom.
- Unplanned tests have an increased risk of Type I errors.

Frequently the conclusions drawn from the null hypothesis of equal treatment population means will not be particularly useful. A conclusion that 'at least one of the population treatment means differs from the overall mean' is not especially informative, although the conclusion that 'there is no difference between population treatment means' will be of merit. Differences between specific treatments or groups of treatments may be of particular interest and these comparisons will be determined from knowledge about the treatments included in an experiment. For example, crop disease levels on untreated control plots are expected to be different from those treated with a fungicide. In these cases it would be useful to formulate additional null hypotheses at the *planning stage*. This will allow more detailed investigation of the anticipated differences in the treatment population means. For example, 'there is no difference between the mean values of the treated and untreated plots' may be an additional null hypothesis.

Hypotheses formulated at the planning stage, before any data are collected, cannot be biased by the data as the data are not available to help determine which hypotheses to test. This means that the chance of rejecting the null hypothesis and stating, incorrectly, that a treatment difference exists, will be equal to the significance level (the Type I error probability). These planned comparisons are known as *contrasts*, occasionally referred to as prior or a priori, and are objective driven. That is, they are formulated to meet the objectives of the experiment and are due to knowledge of the treatments. The number of planned comparisons must be no more than the number of degrees of freedom available.

If the comparisons are not formulated until after the data have been inspected then the comparisons chosen will be data driven, which usually means that only the larger differences between treatments will be selected for testing as treatments which are apparently similar will not be selected for testing. These are unplanned comparisons, occasionally referred to as posterior, post hoc or a posteriori. For these comparisons the stated significance level of the test will not be the same as the Type I error rate. The Type I error rate will be greater than the significance level as the comparisons selected will tend to be in favour of finding a significant difference. It is the problem of controlling this error rate that makes it difficult to compare treatment means.

The probability of making a Type I error also increases as the number of comparisons made increases. The overall Type I error rate is the probability of making a Type I error in a set of comparisons and this will be higher than the Type I error rate for an individual comparison. The overall Type I error rate is sometimes referred to as the experimentwise error. If k pairwise comparisons are tested at a significance level of α, the overall Type I error rate is $1-(1-\alpha)^k$ which is the probability that at least one of the pairwise comparisons will result in erroneously rejecting the null hypothesis. If the comparisons being made are not independent this calculation only gives a rough idea of the true error rate. The other side of this argument makes use of the formula to maintain a constant overall Type I error rate and determines the significance level to operate at to achieve this for a given number of comparisons. When more than one pairwise comparison is to be made, each individual test is made at a lower significance level than $\alpha\%$ in order to maintain the overall Type I error rate at $\alpha\%$.

As an illustration, Table 6.27 indicates the overall Type I error rate when using a significance level (α) of 0.05 for each comparison. For example, if four independent comparisons are being made, each at the 5% level of significance, the probability of making at least one Type I error is 0.19. If the overall Type I error rate is to be restricted to no more than 5% than for four independent pairwise comparisons each comparison needs to be tested at a significance level of 1.3%.

Table 6.27   Overall Type I error rates and significance levels when making different numbers of pairwise comparisons

No. of pairwise comparisons	Overall Type I error rate for individual comparisons made at 5% significance level	Significance level of comparisons required to maintain an overall Type I error rate of 5%
1	0.05	0.05
2	0.10	0.025
3	0.14	0.017
4	0.19	0.013
5	0.23	0.010
10	0.40	0.005

6.1.6.1   Contrasts

In many experiments the objectives or prior knowledge of the treatments will lead the researcher to asking additional questions that will lead to additional hypotheses being formulated. These will be more specific than the overall null hypothesis of no treatment effect and are formulated before any experiment data are collected so they cannot be biased in any way by the data. Such hypotheses which form linear combinations of mean values are referred to as linear contrasts. Details of how to calculate these contrasts and test them can be found in Steel and Torrie (1980) and Montgomery (1991). A contrast will have one degree of freedom and is tested using a t-test, although it can also be compared with the residual mean square and an F-test performed. Since the hypotheses are not data driven then the stated significance level, the chosen value of α, of any test will be equal to the Type I error rate.

If the F-test in the analysis of variance leads to a non-significant result, this leads to the null hypothesis being accepted and the conclusion is that there are no treatment differences. Even if additional hypotheses were formulated when the experiment was planned there will be no point in testing these hypotheses since it has been concluded that there are no treatment differences so there is no point in testing for specific differences. In fact, further testing could lead to misleading results as there will still be a chance of incorrectly concluding that a treatment difference exists. So contrasts should only be used when the analysis of variance table has indicated a significant treatment effect. The experimenter should be aware, however, that an overall non-significant result may be hiding a significant individual treatment effect. Such a scenario may be more likely when treatment factors have many levels. The conclusion of a non-significant treatment effect could be the correct conclusion or a false one that leads to the conclusion of no evidence of a treatment effect when in fact a treatment effect exists. This would suggest the experiment was too insensitive to the presence of treatment differences. The correct non-significant conclusion is made with a probability of $1-\alpha$ (the level of significance) and the false non-significant conclusion with a probability of β, (the probability of making a Type II error).

A special case of linear contrasts occurs when the set of contrasts form an orthogonal set. Orthogonal contrasts are independent and for such contrasts the treatment sum of squares in the analysis of variance can be partitioned into components, one for each contrast. Each contrast will have one degree of freedom and if there are t treatments in an experiment then a maximum of $t-1$ orthogonal contrasts can be tested as there are only $t-1$ treatment degrees of freedom. However, it is not necessary to use all the degrees of freedom. If a contrast (or hypothesis) is regarded as a question being asked of the data and it only makes sense to ask meaningful questions then the number of contrasts tested should be determined by the objectives of the experiment with an upper limit of $t-1$. It is possible for there to be more than one set of orthogonal contrasts for a given set of treatments. The experimenter has to choose the set that is the most appropriate for the experiment objectives.

In some circumstances it may be needlessly restrictive to limit the contrasts to $t-1$ orthogonal contrasts. If non-orthogonal contrasts are to be tested or more than $t-1$ contrasts are to be tested then the overall Type I error rate can be controlled by decreasing the significance level of the test (Box, Hunter and Hunter, 1978); so instead of using the standard 5% level of significance the α value used to reject the null hypothesis will be less than 0.05. The actual value is calculated as $1 - (1 - \alpha^1)^{1/k}$ where α' is the required overall Type I error rate and there are k comparisons. The appropriate values of α to maintain an overall Type I error rate of 5% are indicated in Table 6.27 for a number of comparisons. For example, with five pairwise comparisons, each would need to operate at the 1% level of significance in order to achieve an overall Type I error rate of 5%. This provides some sort of compensation for trying to ask more questions of the data than they can reliably answer.

Another special form of orthogonal contrast is the suite of polynomial contrasts. These contrasts are designed to investigate the existence of polynomial effects in a series of related treatments, such as diets with increasing concentrations of one particular foodstuff; the factor levels must have a natural numerical order. When the levels are equally spaced the contrasts can be calculated quite easily using the method of least squares. An illustrative example is provided in this text in Section 6.1.6.5. The use of orthogonal polynomials is clearly explained and illustrated by example in both Montgomery (1991) and Steel and Torrie (1980), with details of the calculations involved also provided.

The first polynomial contrast examines the linear effect; the response variate increasing or decreasing at a constant rate as the treatment levels increase. The second examines the quadratic effect by testing the statistical significance of adding a squared term to the linear effect, thus allowing the rate of increase or decrease of the response variate to change with increasing treatment levels. The third polynomial contrast investigates the significance of adding a cube term to the quadratic effect. And so on.

Each contrast uses a single degree of freedom so the maximum order of polynomial that can be used is restricted by the number of treatment degrees of freedom and hence by the number of treatment levels in the treatment factor. With a treatment factor of five levels it is possible to investigate polynomials up to the fourth order, i.e. the quartic. With a two-level treatment factor it is only possible to examine the first-order polynomial, the straight line. If a series of polynomial contrasts are incorporated into the analysis of variance table then the shape of the variate response to treatment is determined by the highest order polynomial contrast that is statistically significant. For example, if the linear, quadratic and cubic contrasts were estimated for a four-level treatment factor and only the quadratic contrast was statistically significant then the quadratic curve would be an appropriate polynomial curve for the data.

6.1.6.2 Multiple range tests

If the comparisons to be tested are data driven, i.e. if they are formulated after the data have been investigated, then only apparently large differences will tend to be tested. Treatments which appear to be similar in mean values will not usually be chosen to be tested. This will mean that there will be a greater probability of making a Type I error (mistakenly claiming a difference between treatments is statistically significant). Multiple range tests are used to test such data-driven comparisons. These tests are non-discriminate so any treatment structure will be completely lost as all treatment factor combinations will be tested as if they were independent. For example, multiple range tests are inappropriate when testing a series of dose rates for a particular product.

Multiple comparison test procedures should not be applied if the F-test in the analysis of variance leads to a non-significant result as this leads to the null hypothesis being accepted and the conclusion that there are no treatment differences. In fact, further testing could lead to misleading results as there would be a chance of incorrectly concluding from any further testing that a treatment difference exists.

Views on the merits of multiple comparison testing are diverse, covering the extremes of those who never do it to those who do it as a matter of course. The majority of views lie between these two extremes and can depend on the work discipline, and the needs and objectives of the experiment. The authors recommend that multiple range testing be used as a useful *indication* of where treatment effects may occur which could be used as evidence for further experimentation.

There are a few approaches to multiple range testing. One is to compare each treatment in turn with a control treatment, which is known as Dunnett's test. A second approach, Scheffe's test, tests all possible comparisons, pairwise or otherwise. The third approach is to test all possible pairwise comparisons and there are several tests for this. These tests are discussed below though the test methodologies are not given (for further details, see Steel and Torrie, 1980; Montgomery, 1991).

Comparing each treatment with a control (Dunnett's test)   In some experiments the objective is to compare each treatment in the experiment with a control or standard treatment. Since these comparisons are formulated at the planning stage of an experiment, before the data are collected they can be classed as planned comparisons. This series of tests is collectively called a Dunnett's test but the individual comparisons made in this test are not independent comparisons and do not form a set of orthogonal contrasts. This test is well protected against Type I errors, so the stated significance level is similar to the overall Type I error rate.

Dunnett's test is used to test the differences between treatments and a control. As with all of these tests, if the F-test in the analysis of variance leads to a non-significant result then the null hypothesis is accepted and the conclusion is that there are no treatment differences and any further testing is not valid. It is possible to get non-significant results for comparisons made with Dunnett's test even when the overall treatment effect is significant. This will occur when the control mean lies between the minimum and maximum treatment means. It indicates that the differences are between the actual treatments themselves rather than between a treatment and the control treatment.

It is recommended when designing an experiment where it is planned to use a Dunnett's test to use additional replication of the control, or standard, treatment as it is to be used in a number of comparisons. A useful guide is to have as many control plots in a replicate as the square root of the number of treatments so in an experiment with t treatments and n replicates of treatment there should be $t\sqrt{n}$ units for the control treatment (Montgomery, 1991).

Making all possible comparisons (Scheffe's test)   In some instances the experimenter has only a vague idea, or no idea, where treatment differences may occur but requires more from an experiment than the conclusion that a treatment difference exists. The Scheffe test can be used to test all possible contrasts, orthogonal or not, therefore an infinite number of tests can be chosen but in practice only a finite set is tested.

The test is designed to make all comparisons (contrasts) whilst keeping the overall Type I error rate to the desired level, α. In order to achieve this, each individual comparison must be tested at a lower rate than α. It is possible to use Scheffe tests for both planned and unplanned comparisons but the test is generally used to test unplanned comparisons, since other more appropriate methods exist for testing planned comparisions. The results of the data-driven hypotheses (unplanned comparisons) can then be used to design further experiments. If the comparisons are planned and orthogonal, contrasts can be used to test each comparison at $\alpha\%$, making orthogonal contrasts more sensitive to differences in mean values than Scheffe comparisons operating at a lower Type I error rate per comparison. If any pairwise comparisons are to be tested then one of the following tests would be more appropriate than a Scheffe test as they are more sensitive to such differences.

Making all possible pairwise comparisons   There are many different tests which exist to compare every possible pair of treatments. These tests use similar methodologies but offer different levels of control over the overall Type I error rate. Four commonly used tests are considered here: the least significant difference test (LSD), Duncan's multiple range test, the Newman-Keuls test and the Tukey test.

For each test the difference between each possible pair of means is compared against a critical value. This critical value varies in the Duncan's multiple range test and the Newman-Keuls test according to the relative rank positions of the treatments involved in the pairwise comparison, but for the other tests mentioned the critical value is constant for all pairwise comparisons. The results are then usually presented by listing the treatments in rank order and underlining mean values which are not significantly different from each other; if any two mean values do not have the same line underscoring them then they are considered to be significantly different. In this way groups of means are formed within which no two means differ significantly from each other. Ideally, non-overlapping groups will be formed, but this is rarely the case, leading to difficulties in interpretation.

The least significant difference (LSD) test has very little control over the overall Type I error rate. As the number of comparisons made increases, the overall Type I error rate also increases. Even with relatively few treatments a large number of pairwise comparisons are possible, e.g. with eight treatments there are 28 possible pairwise comparisons. Therefore the more treatments that are tested, the more likely it is that a Type I error will be made and it will be concluded that a treatment difference exists when in fact it does not.

Duncan's multiple range test has more control over the overall Type I error rate than the LSD test and does take into account the number of mean values which lie in magnitude between the two which are being tested. It is a relatively powerful test. The Newman-Keuls test has a smaller overall Type I error rate than Duncan's test and is therefore more conservative, so it is less likely that this test will lead to the conclusion that a pair of mean values are different. The Tukey test takes account of the total number of mean values that are being tested. It has more control over the overall Type I error rate than any of the other tests discussed.

6.1.6.3   Discussion of tests

If multiple range tests are performed when the analysis of variance F-test has found no significant difference between treatment mean values, it is possible that whichever multiple comparison test is used some differences may be detected. This is due to the higher overall Type I error rate associated with the multiple range test. The F-test in the analysis of variance is the most reliable result for means comparisons and further tests should not be performed if the analysis of variance indicates there are no treatment differences. It is preferable to veer to the more conservative tests than the more liberal tests.

The Scheffe test is designed for all possible comparisons so will cater for the Dunnett's and pairwise multiple comparison tests. However, if the required comparisons are only 'each treatment against one other' or 'all possible pairwise comparisons' then it would be better to use the Dunnett's test or a multiple range test, respectively, as they are more sensitive to such differences in mean values. Orthogonal comparisons are better tested using orthogonal contrasts than a Scheffe test.

In terms of Type I error rates, orthogonal contrasts are the only tests that ensure the overall Type I error rate is equal to the significance level of each test. In the Scheffe test each comparison is tested at a lower significance level than $\alpha\%$, the overall Type I error rate. Tukey's test operates at an overall Type I error rate of $\alpha\%$ but as it considers only pairwise comparisons it is more sensitive than the Scheffe test to pairwise differences. The other pairwise comparison procedures operate from the opposite point of view in that each comparison uses an $\alpha\%$ significance level, resulting in an overall Type I error rate that exceeds $\alpha\%$. Tukey's test is the most conservative of the multiple pairwise comparison procedures discussed here, with the lowest overall Type I error rate. The LSD test will produce more significant results than the others but will make more Type I errors in the process. The Duncan and Newman-Keuls tests fall between the two, with Newman-Keuls being the more conservative of the two.

6.1.6.4   Example: contrasts

This example uses the data from the experiment described in Section 4.2.2.6 and analysed in Section 6.1.5.2. The study investigated the effects of nine fungicides, B to J, and an untreated control treatment, A, on thousand grain weight of winter wheat.

At the planning stage it was decided to compare the untreated control treatment, A, against the average of all the fungicide treatments, B to J, and also to compare the five innovative treatments, F to J, against the industry standard treatment, D. Treatments B, C and E were not used in specific comparisons. It can be shown that these two comparisons are mutually orthogonal. The effect of fungicide on thousand grain weight was found to be significant at the 0.001 level therefore the planned comparisons can be made. The analysis of variance table is given below in Table 6.28 and the treatment mean values in Table 6.29. In Table 6.28, the fungicide treatment effect is shown as a whole and also divided into three elements, i.e. the two contrasts and the rest.

The comparison of control treatment against all fungicides was significant at the 0.001 level. The null hypothesis of no difference between the control treatment and all other treatments can be rejected. It can therefore be concluded that the mean thousand grain weight for the control treatment differs from the mean thousand grain weight for all fungicide treatments. The comparison of new treatments against the standard treatment was not significant at the 0.05 level. It can be concluded that there is no evidence of a difference

Table 6.28 Analysis of variance table including two contrasts for the thousand grain weight data (analysed in Section 6.1.5.2)

Source	df	ss	ms	vr	$F_{probability}$
Block (Bl_h)	3	6.093	2.031	1.07	0.380
Fungicides (T_i)	9	190.504	21.167	11.11	< 0.001
Control v. treated	1	119.947	119.947	62.98	< 0.001
New v. standard	1	5.896	5.896	3.10	0.090
Remainder	7	64.661	9.237	4.85	0.001
Residual (e_{hi})	27	51.422	1.905		
Total	39	248.019			

Table 6.29 Treatment mean values for the thousand grain weight (g) data

	Treatment									
	A	B	C	D	E	F	G	H	I	J
Mean value	43.30	47.47	51.27	47.60	51.27	48.35	48.82	47.70	49.22	49.92
N	4	4	4	4	4	4	4	4	4	4

between the two groups of treatments. The test on the sums of squares not accounted for by the two orthogonal contrasts ('Remainder' in Table 6.28) is also significant at the 0.1% level. It can therefore be concluded that there is at least one further difference between the treatment mean values that has not been identified. This last conclusion affects the usefulness of the conclusion for the first contrast, control versus treated. The third contrast concludes that there is at least one difference between the treatment mean values other than 'new versus standard' so taking the average of all fungicides as used in the first contrast is not so sensible as the fungicides are different.

6.1.6.5 Example: polynomial contrasts

The analysis considered in this example is from the experiment designed in Section 4.3.4.6. The experiment investigated the effects on potato yield of two physiological ages (A1 and A2) and two planting dates (P1 and P2) of seed potatoes treated with six different equispaced rates of nitrogen (N1 to N6). The experiment used four randomized blocks with the four planting date/physiological age–treatment combinations randomized to the four main-plots in each block and the six nitrogen rates randomized to the six sub-plots in each main-plot. There were 24 plots in each block. The initial analysis of variance table identified statistically significant nitrogen and planting date effects with a significant interaction between these two treatment factors.

It was suggested that the experimenter would want to examine the yield response to increasing nitrogen application in more detail. One technique for achieving this is to use polynomial contrasts and this is illustrated in the analysis of variance in Table 6.30.

In Table 6.30 polynomial contrasts up to and including the third-order polynomial (cubic) have been calculated for the significant nitrogen effect and the significant

Table 6.30   Analysis of variance table including polynomial contrasts for the potato yield data

Source	df	ss	ms	vr	F$_{probability}$
Block (Bl_h)	3	69.974	23.325		
Age (A_i)	1	18.123	18.123	1.49	0.254
Planting (P_j)	1	620.730	620.730	50.94	<0.001
AP interaction (AP_{ij})	1	3.592	3.592	0.29	0.600
Main-plot residual (em_{hij})	9	109.667	12.185		
Main-plot total	15	822.088			
Nitrogen (N_k)	5	2994.509	598.902	68.72	<0.001
Linear	1	2066.377	2066.377	237.10	<0.001
Quadratic	1	741.790	741.790	85.11	<0.001
Cubic	1	176.884	176.884	20.30	<0.001
Remainder	2	9.458	4.729	0.54	0.584
AN interaction (AN_{ik})	5	26.372	5.274	0.61	0.696
PN interaction (PN_{jk})	5	197.020	39.404	4.52	0.001
Linear	1	174.181	174.181	19.99	<0.001
Quadratic	1	19.148	19.148	2.20	0.144
Cubic	1	1.981	1.981	0.23	0.635
Remainder	2	1.710	0.855	0.10	0.907
APN interaction (APN_{ijk})	5	19.681	3.936	0.45	0.81
Sub-plot residual (e_{hijk})	60	522.914	8.715		
Total	95	4582.583			

interaction term involving the nitrogen effect. The polynomial terms have not been shown for any treatment effects involving nitrogen that are not statistically significant. The first-order polynomial or straight line is labelled 'linear', the second-order polynomial is labelled 'quadratic', and the third-order polynomial is labelled 'cubic'. Each of these uses a single degree of freedom leaving two degrees remaining from the total treatment sums of squares that have not been taken up by the polynomial terms. The rows labelled 'remainder' indicate the portion of the treatment sums of squares that has not been accounted for by the three polynomial terms. As the polynomial contrasts are orthogonal, the sums of squares for these four terms (linear, quadratic, cubic and remainder) sum to the treatment sums of squares from which they are derived. The polynomial contrasts for the nitrogen treatment effect represent the linear, quadratic and cubic response of yield to nitrogen and for the interaction term they represent the interaction of planting date with each of the linear, quadratic and cubic nitrogen effects.

   Considering the interaction term first, the overall interaction term was significant indicating the shape of the nitrogen response curves differed in some respect between the two planting dates. The highest statistically significant interaction polynomial is the linear one, suggesting the response curves differ only in respect of the straight line element of the response. Considering the main nitrogen effect, all three polynomial terms are highly statistically significant implying that, of the polynomial curves, the yield response to nitrogen is best explained by the cubic term. The significant quadratic term implies that the curve fit is significantly improved by adding a quadratic term to the linear one; the significant cubic term implies that the overall fit is significantly improved again by adding a cubic term to the quadratic one. The non-significant 'remainder' term indicates that the complete nitrogen effect can be explained by these three polynomial terms, there is no component of the treatment effect that is unaccounted for.

Investigations do not end at this point. The curves have to be assessed and in agriculture it is not often that a relationship between two variables is sensibly explained by a cubic curve. Rather than immediately estimating the cubic model parameters it may be wiser to investigate a non-linear curve (see Section 6.1.7.4) that can be more easily explained than a cubic curve in terms of crop growth.

6.1.6.6   Example: multiple range tests

In this example the same data set is used to illustrate all of the multiple comparison test procedures mentioned in Sections 6.1.6.1 and 6.1.6.2. To compare treatment mean values in reality only *one* method should ever be used on an individual variable. It is not appropriate to analyse the same data using several methods of means comparisons and if done it is likely to produce conflicting conclusions. The two contrasts, 'control versus treated' and 'new versus standard', can be tested using a Scheffe test. The Scheffe critical values were calculated for the 10%, 5%, 1% and 0.1% significance levels. The contrast of 'control versus treated' was found to be significant at the 0.1% level, and the second contrast, 'new versus standard' was found to be non-significant at the 10% level. When the orthogonal contrast technique was used to test this comparison the result was significant at the 10% level, $p = 0.090$, illustrating the reduced power associated with the Scheffe test in comparison to orthogonal contrasts. It can therefore be concluded that at the 0.1% level of significance the mean thousand grain weight of the control treatment differed from the mean thousand grain weight of the nine fungicide treatments. The second conclusion is that there is no evidence of a difference between the average thousand grain weight of the novel treatments and the average thousand grain weight of the standard treatments.

Rather than using orthogonal contrasts, if at the planning stage it was decided to compare each treatment with the untreated control treatment, A, a Dunnett's test can be performed since the treatment effect was statistically significant. This should not be performed in addition to orthogonal contrasts. This test considers the difference between each treatment mean value and the control mean value and compares this with a value calculated from tables. The differences between the control mean and treatment means are as follows:

Treatment B	$47.47 - 43.30 = 4.17$
Treatment C	$51.27 - 43.30 = 7.97$
Treatment D	$47.60 - 43.30 = 4.30$
Treatment E	$51.27 - 43.30 = 7.97$
Treatment F	$48.35 - 43.30 = 5.05$
Treatment G	$48.82 - 43.30 = 5.52$
Treatment H	$47.70 - 43.30 = 4.40$
Treatment I	$49.22 - 43.30 = 5.92$
Treatment J	$49.92 - 43.30 = 6.62$

Each of these differences lies within the 5% critical region for the test so it can be concluded that the mean thousand grain weight for each of the fungicide treatments differs from the mean thousand grain weight of the untreated control treatment.

If no comparisons had been planned and a treatment difference had been found, the difference could be further investigated using one of the tests that test all pairwise

Table 6.31 Comparison of results from four multiple range tests on the thousand grain weight treatment mean values

Fungicide	Mean value	Pairwise multiple comparison test			
		LSD	Newman-Keuls	Duncan	Tukey
A	43.30				
B	47.47				
D	47.60				
H	47.70				
F	48.35				
G	48.82				
I	49.22				
J	49.92				
C	51.27				
E	51.27				

comparisons. The four tests discussed were applied to the data and the results are illustrated in Table 6.31. It should be noted that only one test should be performed in practice.

Each test has split the treatments into three or four groups. Within each group the treatment mean values can be considered to be the same. Any two treatment mean values joined by the same vertical line are not significantly different. Any two treatments not joined by the same vertical line have mean values that are significantly different.

For the least significant difference test above, it can be concluded that treatment A is significantly different from all the other treatments; treatment B is significantly different from treatments J, C and E; treatment D is significantly different from treatments J, C and E; and treatment H is significantly different from treatments J, C and E. Treatment F is significantly different from treatments A, B, D and H and so are each of treatments G, I, J, C and E.

Similar conclusions can be drawn from the other three multiple range tests though there are slight differences in the composition of each group.

6.1.7 REGRESSION ANALYSIS

Keypoints

- Regression analysis is used to model a relationship between two or more variables.
- Fitted models are only appropriate in the range of independent x-values used to estimate them.
- Residual plots are used to assess whether an appropriate model has been fitted.
- Influential data points can affect which model is fitted.

In most experiments many different variables are measured on each experiment unit, however all analytical techniques covered so far have only considered each variable separately. This section deals with measures of association and relationships between two or more quantitative variables. There are several different ways in which potential relationships between variables can be investigated. There are multivariate methods that consider the relationship between many variables but these can be very complex (see Section 7.1 for a brief overview). The simpler approaches of correlation and regression are considered in this section.

The basics of linear regression are covered, and multiple and non-linear regression are mentioned in brief. No formulae are presented and the interested reader is referred to the texts mentioned below for the details of regression calculations and the formulae required for statistical testing of the goodness-of-fit criteria. Many of the listed references provide a chapter on linear regression analysis (Montgomery, 1990; Steel and Torrie, 1980). Two useful texts devoted entirely to the subject of regression analysis are Weisberg (1985) and Draper and Smith (1981). The former provides a good thorough introduction to regression analysis without being too mathematical and the standard regression formulae are clearly presented and illustrated. The latter is the standard reference for regression analysis but the newcomer to the subject may find it rather heavy going in places.

6.1.7.1 Correlation

Correlation coefficients form a measure of the linear association between two variables. There are a variety of different measures of correlation and their values always lie between -1 and $+1$. The product-moment correlation coefficient is the most commonly used measure and is usually referred to as the correlation coefficient. This product-moment correlation is usually denoted by the population value, ρ, and it is estimated by the sample correlation coefficient, r. The product-moment correlation is appropriate for variables that are continuous and normally distributed.

Scatter plots of the two variables, where the value of one variable is plotted on the vertical axis and the value of the other is plotted on the horizontal axis, are vital for a meaningful interpretation of correlation coefficients. If there is no discernible linear relationship then the correlation coefficient will be zero or close to zero. A correlation coefficient of $+1$ indicates a perfect positive linear association between two variables, illustrated by an upward slope on the plot. A value of -1 indicates a perfect negative linear association between two variables, illustrated by a downward slope on the plot. Such perfect relationships are extremely rare in biological science.

Conclusions about the form of a relationship should never be drawn from a correlation coefficient on its own. It is only by looking at a plot of the data that the possible existence of a relationship may be detected. Anscombe (1973) illustrated this point perfectly by constructing four data sets, each with their own particular characteristics but all with the same correlation coefficients. A sketch of the four raw data plots is provided in Figure 6.48. The correlation coefficient in each case is 0.82, which could be taken to indicate the presence of a strong linear relationship between the two variables, but a brief glance at the plots quickly dispels that idea. The plot in diagram A shows the two variables to have a fairly linear relationship. In diagram B there is very little scatter but the relationship is obviously curvilinear. In diagram C there is one outlying point; the rest of the data show a strong linear relationship and without the outlier a much higher correlation coefficient would be achieved. In diagram D, except for one x-value, all the x-values are identical. Consequently there is no information provided on what the relationship may be. Weisberg (1985) and Chatfield (1995) also discuss these data sets.

The presence of a correlation does not imply a 'cause and effect' scenario. Changes in one variable may not necessarily be due to changes in the other even though the changes

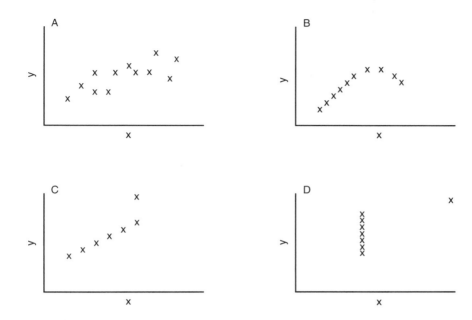

Figure 6.48   Four plots of hypothetical data showing the need to plot data before interpreting a correlation coefficient. Reproduced by permission from Anscombe (1973)

may be related, and it is possible to have a high degree of correlation between two variables that just happen to be changing together. An example cited in Box, Hunter and Hunter (1978) neatly illustrates this. Data, collected over a number of years, on annual human population figures and the number of storks in the same year, were plotted on a scatter graph with human population on the vertical axis and stork population on the horizontal axis. This displayed a straight-line trend that showed as the stork population increased so did the human population. Obviously there is no direct causal effect between the two variables; an increase in stork population does not cause an increase in the human population just as reducing the stork population would not lead to a reduction in the human population.

It may be possible to find a significant correlation between two variables that are associated with each other in that one variable is derived from the other. Such correlations are not necessarily useful. For example, total Nitrogen offtake (N offtake) and N offtake in the grain may show a high correlation, but total N offtake is the sum of grain N offtake and straw N offtake so it may be more useful to correlate grain N offtake with straw N offtake. Thought should be given to sensible interpretation of correlation coefficients when derived variables are used. If a variable is correlated with its ratio to a second variable, or if two ratios with the same denominator are correlated, spurious relationships may well be found.

Correlation coefficients can be tested for statistical significance. A 'large' correlation coefficient is not always significant and a 'small' coefficient may well be. It is the number of pairs of observations used that determines the point at which a correlation coefficient becomes significant. Significant correlations become smaller in absolute magnitude as the number of pairs of observations increases.

6.1.7.2 Simple linear regression

The correlation coefficient will only give a measure of linear association and will not give any indication of which straight line, if a straight line is appropriate, best represents the relationship. Simple linear regression analysis is the means of determining this relationship. 'Simple' refers to the fact that one independent variable is involved in the analysis and not to the complexity of the analysis itself. A plot of the data will reveal whether a straight-line model is appropriate or not.

The two variables involved in a simple regression relationship are not regarded equally; changes in one (the independent variable) are thought of as causing changes in the other (the dependent or exploratory variable). Regressing y on x is not the same as regressing x on y and it is important that this distinction is made. For example, it may make sense to regress yield of a cereal crop (the dependent variable, y) on disease levels in that crop (the independent variable, x) since changes in disease levels could well affect yield. It is unlikely that yield will affect disease levels so a regression of disease on yield would be nonsensical. Regression of y on x and x on y will produce the same model if x and y are perfectly correlated but the interpretation will still be different.

If the independent variable is denoted by x and the dependent variable by y, the simple regression analysis fits a straight-line model of the following form:

$$y = a + bx$$

This model will not fit the data points exactly but the observations will deviate randomly from the straight-line model as illustrated in Figure 6.49. The y-values calculated using

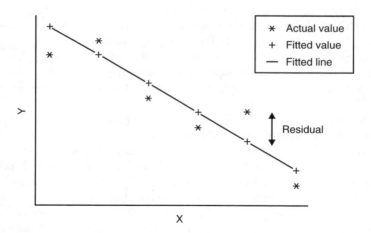

Figure 6.49   An illustration of the concept of fitted values, actual values and the fitted model (in this instance a straight line). After Draper and Smith (1981)

the model are the fitted values and form the straight line. The difference between these fitted values and the observed y-values are the residuals. The regression analysis estimates the values of the model parameters a and b and therefore fits the straight-line model. There are a number of ways of estimating a and b, the most common one being the 'method of least squares' which finds the values of a and b that minimizes the sum of the squared residuals.

The assumptions required for linear regression analysis are the same as for analysis of variance, constant variance, normally distributed and independent errors. As with analysis of variance, plotting the residuals against the fitted values is a useful way of assessing the validity of the model fitted. Again, a random horizontal scatter about the zero residual will indicate whether the appropriate model has been used. There should be no obvious patterns or trends in the plot.

If the initial raw data plot indicates a curved response to the independent variable then this may still be a linear regression. By means of an appropriate mathematical function or transformation it will be possible to fit a model using linear regression. The terminology 'linear regression' can be applied to any model that has mathematically linear parameters and it is not a term that reflects the shape of the function. Both the dependent and independent variables may or may not be transformed in order to fit a model. For example, some ways in which the independent variable, x, in the straight-line model can be transformed are as follows:

$$y = a + bx + cx^2$$

$$y = a + bx + cx^2 + dx^3$$

$$y = a + blnx$$

The straight-line, quadratic and cubic equations just mentioned are the first three in the suite of polynomial equations that are related to the polynomial contrasts mentioned previously in Section 6.1.6.1.

If a data set is to be used for regression analysis then an adequate range of x-values must be available. A line or curve can only be fitted to the range of x-values available and any conclusions can only be drawn about data within this range. A model is fitted to values from a given range and must not be used for predicting or estimating values outside the given range as it will not be known whether the model alters outside the range as no information is provided. The fitted line will be estimated with more certainty in areas of greater intensity of data pairs. Thus if it is possible to measure the y-values for chosen x-values the selection of these x-values will vary depending on the objectives of the study. If all points are of equal interest, which is often the case with a straight line, an even spread of x-values throughout the range is desirable. If the objective of the experiment is to fit a mathematical model to the data set and there is no prior knowledge of the form of the model then an even distribution of x-values is required. If some particular value is of interest, for example an optimum value, and it is known roughly where this value might lie on the x-axis, the true location of this value will be found with more certainty if there are more x-values around the expected location with few x-values at the extremes.

An adequate number of data points is required in the range of x-values to have any certainty that the correct model is fitted. To a certain extent the number of different x-values available determines the complexity of the model that can be sensibly fitted (Draper and

Smith, 1981). The same straight line will always be drawn through two data points but an infinite number of more complicated curves can be drawn through the same two points. To fit a straight-line model, two distinct x-values are an absolute minimum requirement in order to estimate the values of the two model parameters. For a quadratic curve the absolute minimum is three because there are three model parameters. In practice, a few more x-values than there are model parameters to be estimated should be used to allow testing for lack of fit. Obviously the number of different x-values that can be used will be limited by other factors in the experiment and by resources. Replication is required if it is required to check the model is repeatable. The balance between the number of replicates and the number of data points should be carefully considered. If the only objective is to fit a model then more data points with x-values at different locations rather than extra replication may be desirable. If an independent or dependent observation is missing for some reason then the corresponding value is also excluded from the regression analysis. It is not necessary to estimate the missing values.

6.1.7.3   MULTIPLE LINEAR REGRESSION

When more than one independent variable is being studied, a multiple linear regression may be appropriate. The model is a straightforward extension of the simple linear regression. The simplest form of the multiple linear regression model for k independent variables, $x_1, \ldots x_k$, is

$$y = a + b_1 x_1 + b_2 x_2 + \ldots + b_k x_k$$

where $a, b_1, b_2, \ldots b_k$ are the model parameters that need to be estimated. Usually with such models there is an element of model development when fitting them in the sense that checks can be made to assess the contribution of each term to the overall fit of the multiple regression model. Ideally the final multiple regression model would be one where all the independent terms are considered to be making a significant contribution to the overall model, but more importantly the final model should make sense in terms of the variables used and in its use.

More complex models can be produced that allow each independent variable to have a curvilinear response. Each variable is not constrained to be included in the model in the same form so one variate could be included in a quadratic form and another in the straight-line form, for example. The multiple regression equation would then be

$$y = a + b_1 x_1 + b_2 x_2 + b_3 x_2^2 + \text{other terms}$$

where a, b_1, b_2, b_3, etc., are the model parameters that need to be estimated.

If different polynomial terms are to be included in the model it is common practice to keep all polynomial terms for that variate in the model, so, for example, both the linear and the square term in a quadratic would be retained in the full model even if the linear term was found not to be making a significant contribution to the overall model (Draper and Smith, 1981). The argument can be presented from the other side which says that only terms making a significant contribution to the model should be retained when developing a multiple regression model. However, the final model should have some practical relevance and the latter approach may lead to a model that is too specific and related only to the particular data set from which it was derived. It is unlikely that models with high-order polynomials are realistic and they may well over-fit the data and be poor predictors. It would be advisable to try another form of model, such as the non-linear models mentioned briefly in Section 6.1.7.4, before choosing a high-order polynomial.

More complex models again can be produced including interactions in the model by using product terms of the independent variables. The form of the multiple linear regression model for k independent variables, $x_1, \ldots x_k$, could then be

$$y = a + b_1x_1 + b_2x_2 + b_3x_1x_2 + b_4x_3 + b_5x_5 + \text{other terms}$$

where $a, b_1, b_2, \ldots b_5$, etc., are the model parameters that need to be estimated.

In all these instances, as with simple linear regression, the difference between the observed y-values and the fitted y-values will give the residual terms. If any of the independent or dependent observations are missing they are not estimated and the whole set of observations for that point is excluded. All the assumptions and principles of simple linear regression hold for multiple linear regression. The reader is referred to Weisberg (1985) and Draper and Smith (1981) for further details on model building.

6.1.7.4 Non-linear regression

Non-linear regression models are defined to be non-linear in the parameters and can take many different forms. They are often a combination of parameters that are linear and non-linear. Exponential models are non-linear models, as are many other forms of growth model. There are some 'standard' forms of non-linear models such as $y = a + be^x$ (exponential) or $y = a + br^x + cx$ (linear plus exponential). Many other forms are non-standard. Biological models are frequently non-linear.

Residuals for these models can be calculated in the same way as for linear models. These models are more complex to handle in terms of estimating the model parameters as iterative procedures have to be employed and initial parameter estimates have to be provided. These initial estimates can have quite an influence on the final estimates produced by the iteration procedure. Some computer packages will do this automatically, though knowledge of how these initial values are determined and some idea of expected parameter values will be necessary for sensible interpretation of the results. The texts referenced in this section mostly concentrate on the easier to use linear regression technique, but Draper and Smith (1981) do provide an introductory chapter to non-linear regression.

6.1.7.5 Comparing linear regression lines

When a number of equations are fitted to subsets in a data set it is frequently of interest to compare the equations to see if they are 'the same'. Are the parameter estimates for the models so similar that they cannot be considered significantly different? If they are then a single model may explain the data just as effectively as a series of models. Details of the statistical tests available to determine this are not provided here; only an outline of the techniques is given below. For more details the reader is referred elsewhere, particularly Weisberg (1985) who gives an example of model comparisons in parallel curve analysis.

There is usually some sort of structure in the data that encourages the researcher to consider different subsets of the whole data sets separately. For example, the relationship between yield of a wheat crop and amount of fertilizer applied to the crop may be investigated for a number of varieties of the same wheat crop. It would be of interest to know not only whether the yield response to nitrogen could be modelled, but whether the models are the same for the different varieties. If there are only two varieties under investigation then, assuming the same form of model has been fitted to each variety, the model parameters could be tested using t-tests. For each parameter a significant test result would indicate that the varieties required different estimates of the model parameters.

However, this is not a particularly efficient way of testing for similarity/equality of models, particularly when each model has a number of parameters to be estimated, and is even less so when there are more than two models to be compared. Parallel curve analysis is an effective means of identifying whether a set of equations is completely separate, parallel, concurrent or coincident. A single curve would indicate that the yield responses of the varieties are the same (all varieties would have the same model intercept and model slope). Parallel curves would indicate the variables respond in the same way to nitrogen application but at different overall yield levels for each variety (the model intercepts would be different for each variety but the slopes would be the same). Concurrent curves would indicate that all varieties gave the same yield with no nitrogen fertilizer application but then responded differently to application of fertilizer (the model slopes would be different for each variety but the intercepts would be the same). Separate curves would indicate that each variety responded in a completely different way to the application of fertilizer (all model parameters would be different for each variety). An F-test is used to compare each of the models against the full model of separate curves for each variety. If the F-test is significant the full model is required because its residual mean square is significantly smaller than that for the alternative model. The same sort of F-test can be used to compare two hierarchical models, e.g. to compare a quadratic curve with a straight line. The simpler model can be regarded as being just as adequate as the more complex one if the F-test is not statistically significant but the more complex model is required if the F-test is significant.

6.1.7.6   Goodness-of-fit

There are several ways of assessing how well a model has been fitted. The methods detailed here are suitable for linear regression models. Although they are frequently also used with non-linear models, the assumptions behind some of the tests may not be strictly valid with these models so these goodness-of-fit checks are not as effective with non-linear models but can be a useful guide. No one method should be used in isolation but results from each method should be considered in conjunction with the others. These methods include examination of residuals, calculating the correlation, testing the significance of the coefficients, checking for outliers and possibly calculating prediction confidence limits if appropriate. If several different models have been fitted and there is little to chose between them in terms of goodness-of-fit then the simplest model should be chosen unless there are sound scientific reasons for choosing another.

Residual plots   Many problems can be detected from the plot of the residual values versus the fitted values. If the model is valid the plot should display a horizontal band of randomly scattered points about the zero residual, as shown in Figure 6.50.

Discussed below are the most common problems and the residual patterns associated with them. Examples of the residual patterns are given although variations of these patterns may well be encountered.

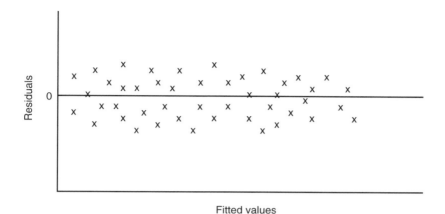

Figure 6.50   A residual plot that looks acceptable, i.e. a random scatter of points lying in a horizontal band around the zero residual. Reproduced by permission from Weisberg (1985)

Non-constant variance can be indicated by any of the three patterns shown in Figure 6.51. Note that the dashed lines are not part of the plot but have been used to emphasize the shape indicated by the points. If non-constant variance is detected this may be remedied by transforming the response variable, y, with a variance-stabilizing transformation such as those mentioned in Section 6.1.4.3 on analysis of variance. There are other more complex methods of regression analysis which may remedy the situation, such as weighted regression, but these are beyond the scope of this text.

If an inappropriate model has been fitted the residual plot may show patterns such as those in Figure 6.52. Transforming the data by adding polynomial terms or fitting a non-linear model may resolve the problem. Figure 6.53 below shows the patterns obtained by the combination of the two problems discussed above.

If a model has been fitted and an independent variable has been omitted from the model the residual pattern may be as shown in Figure 6.54. This can be corrected for by fitting a revised model with all independent variables, though this may not always be possible if the appropriate variable has not been measured.

Residual plots can be used for detecting outlying data points. If any such points are found, the original data must be checked to ensure this is a true value. If no other problems exist the residual plot will show a horizontal band around zero with a single isolated point, as in Figure 6.55. Obviously such isolated points can also occur in residual plots that indicate the other problems mentioned above.

All of the residual plots shown so far have been theoretical plots and show a very clear picture which is easy to interpret. In reality, residual plots are rarely like this and the interpretation can be quite difficult and subjective. Is this an indication of non-constant treatment variability or is it caused by outliers? Is there a curvature to the plot or not? It may

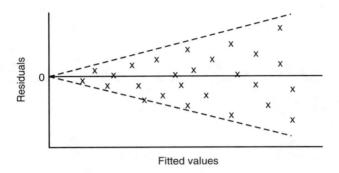

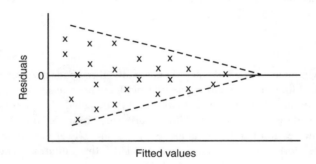

Figure 6.51   Residual plots indicating non-constant variance; a variance stabilizing transformation may remedy the problem. Reproduced by permission from Weisberg (1985)

be unclear exactly what is happening in the plot. Figure 6.56 shows a residual plot of data taken from a dairy study. There could be several interpretations put onto this plot. The point on the bottom right may be regarded as an outlier with the rest of the points being a random scatter. Alternatively, the two points in the top left corner could be regarded as outliers and the remaining points considered to indicate non-constant variance. Experience and knowledge of the data will help most in interpreting these plots.

Coefficient of determination   The coefficient of determination, R^2, is probably the most commonly used, and abused, goodness-of-fit criteria. Models are often described as being adequate based solely on this measure. However, it is a useful measure that is appropriate for all linear models. It is usually defined as the ratio of the sum of squares explained by the regression to the total sum of squares. It takes values between 0 and 1 inclusive, and

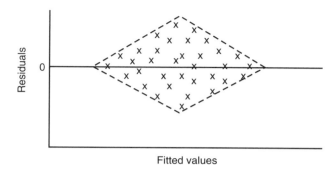

Figure 6.51   (continued)

the closer the value is to 1 the better the fit of the model. In simple linear regression the coefficient of determination is identical to the square of the correlation coefficient. It is actually a measure of the square of the correlation between the observed and fitted values and is also known as the multiple correlation coefficient.

The value of R^2 can be increased by simply adding more variables to the model, regardless of whether the variables contribute anything useful to the model. For this reason it is preferable to use an adjusted value for R^2 which takes into account the number of variables included in the model and therefore the number of parameters the model is fitting. This gives a measure of the variability accounted for by the regression, with a high value of R^2 being taken to mean a good fit. However, high values can occur even with inappropriate models and account should always be taken of the other goodness-of-fit tests.

Testing the regression   Having fitted a model, an analysis of variance divides the total variability in the y-values into the portion explained by the model and the unexplained variation. This unexplained variation is the residual mean square. The residual degrees of freedom is affected by the number of model parameters that need to be estimated. A variance ratio can be calculated to test the significance of the model in the same way as for any other analysis of variance method. The test should be statistically significant for a good model.

Testing regression coefficients   The calculated regression coefficients can be tested to ascertain whether they are significantly different from zero or any other sensible value. This involves performing a t-test for each coefficient estimated. This would normally be done when building a model to be used for prediction purposes. For a good model the parameters should be statistically significant.

Influential data points   Influential data points are ones which, when removed from the regression analysis, cause a different model to be fitted. This can mean the coefficients of the model changing or the terms included in the model changing. Such points are sometimes said to have high leverage. These values must not be discarded. They must be treated in the

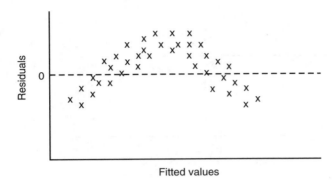

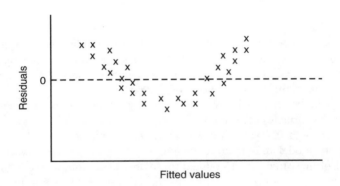

Figure 6.52   Residual plots indicating an inappropriate model has been fitted; adding terms or fitting a
non-linear model may remedy the problem. Reproduced by permission from Weisberg (1985)

same way as outliers in analysis of variance (see Section 6.1.4.2), but it must be noted that
outliers are not necessarily influential data points. An influential data point could cause a
curvilinear model to be fitted to the data rather than a straight-line model which may be
indicated by the rest of the points. Other influential data points may simply cause a vertical
shift in the model to be fitted, so the model remains the same but the intercept changes, or
they may change all the estimates of the model parameters but keep the same model by
changing the slope.

Confidence intervals   The fitted equation is often used to predict values of the response
variable and to estimate fitted values. Confidence intervals can be calculated for both these
situations.
     When regression techniques are used for prediction purposes a response value which
was not used in the model estimation is predicted for a given value of the independent

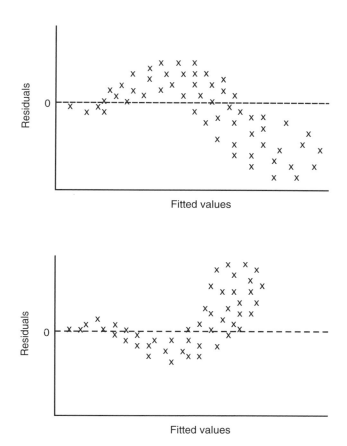

Figure 6.53   Residual plots indicating non-constant variance and an inappropriate model has been fitted. Reproduced by permission from Weisberg (1985)

variable. Such predictions can be made for all values of the independent variable. Confidence limits on these predictions can be plotted and the vertical difference between these limits will not be constant but will diverge at the extremes of the independent variable. This makes intuitive sense because in the middle of the relationship there is more 'surety' about what is happening because the points around the middle provide support to the model fitted. At the extremes of the range of x-values there is less information available, and outside of the range of x-values there is none, so there is less certainty in the relationship. Values should not be predicted outside the range of independent values used to estimate the model.

When regression techniques are used for estimation purposes an average value for the response variable at a given value of the independent variable is estimated. The fitted value is the best estimate of this average value. Confidence intervals can be calculated for each

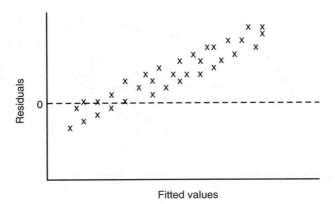

Figure 6.54   Residual plot indicating that an independent variable has been omitted from the model.
After Draper and Smith (1981)

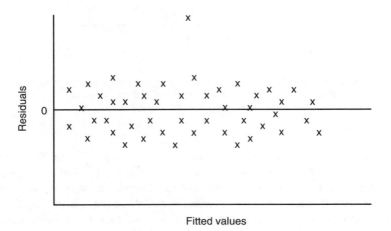

Figure 6.55   A residual plot with an outlier

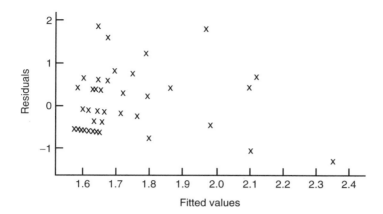

Figure 6.56   A residual plot of data from a dairy study showing the difficulty in interpreting residual
plots from actual data. After Weisberg (1985)

fitted value. These can be shown on a plot as a symmetrical vertical line drawn through each
point. Each line does not have to be the same size but will be dependent on the variability
associated with each point.

6.1.7.7   Example: linear regression

This is an experiment in which the aim is to find what, if any, relationship exists between the
yield of sugar beet and the amount of irrigation water applied. Seven levels of irrigation are
used, ranging from zero to over 300 mm, and each treatment is replicated four times. The
experiment design is a single-treatment factor randomized blocks design with four blocks.
The actual amount of irrigation water applied to each plot for a specific treatment varied
slightly from the target level, and it is the actual amount applied that is used in the following
regression analysis. A graph of the raw data is given in Figure 6.57.

The graph suggests that there is a reasonably well-defined relationship between the two
variables. The correlation coefficient is 0.85 but the graph indicates the relationship is not a
straight-line one. There are two points at the lower irrigation levels (0 and 50 mm) that have
slightly higher root weights than the other replicate points at these levels. It is worth
checking that they are genuine observations before proceeding any further; they could be
errors or they could be indicating increased variability between observations at the lower
irrigation rates. However, the following analysis is conducted with all observations present.
The simplest curvilinear model, a quadratic curve, is tried first to see how well it fits.

The quadratic equation is estimated as

$$R = 11.4 + 0.00579W - 0.00013W^2$$

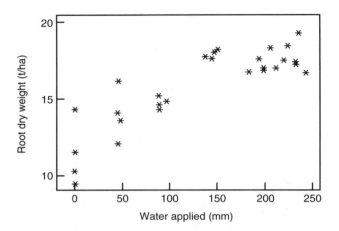

Figure 6.57   Relationship between sugar beet yield (t/ha) and water applied (mm)

where R is the dry weight of roots (t/ha), and W is the amount of irrigation water applied (mm).

The overall fit of the model as judged by the F-test on the regression mean square is statistically significant, $p < 0.001$. Each of the model coefficients, 11.4, 0.00579 and -0.00013, is tested with a t-test which indicates they are significantly different from zero ($p < 0.005$ at least). So they are all judged to be making a significant contribution to the model. The coefficient of determination is calculated as 0.782. The graph of the model residuals versus the fitted values is drawn in Figure 6.58. The two observations on the raw data plot that are slightly high are clearly indicated in the graph. The overall impression of the graph is not quite of a random scatter of points in a horizontal band around the zero residual, but it is not too far removed from that and the two 'slightly high' points are influencing the shape to some extent. (In fact, if those two points are ignored in the regression analysis there is no indication of any problems with the plot of residuals, but they should not be removed from the analysis for that reason.) All the indications point to the quadratic model providing a reasonable fit to the data but it is possible that some other model would fit just as well.

The correlation coefficient for these data is high and in fact is statistically significant, $p < 0.001$, and if the data had not been plotted first it might have been very tempting to just fit a straight line to the data. If this model is fitted the following regression analysis is obtained.

The straight-line equation is estimated as

$$R = 12.4 + 0.0255W$$

where R is the dry weight of roots (t/ha), and W is the amount of irrigation water applied (mm).

As with the quadratic equation the overall statistical significance of the regression model is high, $p < 0.001$ and each of the model parameter estimates are significantly different from

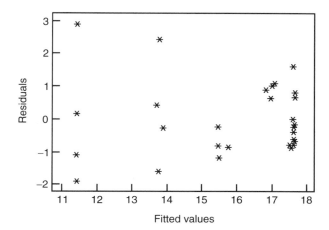

Figure 6.58   Residual plot from fitting a quadratic model to the sugar beet data. The raw data are shown in Figure 6.57

zero, $p < 0.001$. The coefficient of determination is 0.711 which is not much less than that for the quadratic model. All the indications so far are that the straight-line model is a reasonably good fitting model. However, the graph of the model residuals plotted against the fitted values (Figure 6.59) quickly dispels that idea. The points are lying on quite a marked curve indicating that either a polynomial term has been missed from the model or that a non-linear model is more appropriate. This plot is clearly indicating that the straight-line model is not an appropriate model for these data, even with the two slightly high points in the lower fitted value range.

Using the approach of keeping the model as simple as possible, the quadratic model would have been the next model to try had the straight-line model actually been the first model fitted. It is possible that the curve fitting would stop at the quadratic model although other models that fit as well or better than that model may possibly be found. Depending on the final use to which the model is put, it may be more appropriate to try to find a function that is not symmetrical; an exponential model would be a good starting point for non-linear curve fitting in this instance. The calculation of confidence limits on model predictions may assist in the choice of a model should two or more models be found that fit the data equally appropriately.

6.2   NON-PARAMETRIC TECHNIQUES

Non-parametric techniques make few assumptions about the nature of the data or the populations from which the data are drawn. They are used for comparing distributions and testing hypotheses concerning the distributions. In many of these tests distributions have to be specified. This can either take the form of an equation, or be described by name, as in the binomial distribution, or be in terms of frequencies or probabilities.

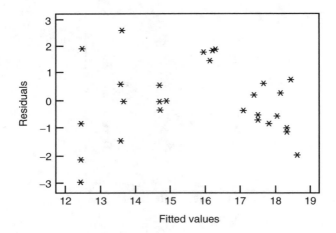

Figure 6.59   Residual plot from fitting a straight line to the sugar beet data. The raw data are shown in Figure 6.57

They can be used on all types of data provided the assumptions of the particular test being used are valid. The minimum assumptions made by non-parametric techniques are that the observations are independent of one another, thus it is still important that a proper randomization is used and that careful consideration is given to experiment procedure.

The scale of measurement is determined by the experiment material and objectives. The nominal scale is the weakest level, where the numbers provide only a classification. The ordinal scale, or ranks, is the next scale of measurement. Here numbers are classified such that the classes can be placed in one and only one order. The interval scale is the second strongest scale and this is for continuous data that have a natural numerical order. The strongest scale of measurement is the ratio scale. It has the same characteristics as the interval scale but the zero value is absolute. Further details on scales of measurement are provided in Section 5.2.2.

A selection of non-parametric techniques are discussed in the following sections. The list of tests is not comprehensive though it is hoped that most situations will be covered. Only a brief discussion of each test is provided and further details will be needed before performing the tests. The required formulae for all of the tests mentioned in the following sections can be found in Siegel and Castellan (1988) who also provide a thorough discussion of each test with illustrative examples. Steel and Torrie (1980) and Sokal and Rohlf (1995) also cover a variety of these tests; both give clear examples of how to conduct the tests but neither provide as detailed coverage as that found in Siegel and Castellan (1988).

For some of the tests the actual method used to perform the test can vary. The tables required to determine statistical significance of the test will depend on the method used. It is therefore very important that the instructions for performing a test are followed exactly and that the tables used are from the same source. When a scale of measurement is specified in this text for a particular test, it is the minimum scale required. For example, if a test requires ordinal data it cannot be used on nominal data but can be used for continuous data, though there may be a more appropriate test for the higher scale of measurement.

Note: Some of these tests are known by different names although they are the same test. Other tests, most notably the chi-square tests, share the same name but are used in different situations.

6.2.1 COMPARISON OF SINGLE SAMPLES

Keypoints

- *Runs test*: tests if a sequence of observations is random; uses dichotomous data; can be used with small sample sizes.
- *Chi-square goodness-of-fit test*: compares observed frequencies with hypothesized frequencies; uses nominal data; there is a limitation on the minimum size of hypothesized frequencies.
- *Kolmogorov-Smirnov test*: compares the observed cumulative distribution with the expected cumulative distribution; uses ordinal data; data are from a continuous distribution; can be used with small sample sizes.

6.2.1.1   Runs test

The runs test is a method of testing the hypothesis that a sequence of observations in a sample is random. There is no parametric equivalent to this test. The test uses the total number of runs in a sample to indicate whether the sample is random, where a run is a succession of identical items preceded and followed by a different item. A few long runs in a sample indicate some clumping of observations and many short runs imply a systematic element to the observations. Both are indicators of non-randomness. The runs test can be used for small sized samples. The data must be binary or reduced to binary type data, for example whether the values lie above or below a specified value. Binary data are also known as dichotomous data.

Assumptions   Data are collected from a single population.

Limitations   The test can only be used on dichotomous data.

Hypotheses

H_0: The sequence of observations occur in random order.

There are three possible alternative hypotheses:

H_1: The sequence of observations does not occur in a random order.
H_1: There are too many runs.
H_1: There are too few runs.

Conclusions   If the test result was not statistically significant the null hypothesis would be accepted. It would be concluded that there is no evidence to suggest that the observations do not occur in random order. If the test result was statistically significant the null hypothesis would be rejected. In a two-tailed test it would be concluded that the sequence of observations is not random. In a one-tailed test it would be concluded that the sequence of observations deviates from randomness in the manner stated for the alternative hypothesis.

Example Analysis of variance was performed on a set of data. It was decided to formally test whether or not the sign (positive or negative) of the residual was occurring at random once the residuals had been arranged in order of the experiment units. (This is one formal method which can be used for checking the assumption that errors are independent for analysis of variance; see Section 6.1.3.) The residuals obtained from the analysis are given in Table 6.32.

The number of runs of plus signs and minus signs can then be counted. In these data there is a run of two plus signs followed by two minus signs and then a plus sign. There are 13 runs in total; these are shown below by underlining each run.

Unit	1	2	3	4	5	6	7	8	9	10	11	12	13	14	15	16	17	18	19	20
Sign	+	+	−	−	+	−	+	+	+	−	+	−	−	+	−	+	−	−	−	+
Run	1		2		3	4	5			6	7	8		9	10	11	12			13

H_0: The + and − signs occur in a random order
H_1: The order of the + and − signs is not random

Significance level, $\alpha = 0.05$
Number of observations $= 20$
Number of + signs $= 10$
Number of − signs $= 10$
Critical region is 'number of runs ≤ 6 and ≥ 16'
Observed number of runs $= 13$

The observed number of runs lies outside the critical region and therefore the test is not significant so the null hypothesis is accepted. It is concluded that there is no evidence that the + and − signs are not in random order. The assumption that the errors are occurring in a random order is valid.

6.2.1.2 Chi-square test for goodness-of-fit

The chi-square test is used to compare observed frequencies with hypothesized frequencies. The hypothesized frequencies are based on the null hypothesis that the observations are

Table 6.32 Residual values obtained from an analysis of variance

Unit no.	1	2	3	4	5	6	7	8	9	10
Residual	5.1	4.1	−7.9	−1.7	0.4	−18.9	4.1	14.1	1.4	−0.7

Unit no.	11	12	13	14	15	16	17	18	19	20
Residual	0.5	−6.5	−4.5	17.8	−7.3	13.3	−1.7	−1.6	−17.5	7.6

drawn from a specified distribution. Observations need to be categorized into mutually exhaustive and exclusive groups.

Assumptions   The observations must be independent. Data are collected from a single population.

Limitations   The data must fall into at least two categories and are measured on at least the nominal scale of measurement. This test takes no account of any order implicit in the categories. The chi-square test is inappropriate if the hypothesized frequencies are small. This means that the test should only be used on data where all hypothesized frequencies are at least one, and no more than 20% of the cells have hypothesized frequencies less than five. If the hypothesized frequencies do not meet this requirement then adjacent categories can be combined if they still make sense.

Hypotheses

H_0: The proportion of observations falling into each of the categories is as hypothesized.

There are numerous alternative hypotheses. It is only possible to infer if the null hypothesis is not accepted. Therefore the general alternative hypothesis is as follows:

H_1: The proportion of observations falling into each of the categories is not as
hypothesized.

Conclusions   If the test result was not statistically significant the null hypothesis would be accepted. It would be concluded that there is no evidence to suggest that the observations do not follow the hypothesized distribution. If the test result was statistically significant the null hypothesis would be rejected. It would be concluded that the observations do not follow the hypothesized distribution.

Example   A set of data were received for analysis from a experiment. It was believed that these data were continuous and had, according to the accuracy of the data received, been measured to an accuracy of three decimal places. If this were true then the distribution of the last digit of the data (that in the third decimal place) would be expected to be uniformly distributed. The actual data values are listed below and the real and hypothesized frequencies are provided in Table 6.33.

23.288, 24.120, 30.988, 23.444, 37.120, 21.432, 22.332, 27.088, 27.356, 25.372, 26.568,
18.772, 24.020, 26.992, 21.052, 20.700, 20.208, 21.980, 24.556, 22.164, 21.080, 26.216,
29.564, 25.752, 32.576, 25.512, 26.300, 26.892, 32.084, 28.496, 28.440, 33.436, 17.800,
25.808, 31.832, 24.372, 23.796, 24.584, 27.116, 38.444, 19.912, 18.420, 22.164, 22.488,
21.756, 23.740, 19.068, 25.400, 20.968, 28.552, 15.464, 32.772, 26.664, 24.976, 26.244,
27.904, 30.352, 27.848, 31.704, 35.320, 29.312, 27.692, 22.360, 34.912, 26.048, 22.572,
14.212, 29.832, 31.760, 27.652, 38.740, 29.988, 25.668, 18.672, 29.524, 24.120, 33.208,
23.088, 23.412

The hypothesized frequencies are calculated as the product of the total number of observations and the theoretical probability.

Table 6.33   Frequency of values of the final digit of data believed to be continuous

Final digit	Theoretical probability	Observed frequency	Hypothesized frequency
0	1/10	17	7.9
1	1/10	0	7.9
2	1/10	24	7.9
3	1/10	0	7.9
4	1/10	13	7.9
5	1/10	0	7.9
6	1/10	10	7.9
7	1/10	0	7.9
8	1/10	15	7.9
9	1/10	0	7.9

H_0: The distribution of the final digit follows a uniform distribution.
H_1: The distribution of the final digit does not follow a uniform distribution.

Significance level, $\alpha = 0.05$
Degrees of freedom $= 9$
Critical region is '$\chi^2_{9, 0.05} \geq 16.919$'

Calculated statistic, $\chi^2 = 93.025$

The calculated χ^2 statistic lies inside the critical region and therefore the test is significant at the 5% level so the null hypothesis is rejected. The conclusion is that the distribution of the final digit does not follow a uniform distribution.

This result seems obvious from the frequencies and not worth testing; however, until the frequencies are written out in this way it would be easy to accept that the data were recorded to three decimal places. In fact, the data were recorded to two decimal places and the extra decimal place was obtained when applying a conversion factor to convert the yields from kilograms per plot to tonnes per hectare.

6.2.1.3   Kolmogorov-Smirnov test

The one sample Kolmogorov-Smirnov test is a goodness-of-fit test which determines whether observations follow a specified theoretical continuous distribution. It is a more powerful test than the chi-square test and is particularly useful with small sample sizes.

Assumptions   The data must be from a continuous distribution. The observations must be independent. The data are collected from a single population.

Limitations   The data must be measured on the ordinal scale at least.

Hypotheses

   H_0: The observations follow the theoretical distribution.

There are numerous alternative hypotheses. It is only possible to infer if the null hypothesis is not accepted. Therefore the general alternative hypothesis is as follows:

H_1: The observations do not follow the theoretical distribution of frequencies.

Conclusions   If the test result was not statistically significant the null hypothesis would be accepted. It would be concluded that there is no evidence to suggest that the observations do not follow the theoretical distribution. If the test result was statistically significant the null hypothesis would be rejected. It would be concluded that the observations do not follow the theoretical distribution.

Example   Analysis of variance was performed on a set of data. It was decided to formally test whether or not the standardized residuals followed a normal distribution with a mean of 0 and variance of 1. (This is one formal method which can be used for checking the assumption that errors are normally distributed for analysis of variance; see Section 6.1.3.) The residuals obtained from the analysis are given below:

   0.289, −1.763, −0.757, 1.057, −0.375, −0.198, −0.824, −0.650, 0.365, 1.262, −0.851, 1.078, −0.090, 0.099, 0.399, −0.023, −0.019, 0.270, −1.647, 0.040,

   H_0: The residuals follow a normal distribution with a mean of 0.000 and variance of 1.0000.
   H_1: The observations do not follow a normal distribution with a mean of 0.000 and variance of 1.0000.

Significance level, $\alpha = 0.05$
Number of observations $= 20$
Critical region is '$D \geq 0.294$'
Calculated statistic, $D = 0.195$

The calculated statistic, D, lies outside the critical region and therefore the test is not significant so the null hypothesis is accepted. The conclusion is that there is no evidence to reject the hypothesis that the standardized residuals follow a normal distribution with a mean of 0.000 and standard deviation of 1.0000.

6.2.2   Comparison of two paired samples

Keypoints

- *Sign test*: compares the populations from two paired samples; uses ordinal data; uses two related samples.
- *Wilcoxon signed rank test*: compares the populations from two paired samples; uses ordinal data; can be used on ordered data; uses two related samples.

6.2.2.1   Sign test

The sign test determines whether two populations are different or not when the samples taken from them are paired. The direction of the difference between the two populations is tested rather than the size of the difference. It is a non-parametric equivalent to the paired t-test and the data must be at least ordinal. It is useful where one half of the pair, X, can be considered in some way to be 'better' or 'greater' than the other member of the pair, Y. Consequently, the test can be used when it is not possible to take quantitative measurements.

Assumptions   It assumes that the paired sample has come from a population with an underlying continuous distribution. Observations must be independent. Differences between pairs should be independent.

Limitations   The test only looks at the direction of the differences and takes no account of the size of the differences. Data measurements must be made on at least the ordinal scale. Experiment units must be matched or paired as described in Section 6.1.1 (paired t-test).

Hypotheses

H_0: Prob $(X > Y) =$ Prob $(X < Y) = 0.5$
The null hypothesis is that the population X is the same as the population Y. That is, for quantitative data the median difference between X and Y is zero.

There are three alternative hypotheses:

H_1: Prob $(X > Y) \neq 0.5$
The median of X is not equal to the median of Y for quantitative data or the two populations are considered to be equal for qualitative data.

H_1: Prob $(X > Y) > 0.5$
The median of X is greater than the median of Y for quantitative data or population X is 'superior' to population Y for qualitative data.

H_1: Prob$(X > Y) < 0.5$
The median of X is less than the median of Y for quantitative data or population X is 'worse' than population Y for qualitative data.

Conclusions   If the test was not statistically significant then the null hypothesis would be accepted. The conclusion would be that there is no evidence to say that the two populations are not the same. If the test was significant then the null hypothesis would be rejected. In a two-tailed test the conclusion would be that the two populations are not equal. In a one-tailed test the conclusion would be that the populations differ in the direction tested.

Example   In this study cattle were scored on the condition of their front and back teats and so the samples are paired as each animal is used to provide its own matched sample. The objective was to find if any difference in the condition of the teats could be related to a condition such as mastitis. The condition of the front teats was scored as either being worse than, better than or equal to the condition of the back teats. This assessment method allows more cattle to be assessed in a minimal time period. Results are given in Table 6.34.

H_0: There is no difference in condition on the front and back teats.
H_1: The condition of the front and back teats is different.

Significance level, $\alpha = 0.05$
Number of pairs resulting in a '+' or '−' $= 14$
The number of times the most frequent sign occurs $= 10$
Probability of observing the number of times the most frequent sign occurs $= 0.190$

Table 6.34   Indication whether the condition of the front teats of 20 cattle was worse than (−), better than (+) or equal to (0) the condition of the back teats

Cow	1	2	3	4	5	6	7	8	9	10
Sign of score difference	+	0	0	+	−	+	−	0	+	+

Cow	11	12	13	14	15	16	17	18	19	20
Sign of score difference	+	0	−	+	0	+	0	+	−	+

The probability of observing the number of times the most frequent sign occurs is greater than 0.05 and therefore the test is not significant so the null hypothesis is accepted. The conclusion is that there is no evidence to reject the hypothesis that the condition of the front and back teats is the same.

6.2.2.2   Wilcoxon signed rank test

This test is used with two paired samples to determine if the two populations are different or not and is another non-parametric alternative to the paired t-test. This test is more powerful than the sign test because the magnitude of the difference between the two populations is tested. It requires at least ordinal data where both the score for one half of the pair, X, can be ranked as being 'greater than or equal to' the other half, Y, and the differences between the two scores for a pair can be ranked.

Assumptions   The differences between paired observations must come from symmetrical distributions. Observations are independent.

Limitations   Ordinal measurement is required of both the observations and the differences between them. The power of the test deteriorates for large samples. Experiment units must be matched or paired as described in Section 6.1.1 (paired t-test).

Hypotheses

H_0: median of differences $(d_i) = 0$
The null hypothesis is that the samples are equivalent, i.e. they are samples from populations with the same continuous distribution and hence have the same median.

There are three alternative hypotheses:

H_1: median $d_i \neq 0$
The median of X is not equal to the median of Y so the two populations are different.

H_1: median $d_i > 0$
The median of X is greater than the median of Y.

H_1: median $d_i < 0$
The median of X is less than the median of Y.

Table 6.35   Scores of foot damage to cows before moving to the new housing system and after they had been in the new system for a set period

Cow no.	1	2	3	4	5	6	7	8	9	10	11	12	13	14	15
Before	27	20	6	14	12	14	9	22	36	6	0	35	40	18	20
After	61	35	61	36	69	22	24	58	36	16	33	10	21	22	16
d_i	−34	−15	−55	−22	−57	−8	−15	−36	0	−10	−33	25	19	−4	4
Rank	11	5.5	13	8	14	3	5.5	12	−	4	10	9	7	1.5	1.5

Conclusions   If the test was not statistically significant then the null hypothesis would be accepted. The conclusion would be that there is no evidence to say that the two populations are not the same. If the test was significant then the null hypothesis would be rejected. In a two-tailed test the conclusion would be that the two populations were not equal. In a one-tailed test the conclusion would be that the populations differ in the direction tested.

Example   An investigation was conducted into a new housing system for cattle. To establish whether the new system increased the amount of damage to the cows' feet, a scoring system was used which scored various aspects of lesions on the cows' feet from which an overall score for each animal was obtained. Each animal was scored as soon as they were moved to the new housing system and then after they had been in the new system for a set period. The scores obtained are tabulated in Table 6.35.

H_0: median $d_i = 0$
There is no difference in the lesion scores at the beginning and end of the study.

H_1: median $d_i < 0$
The lesion scores increased during the study.

Significance level, $\alpha = 0.05$
Number of ranks $= 14$
Sum of the positive ranks, $W = 17.5$
Probability of obtaining the sum of positive ranks $= 0.015$

The probability of observing the sum of positive ranks is less than 0.05 and therefore the test is statistically significant at the 5% level so the null hypothesis is rejected. The conclusion is that the lesion score has increased.

6.2.3   Comparison of two independent samples

Keypoints

- *Fisher's exact test*: compares the proportion of observations in one category; independent observations are required; uses dichotomous data; is used with small sample sizes.
- *Chi-square test*: tests for independence of rows and columns; compares observed frequencies with expected frequencies; independent observations are required; uses nominal data.

- *Median test*: compares the medians of two independent samples; independent observations are required; uses ordinal data; can be used with truncated data; unequal sample sizes are possible.
- *Mann-Whitney/Wilcoxon rank test*: compares the medians of two independent samples; independent observations are required; uses ordinal data; data are sampled from continuous distributions; unequal sample sizes are possible.
- *Kolmogorov-Smirnov test*: compares the cumulative distributions of two independent samples; independent observations are required; uses ordinal data; data are from continuous distributions; unequal sample sizes are possible; can be used with small sample sizes.

6.2.3.1 Fisher's exact test for 2 × 2 contingency tables

This is a test of location which determines whether two groups differ in the proportions in which they fall into two mutually exclusive categories. It will determine if the rows (treatments) and columns (scores) of the contingency table are independent by testing the proportion of values of group 1 that falls into the first category (p_1) and the proportion of group 2 that falls into the same category (p_2). The test is used for analysing dichotomous data for two small samples. It can be used on frequency data.

Assumptions   The observations must be independent.

Limitations   The test is performed on a 2 × 2 contingency table where the sample size is small. The scores should be measured on the nominal or ordinal scales.
Hypotheses

   H_0: $p_1 = p_2$ The rows and columns of the contingency table are independent.

There are three alternative hypotheses:

   H_1: $p_1 \neq p_2$ The proportion of group 1 is not equal to the proportion of group 2 in the first category; the rows and columns are not independent.
   H_1 : $p_1 > p_2$ The proportion of group 1 is greater than the proportion of group 2 in the first category.
   H_1 : $p_1 < p_2$ The proportion of group 1 is less than the proportion of group 2 in the first category.

Conclusions   If the test was not statistically significant then the null hypothesis would be accepted. The conclusion would be that there is no evidence to say the rows and columns of the table are not independent. If the test was statistically significant then the null hypothesis would be rejected. In a two-tailed test the conclusion would be that the two groups are not independent; the proportions falling into the first category differ. In a one-tailed test the conclusion would be that the proportions falling into the first category differ in the direction stated.

Example   In a study assessing the effect of two different diets on the level of heel erosion suffered by cattle, each animal was fed one of the diets and after a period of time the amount of heel erosion was assessed and scored. Few distinct scores were recorded so the data were converted to presence/absence of heel erosion before analysis. The resultant contingency table is given in Table 6.36.

Table 6.36   Contingency table indicating the presence or absence of heel erosion for cattle fed on two different diets

	Heel erosion	
	Absent	Present
Diet 1	8	4
Diet 2	2	6

p_1 = proportion of diet 1 samples in the heel erosion absent category.
p_2 = proportion of diet 2 samples in the heel erosion absent category.

H_0: $p_1 = p_2$ The presence or absence of heel erosion is independent of the two diets.
H_1: $p_1 > p_2$ There is an increase in the presence of heel erosion for diet 1.

Significance level, $\alpha = 0.05$
Number of observations $= 20$
$p_1 = 8/12$
$p_2 = 2/8$
Probability of observing p_1 and $p_2 = 0.075$

The probability of observing p_1 and p_2 is greater than 0.05 and therefore the test is not statistically significant at the 5% level so the null hypothesis is accepted. The conclusion is that the diet has not affected the presence or absence of heel erosion. However, the result is not really conclusive as the test was significant at the 10% level.

6.2.3.2   Chi-square test

The chi-square test is used to detect any sort of difference (location or dispersion) between two independent groups. It is used with frequency data falling into discrete categories. The test ignores any ranking implicit in the categories. The data are summarized in a contingency table where usually the columns represent the two groups and the rows represent the categories. The expected frequencies in each cell of the contingency table are then compared with the observed frequencies. If necessary, the data can be artificially classified into categories but information, and therefore power, are lost doing this.
Assumptions   The observations must be independent.

Limitations   The data must fall into at least two categories and are measured on at least the nominal scale of measurement. The chi-square test is inappropriate if the expected frequencies in the contingency table are small. This means that the test should only be used on data where all expected frequencies are at least one, and no more than 20% of the cells have expected frequencies less than five. If these criteria is not satisfied then adjacent categories can be combined if they still make sense. The expected frequency for a cell in a contingency table is calculated as the product of the sum of the row frequency and the column frequency in which the cell lies, divided by the total frequency.
   A special case of the chi-square test is for a 2×2 contingency table. This test includes a continuity correction which markedly improves the accuracy of the expected frequency

calculations. The test with the continuity correction can be used when there are more than 40 observations. The chi-square test without the continuity correction can be used if all expected frequencies are greater than five when there are between 20 and 40 observations in total. Fisher's exact test should be used if there are 20 or fewer observations or if there is an expected frequency of less than five.

Hypotheses

H_0: There is no interaction between the contingency table rows and columns; the proportions in the first group categories are the same as the proportions in the second group categories.

There is only one alternative hypothesis:

H_1: There is an interaction between the contingency table rows and columns; the proportions in the categories differ between the two groups.

Conclusions

If the test was not statistically significant then the null hypothesis would be accepted. The conclusion would be that there is no evidence to say that there is an interaction between the rows and columns of the table. If the test was statistically significant then the null hypothesis would be rejected. The conclusion would be that there is an interaction, i.e. the two groups are not independent of the categories.

Example In an experiment looking at various factors affecting newly planted woodland the percentage of saplings which survived was assessed on different plots for two species. The percentages were banded into six categories. The resultant frequency table is shown in Table 6.37.

H_0: There is no difference between the distribution of survival rates of the two species.
H_1: There is a difference between the distribution of survival rates of the two species.

Significance level, $\alpha = 0.01$
Degrees of freedom $= 5$
Critical region is '$\chi^2_{5,0.01} \geq 15.086$'

Calculated statistic, $\chi^2 = 26.86$

The calculated statistic χ^2 lies within the critical region and therefore the test is statistically significant at the 1% level so the null hypothesis is rejected. The conclusion is that there is a difference in the distribution of survival rates of the two species.

Table 6.37 Frequency table showing the percentage survival of two tree species

Survival (%)	Sycamore	Birch
0–9	9	30
10–24	10	17
25–49	12	13
50–74	21	11
75–99	10	5
≥ 100	18	4

6.2.3.3   Median test

The median test is used to test whether two independent groups have been sampled from continuous populations, X and Y, with the same median. The groups do not have to be the same size. The test is performed by calculating the overall median and then categorizing the data as to whether they fall above or below this value. For this reason it is not necessary to measure all observations, only sufficient to determine the median value and the relative magnitude of the other values. The test is useful, therefore, with truncated data.

Assumptions   The observations must be independent.

Limitations   The data must be measured on at least the ordinal scale of measurement. The median test is one of the least powerful two-sample tests.

Hypotheses

   H_0: $X = Y$ The two distributions have the same median.

There are three possible alternative hypotheses:

   H_1: $X \neq Y$ The medians of the two distributions are not the same.
   H_1: $X > Y$ The median of one distribution, X, is greater than that of the other distribution, Y.
   H_1: $X < Y$ The median of one distribution, X, is less than that of the other distribution, Y.

Conclusions   If the test was not statistically significant then the null hypothesis would be accepted. The conclusion would be that there is no evidence to say the distributions do not have the same median. If the test was statistically significant then the null hypothesis would be rejected. In a two-tailed test the conclusion would be that the two samples have come from distributions with different medians. In a one-tailed test the conclusion would be that the samples have come from distributions with medians differing in the direction stated.

Example   An experiment was conducted to investigate the effect of fertility treatments on the length of time between a dairy cow calving and conceiving. The fertility treatment was tested against an untreated control. The length of time between calving and next conception was measured in days. After 120 days not all of the cows had conceived but it was decided to analyse the data available. The data were as follows:

   Treatment: 45 65 108 70 >120 58 >120   106 >120 >120 65 >120 61   88   109
   Control: 69 64   56 92   103 96   71 >120 >120   92 64   82 59 >120 >120

   H_0: There is no difference in the length of time between calving and conception for either treatment.
   H_1: The length of time between calving and conception for either treatment is different for each treatment.

Significance level, $\alpha = 0.05$
Number of observations for treatment $= 15$
Number of observations for control $= 15$

Degrees of freedom $= 1$
Critical region is '$\chi^2_{1,\,0.05} \geq 3.841$'
Calculated statistic $= 0.54$

The calculated χ^2 statistic lies outside the critical region and therefore the test is not significant so the null hypothesis is accepted. The conclusion is that there is no evidence that the treatment has any effect on the length of time between calving and conception.

6.2.3.4   Mann-Whitney/Wilcoxon rank test

The Mann-Whitney/Wilcoxon rank test is used to determine whether two independent groups have been drawn from two populations, X and Y, with the same medians. It is one of the most powerful non-parametric tests. The test considers the rank value of each observation rather than just its location with respect to the sample median, thereby producing a more powerful test than one that just considers differences in median values. The groups can have small numbers of observations and do not have to be of equal size. It is an alternative to the parametric two-sample t-test.

Assumptions   The test assumes that data are sampled from a continuous distribution and that the observations are independent.

Limitations   The data must be measured on the ordinal scale at least. The test should only be used with large sample sizes because it loses power when sample sizes are small.

Hypotheses

H_0: $P(X > Y) = 0.5$ The two continuous distributions have the same median.

There are three possible alternative hypotheses:

H_1: $P(X > Y) \neq 0.5$ The medians of the two continuous distributions are not the same.
H_1: $P(X > Y) > 0.5$ The median of one distribution, X, is greater than that of the other distribution, Y.
H_1: $P(X > Y) < 0.5$ The median of one distribution, X, is less than that of the other distribution, Y.

Conclusions   If the test was not statistically significant then the null hypothesis would be accepted. The conclusion would be that there is no evidence to say the distributions do not have the same median. If the test was statistically significant then the null hypothesis would be rejected. In a two-tailed test the conclusion would be that the two samples have come from distributions with different medians. In a one-tailed test the conclusion would be that the samples have come from distributions with medians differing in the direction stated.

Example   An investigation was undertaken to study the various factors that may be used to predict whether or not a field is infested with wireworms. Fields that were known to be infested and not infested were studied and factors assessed on each field. The factors were then tested to find if there was any difference between infested and non-infested fields. The percentage of stone in the soil was one of the factors investigated and the percentages measured are as detailed in Table 6.38.

Table 6.38 Percentage stone content of soil in fields infested and not infested with wireworms

Non-infested fields (% wt)	Rank	Infested Fields (% wt)	Rank
20.4	25	0.4	8
0.1	3	1.1	11
10.5	19	0.1	3
0.2	6.5	0.2	6.5
1.4	12	0.1	3
5.3	17	8.4	18
0.9	10	16.6	24
11.2	22	3.7	14
4.7	16	10.3	20
4.5	15	0.1	3
		10.8	21
		1.8	13
		13.0	23
		0.7	9
		0.1	3

H_0: There is no difference between the percentage of stone in infested and non-infested fields.

H_1: The percentage of stone in infested and non-infested fields is different.

Significance level, $\alpha = 0.05$
Number of observations for non-infested fields $= 10$
Number of observations for infested $= 15$
Smallest sum of ranks $= 145.5$
Largest sum of ranks $= 179.5$
Probability of observing the sums of ranks $= 0.373$

The probability of observing the sums of ranks is greater than 0.05 and therefore the test is not statistically significant at the 5% level so the null hypothesis is accepted. The conclusion is that there is no evidence of any difference in the percentage of stone in wireworm infested and non-infested fields.

6.2.3.5 Kolmogorov-Smirnov test

The Kolmogorov-Smirnov test assesses whether two independent samples are from populations having the same distribution. When a two-tailed alternative hypothesis is used the test looks for any sort of difference, such as location, dispersion, skewness, etc., but when a one-tailed alternative hypothesis is used the test only looks for differences in location. Both the one- and two-tailed alternatives can be used with very small sample sizes. Equal sample sizes are not required. The two-tailed test can be used when the population is not continuous but the test then becomes more conservative than when the continuity assumption holds and there will be a greater chance of making a Type II error.

Assumptions The data should be from a continuous distribution. The observations must be independent.

Limitations Data must be measured on at least the ordinal scale.

Hypotheses

H_0: The two samples follow the same distribution.

There are three alternative hypotheses:

H_1: The two samples are drawn from distributions that differ in at least one characteristic.

H_1: The first sample is drawn from a distribution with some larger characteristic value than the second sample.

H_1: The first sample is drawn from a distribution with some smaller characteristic value than the second sample.

Conclusions If the test was not statistically significant then the null hypothesis would be accepted. The conclusion would be that there is no evidence to say the distributions are not the same. If the test was statistically significant then the null hypothesis would be rejected. In a two-tailed test the conclusion would be that the two samples have come from distributions that differ with respect to at least one characteristic. In a one-tailed test the conclusion would be that the samples have come from distributions that differ in their location in the specified direction.

Example In an investigation into the somatic cell counts in dairy cows, the cell counts of animals in their first and second lactation were measured and the results are tabulated in Table 6.39. The aim was to establish whether the distribution was the same in each lactation.

H_0: The distribution of the somatic cells is the same in both lactations.
H_1: The distribution of somatic cells is different in the two lactations.

Significance level, $\alpha = 0.05$
Number of observations in lactation $1 = 10$
Number of observations in lactation $2 = 8$
Critical region is '$D \geq 0.533$'
Calculated statistic, $D = 0.675$

Table 6.39 Somatic cell counts of cows in the first and second lactations

Lactation 1	Lactation 2
81.4848	56.7318
57.1718	36.8624
49.5532	46.4822
48.1518	39.0318
43.8125	43.3310
47.3996	47.3938
42.9837	44.0427
51.6716	43.1454
48.1547	
50.1167	

The calculated statistic, D, lies inside the critical region and therefore the test is statistically significant so the null hypothesis is rejected. The conclusion is that there is a difference in the distributions of the somatic cell counts in the first and second lactations.

6.2.4   Comparison of more than two matched samples

Keypoints

- *Friedman's test*: compares the medians of more than two matched samples; data observations must be independent; uses ordinal data; uses data from continuous distributions; can be used with small samples.

Friedman's test

Friedman's test is used to test the hypothesis that many matched samples could have come from populations with the same median. The samples must be matched so equal numbers of observations are required for each sample. Matching is achieved in a similar way as for two related samples (see Section 6.1.1.2). The test can be used with small sample sizes. Each sample represents a treatment or condition under investigation. The samples are independent.

Assumptions   The test assumes that the data for each sample come from the same continuous distribution.

Limitations   The samples must be matched and equal sample sizes are required. The data must be measured on the ordinal scale at least.

Hypotheses

H$_0$: $X_i = X_j$ for all $i \neq j$
The population medians are the same for all samples.

There is only one alternative hypothesis:

H$_1$: $X_i \neq X_j$ for at least one pair of $i \neq j$
At least one pair of medians is different.

Conclusions   If the test was not statistically significant then the null hypothesis would be accepted. The conclusion would be that there is no evidence to say the medians are not the same. If the test was statistically significant then the null hypothesis would be rejected. The conclusion would be that at least one of the samples has a different median value from the rest.

Note: When the test result is statistically significant the conclusion is that at least one of the samples is different. Multiple comparisons are possible which may indicate which sample or group of samples are different. The tests available are the equivalent of the parametric pairwise comparisons test and Dunnett's test. With these tests the chances of making a Type I error are increased so the test results should be treated with caution.

Example   In a study investigating the quality of four different oat varieties, one of the assessments made was a score of grain ripeness. The study was run in four randomized

Table 6.40   Grain ripeness scores in four varieties of oats

	Grain Ripeness Score			
	Block 1	Block 2	Block 3	Block 4
Variety 1	9	8	9	9
Variety 2	7	6	4	7
Variety 3	9	8	9	8
Variety 4	6	6	4	6

blocks so these constituted the four related samples and the data in Table 6.40 were collected.

H_0: The median ripeness score is the same for each variety.
H_1: The median ripeness score is different for at least one of the four varieties.

Significance level, $\alpha = 0.05$
Number of related samples $= 4$
Number of observations per sample $= 4$
Critical region is 'calculated statistic ≥ 7.8'
Calculated statistic $= 11.4$

The calculated statistic lies within the critical region and therefore the test is statistically significant at the 1% level so the null hypothesis is rejected. It is concluded that the median ripeness score is different for at least one of the four varieties.

6.2.5   Comparison of more than two independent samples

Keypoints

- *Chi-square test*: tests for independence of rows and columns; compares observed frequencies with expected frequencies; independent observations are required; uses nominal data.
- *Median test*: compares the medians of more than two independent samples; independent observations are required, uses ordinal data; can be used with truncated data; unequal sample sizes are possible.
- *Kruskal-Wallis test*: determines whether more than two independent samples have been drawn from populations with the same median; independent observations are required; uses ordinal data; data are from continuous distributions; unequal sample sizes are possible.

6.2.5.1   Chi-square test

The chi-square test is testing for any sort of differences (location or dispersion, etc.) between more than two independent groups. It is appropriate for frequency data in discrete categories. The data are summarized in a contingency table where the columns represent the groups and the rows represent the categories. The expected frequencies in each cell of the contingency table are then compared with the observed frequencies. If necessary, the data can be artificially classified into categories but information, and therefore power, are lost doing this.

Assumptions   The observations must be independent.

Limitations   The data must fall into at least two categories and should be measured on at least the nominal scale of measurement. The test takes no notice of any ordering implicit in the categories. The chi-square test is inappropriate if the expected frequencies in the contingency table are small. This means that the test should only be used on data where all expected frequencies are at least one and no more than 20% of the cells have expected frequencies less than five. If these criteria is not satisfied then adjacent categories can be combined if they still make sense. The expected frequency for a cell in a contingency table is calculated as the product of the sum of the row frequency and the column frequency in which the cell lies divided by the total frequency.

Hypotheses

   H_0:   The rows and columns of the table are independent, i.e. the group frequencies have
          come from the same populations.

There is only one alternative hypothesis:

   H_1:   The rows and columns of the table are not independent.

Conclusions   If the test was not statistically significant then the null hypothesis would be accepted. The conclusion would be that there is no evidence to say that the rows and columns are not independent. If the test was statistically significant then the null hypothesis would be rejected. The conclusion would be that the group frequencies are not independent of the categories.

Example   A study was carried out to investigate the effectiveness of five different herbicides (A, B, C, D and E) in controlling weeds in new tree plantations. The toxic effect of the herbicides was scored on various criteria and the scores summed to give an overall score between 0 and 50. These scores had to be combined into categories because there would be 50 categories if each score was a category and the expected frequencies would have been too low. The frequency data in Table 6.41 were collected.

   H_0:   There is no difference between the distribution of toxicity scores for the five
          herbicides.
   H_1:   There is a difference between the distribution of toxicity scores for the five
          herbicides.

Significance level, $\alpha = 0.05$
Degrees of freedom $= 12$
Critical region is '$\chi^2_{12,\,0.05} \geq 21.03$'

Calculated statistic, $\chi^2 = 23.34$

The calculated χ^2 statistic lies within the critical region and therefore the test is statistically significant at the 5% level so the null hypothesis is rejected. The conclusion is that there is a difference in the distribution of toxicity scores between the five herbicides.

Table 6.41   Toxicity scores on trees receiving one of five herbicides

Toxicity score	Herbicide				
	A	B	C	D	E
0	76	72	74	67	83
1–5	10	9	11	9	12
6–10	7	9	4	9	1
> 10	3	6	7	11	0

6.2.5.2   Median test

The median test is used to test whether more than two independent groups have been sampled from two populations, X and Y, with the same median. The groups do not have to be the same size. The test is performed by calculating the overall median and then categorizing the data as to whether they fall above or below this value. For this reason it is not necessary to measure all observations, only sufficient to determine the median value and the relative magnitude of the other values. The test is useful, therefore, with truncated data.

Assumptions   The observations must be independent. It is assumed the groups come from the same continuous distribution.

Limitations   The data must be measured on at least the ordinal scale of measurement.

Hypotheses

H_0: $X_i = X_j$ for all $i \neq j$
The population medians are the same for all samples.

There is only one alternative hypothesis:

H_1: $X_i \neq X_j$ for at least one pair of $i \neq j$
At least one pair of medians is different.

Conclusions   If the test was not statistically significant then the null hypothesis would be accepted. The conclusion would be that there is no evidence to say the medians are not the same. If the test was statistically significant then the null hypothesis would be rejected. The conclusion would be that at least one of the samples has a different median value from the rest.

Example   In an experiment there were three different types of cover crop grown before growing a cereal crop. There were six replicates of each cover crop. The total soil mineral nitrogen content of the soil was measured at harvest. Total soil mineral nitrogen is a derived variable and one of the variables that is part of the calculation can only be measured to a specified lower limit. This means that soil mineral nitrogen can only be calculated to a specified limit; values less than 2.44 cannot be detected so the data are truncated. The data are tabulated in Table 6.42.

H_0: There is no difference in the total soil mineral nitrogen for each cover crop.
H_1: There is a difference in the total soil mineral nitrogen for each cover crop.

Table 6.42   Soil mineral nitrogen results after the harvest of three cover crops

	Replicate no.					
	1	2	3	4	5	6
Crop 1	<2.45	3.26	4.08	9.80	3.67	10.20
Crop 2	8.16	<2.45	3.67	4.90	3.67	4.90
Crop 3	<2.45	10.61	3.67	3.67	4.89	3.62

Significance level, $\alpha = 0.05$
Degrees of freedom $= 2$
Critical region is '$\chi^2_{2,0.05} \geq 5.99$'

Calculated statistic, $\chi^2 = 0.45$

The calculated χ^2 statistic lies outside the critical region and therefore the test is not significant so the null hypothesis is accepted. The conclusion is that there is no evidence of a difference in total soil mineral nitrogen for each cover crop.

6.2.5.3   Kruskal-Wallis test

The Kruskal-Wallis test is used to determine whether the samples could be from populations with the same median. The test does not require equal sized samples and sample sizes can be very small. It is the non-parametric equivalent to the one-way analysis of variance as any structure in the treatment list will be ignored by the test.

Assumptions   This test assumes that the independent samples are taken from a continuous distribution.

Limitations   The data must be measured on at least the ordinal scale. Any structure implicit in the treatments will be lost.

Hypotheses

   H_0: $X_i = X_j$ for all $i \neq j$
   The population medians are the same for all samples.

There is only one alternative hypothesis:

   H_1: $X_i \neq X_j$ for at least one pair of $i \neq j$
   At least one pair of medians is different.

Conclusions   If the test was not statistically significant then the null hypothesis would be accepted. The conclusion would be that there is no evidence to say the medians are not the same. If the test was statistically significant then the null hypothesis would be rejected. The conclusion would be that at least one of the samples has a different median value from the rest.
   Note: When the test result is statistically significant the conclusion is that at least one of the treatments is different. Multiple comparisons are possible which may indicate which treatments or group of treatments are different. The tests available are the equivalent of the

Table 6.43   Above-ground dry matter results for a crop sown on three different dates

	Replicate no.				
	1	2	3	4	5
Sowing date 1	944	669	160	316	218
Sowing date 2	543	462	441	421	450
Sowing date 3	10	8	46	14	8

parametric pairwise comparisons test and Dunnett's test. As with these tests the chances of making a Type I error are increased so the test results should be treated with caution.

Example   In a study looking at different sowing dates for a crop, the above-ground dry matter was assessed throughout the growing season. There were five replicates in the experiment and the data are given in Table 6.43. In the early stages the final sowing date was producing much smaller values than the other two. This led to highly skewed data for which no transformation could be found that left the data satisfying the analysis of variance assumptions. An alternative analytical technique is required.

H_0: There is no difference between the above-ground dry matter for the three sowing dates.

H_1: At least one of the above-ground dry matter median sowing dates is different.

Significance level, $\alpha = 0.05$
Degrees of freedom $= 2$
Critical region is '$\chi^2_{2,\,0.05} \geq 5.991$'

Calculated statistic, $\chi^2 = 9.52$

The calculated χ^2 statistic lies inside the critical region and therefore the test is statistically significant so the null hypothesis is rejected. The conclusion is that at least one of the sowing dates has a different median above-ground dry matter.

6.2.6   Comparison of tests

In the following sections the non-parametric tests discussed are briefly summarized. Where there is a choice of tests, guidance is provided on which test is likely to be the most appropriate. It should be remembered that not all non-parametric tests have been included in these notes; other tests may be more appropriate for the given circumstance.

6.2.6.1   Summary of single-sample tests

The tests can be used to ascertain if samples have been drawn from specific populations, if there are differences in location between the sample and the population, if samples are random or if observed frequencies are as expected.

The runs test is the only test for randomness and looks for clustering or dispersion data. The other two tests are both goodness-of-fit tests based on hypothesized frequency distributions. The chi-square test compares observed frequencies with hypothesized frequencies on categorical data. The Kolmogorov-Smirnov test is used to compare observed cumulative frequency distributions with expected cumulative frequency distributions. The Kolmogorov-Smirnov test is generally more powerful than the chi-square test when it can be used, particularly when sample sizes are small.

6.2.6.2 Summary of two related sample tests

These tests are used to determine whether two paired samples have different medians. Both the sign test and the Wilcoxon signed rank test have the same basic hypotheses but the Wilcoxon signed rank test should be used when the differences between the elements of the paired samples can be ranked. If the samples can only be scored on a 'greater than/less than' basis then the sign test should be used.

6.2.6.3 Summary of two independent sample tests

These test the hypothesis that two independent samples come from the same population. Different tests are used in various circumstances. Some tests may be used to test whether two samples represent populations that differ in their measures of location. Others are not so specific and are used to determine whether the samples are from populations that differ in any respect at all, such as dispersion, location or skewness.

Two of the tests considered, Fisher's exact test and the chi-square test, assess the proportion of observations falling into two or more categories. The Fisher's exact test can only be used on data falling into two categories whereas the chi-square test has no such restrictions. The chi-square test cannot be used with small sample sizes though Fisher's exact test can.

Three of the tests considered – the median test, the Mann-Whitney/Wilcoxon test and the one-tailed Kolmogorov-Smirnov test – can be used to detect differences in location. The median test is useful for truncated data or if the data can only be classified as above or below the median. The Mann-Whitney/Wilcoxon test is a more powerful test so should therefore be used when possible. The one-tailed Kolmogorov-Smirnov test is slightly more powerful than the Mann-Whitney/Wilcoxon when used with small sample sizes. If sample sizes are large the Mann-Whitney/Wilcoxon test is more powerful than the one-tailed Kolmogorov-Smirnov test. For all sample sizes the one-tailed Kolmogorov-Smirnov test is more powerful than the median test.

Two of the tests considered – the two-tailed Kolmogorov-Smirnov test and the chi-square test – can be used to detect any differences in the sample populations, such as location, dispersion, skewness, etc. A two-tailed Kolmogorov-Smirnov test should be used in preference to the chi-square test if the data are continuous. The two-tailed Kolmogorov-Smirnov test is the more powerful test but is more conservative if data are not continuous.

Note: The Kolmogorov-Smirnov test can be used in two different ways depending on whether it is used as a one-tailed test or a two-tailed test.

6.2.6.4 Summary of more than two related sample tests

Friedman's test is used to determine whether three or more matched samples have different medians.

6.2.6.5 Summary of more than two independent sample tests

The three tests considered in Section 6.2.5 all test for differences in location between several independent samples. The chi-square test can only be used with frequencies or categorical data. If data are not of this form they can be artificially categorized though information will be lost.

Two of the tests – the median test and the Kruskal-Wallis test – can be used to determine differences in location of the independent samples. The median test is useful for truncated data or if the data cannot be ranked but can only be classified as above or below the median. The median test is less affected by outliers than the Kruskal-Wallis test. When applicable the Kruskal-Wallis test should be used in preference as it is a more powerful test.

6.3   COMPARISON OF PARAMETRIC AND NON-PARAMETRIC TECHNIQUES

Keypoints

	Parametric	*Non-parametric*
Assumptions	Many	Few
Power	Strong	Weak
Hypotheses	Parameter based	Distribution based
Data types	Quantitative	All types
Interpretation of test results	Relatively straight-forward	Can be difficult
Data structure	Any	Limited
Population	Must be known	Can be unknown
Sample size	Limitations on minimum size	Can be small

Parametric techniques make many assumptions about the population from which the data are drawn. They are used to estimate and compare population parameters and the distribution of the population must be known. Non-parametric techniques do not make as many assumptions about the data and those that are made tend to be weaker. Non-parametric techniques are used to compare characteristics of the population distributions and it is not always necessary to know the form of the population distributions.

Non-parametric techniques are more robust than parametric techniques because they make fewer assumptions about the nature of the data. They are less powerful than parametric techniques if the underlying distribution of the data is reasonably close to a standard distribution because they extract less information from the data and so are less likely to detect a difference if one exists. If the data can be transformed to a distribution that enables parametric techniques to be used, more information will be gained than by using non-parametric statistics on the untransformed data. If all the assumptions of a parametric test are valid then the non-parametric test would require more observations to be as effective. If all the assumptions of parametric techniques are met then these techniques are always more powerful than the non-parametric equivalents although non-parametric techniques come very close to matching parametric techniques for small sample sizes.

The results from non-parametric tests are not always so easy to interpret as those from parametric tests so they should only be employed when parametric techniques are not appropriate. For example, they are less appropriate for testing complex models, particularly those involving a hierarchical data structure or interactions. However, these techniques are generally robust and cover a wide range of distributions whereas many parametric techniques require the data to be normally distributed. Consequently non-parametric techniques are useful for analysing data with non-normal distributions or distributions that are not easily specified or are unknown.

Non-parametric statistics can be used on a variety of data types. They should be used on nominal data since such data, by their nature, usually violate the assumptions of parametric techniques. They can be used on count data and rank or ordered data, such as infestation scores. If it is only possible to collect data by category (e.g. type of ground cover could be categorized as grassland or forest), then non-parametric techniques must be used as parametric techniques are inappropriate. If sufficient categorical data can be collected then the power of the non-parametric test may well be adequate despite the weak measurement scale. If continuous data are not suitable for analysis via parametric techniques, non-parametric statistics can be used. This may occur when sample sizes are small or when the data have a very skewed distribution.

REFERENCES

Anscombe FJ (1973) Graphs in statistical analysis. *Am. Statist.* **27**: 17–21.

Box GEP, Hunter WG and Hunter JS (1978) *Statistics for Experimenters: An Introduction to Design, Data Analysis, and Model Building.* John Wiley, New York, Chapters 1, 9 and 15.

Chatfield C (1995) *Problem Solving: A statistician's guide*, second edition. Chapman and Hall, London, Chapter 6 and Appendix C2.

Cochran WG and Cox GN (1957) *Experimental Designs*, second edition. John Wiley, New York, Chapters 1, 3, 4, 5 and 7.

Draper N and Smith H (1981) *Applied Regression Analysis*, second edition. John Wiley, New York, Chapters 1–3, 5, 7 and 10.

Genstat 5 Committee (1993) *Genstat 5 Release 3 Reference Manual.* Oxford University Press, Oxford, Chapter 9.

Montgomery DG (1991) *Design and Analysis of Experiments*, third edition. John Wiley, New York, Chapters 3–7 and 14.

Siegel S and Castellan NJ, Jr (1988) *Nonparametric Statistics for the Behavioral Sciences*, second edition. McGraw-Hill, New York, Chapters 1, 3–8.

Sokal RR and Rohlf FJ (1995) *Biometry*, third edition. Freeman, New York, Chapter 13.

Snedecor GW and Cochran WG (1989) *Statistical Methods*, eighth edition. Iowa State University Press, Ames, Chapter 13.

Steel RGD and Torrie JH (1980) *Principles and Procedures of Statistics: A Biometrical Approach*, second edition. McGraw-Hill, New York, Chapters 7–9, 15, 16 and 24.

Weisberg S (1985) *Applied Linear Regression*, second edition. John Wiley, New York, Chapters 1, 3, 5, 6, 7 and 12.

CHAPTER 7

Other Statistical Techniques

Previous chapters have concentrated on the design of experiments, their analysis and interpretation of results, and general tools for data exploration have been presented. There are many more statistical topics such as multivariate statistics, time series analysis, survey methods, clinical trials, survival analysis, spatial statistics, demographic statistics, and process control, which cover a wide variety of statistical techniques not mentioned so far in this text. Each of these topics requires the use of standard data exploration techniques and their own specialist techniques to examine, investigate, explore and form conclusions from data.

In this chapter we give the reader a brief insight into some of these other topics. The aim here is to whet the reader's appetite. Some of the statistical techniques particularly associated with each topic are mentioned. We indicate what the techniques aim to do, the assumptions that need to be made and the sort of data they can be used with. No formulae are presented and only the briefest information, if any, on the actual procedures is given. It is left to the interested reader to refer to more detailed texts, such as those referenced, which provide a much fuller coverage of the topics.

7.1 MULTIVARIATE ANALYSIS

In many instances with experimentation and research, measurements are taken on a whole range of different variables which may or may not be related, directly or indirectly, to each other in some way. Multivariate analysis can be used to investigate the structure of the data and to identify the main features of the data. This is mainly achieved by hypothesis testing, by fitting models to the data, by classifying the variables into groups or clusters or dividing the individual observations into groups. Although each multivariate technique has its own specific function or use, few of the techniques are used in complete isolation. It is more usual to use a few multivariate techniques to explore the data though some situations do lend themselves to the use of one specific technique. Many of these techniques are used as stepping stones to point to the next stage of analysis.

Many of the techniques assume the data are from a multivariate normal distribution. However, in reality it is unlikely that all variables under consideration will follow a normal distribution. Even if they do it is unlikely that the researcher will be in a position to test this assumption on each variable individually; the sheer number of variables being used may make this too time-consuming a task. In practice, it is usually assumed that the normality assumption holds and the multivariate analysis is conducted on the data without any formal normality testing or transformation to normal distributions being carried out (Chatfield and Collins, 1980). In this section some of the main techniques used are briefly outlined and any assumptions associated with the technique are noted. A few references have been provided from a selection of texts the authors have found useful.

7.1.1 Multivariate t-test

The one-sample and two-sample t-tests mentioned previously in Section 6.1.1 look for differences between a single mean value and a specified mean value or between two mean values, for a single variable so only one comparison is made with each test. A second test would be required if another variable was to be tested. The multivariate t-test is called Hotelling's T^2 and is an extension of the univariate t-test as it tests for differences between two or more variables simultaneously so two or more comparisons are made with each test. Each comparison is of the same form as the univariate t-test; it either tests for a difference between the mean value and a specified value (the one-sample case) or tests for a difference between a mean value from one group and a mean value from a second group (the two-sample case). There will be as many comparisons made as there are variables under consideration. For example, if the milk yield, weight and food intake of cows and heifers are measured then Hotelling's T^2 can be used to make a multiple comparison of cows and heifers. The three mean values of milk yield, weight and feed intake for the first group, cows, can be compared simultaneously with the three corresponding mean values for the second group, heifers.

In both cases a calculated test statistic is compared against a critical value to determine if the null hypothesis of no difference is to be rejected or accepted. The test is structured so that, although many simultaneous comparisons are being made, there is only an $\alpha\%$ chance of rejecting the null hypothesis when it is correct; the test does control the probability of making Type I errors.

If the test proves to be statistically significant, the null hypothesis is rejected and the alternative hypothesis is accepted. It is not possible to say, from the Hotelling's T^2 alone, which of the mean values is causing the difference. Simultaneous confidence limits on the mean value for the one-sample case or on the differences between the two mean values for the two-sample case can be calculated to determine which variables are contributing towards a significant result. If the interval for a variable contains the specified value (one-sample case) or the value zero (two-sample case) then the test is not significant for that variable and the difference found in the Hotelling's T^2 test is not due to that variable. It only makes sense to calculate these after the Hotelling's T^2 has been found to be statistically significant.

Assumptions   For the single-sample case the test assumes the observations have been randomly sampled and are from a multivariate normal distribution.

For the two-sample case the test assumes that the observations in each group have been randomly sampled, that the groups are from a multivariate normal distribution and that the population variance–covariance matrices of the two groups are equal. It is not necessary to have equal numbers of observations in each group. For large numbers of observations in each sample the T^2 test is quite robust against departures from equal variance–covariance matrices and multivariate normality.

A full discussion of these tests with examples and formulae can be found in Chatfield and Collins (1980), which provides a useful introduction to multivariate statistics.

7.1.2 Multivariate analysis of variance

This extension of the multivariate t-test allows comparisons between mean values from more than two groups. The comparisons are being made simultaneously on multiple response variables where the response variables are a series of different characteristics being measured. It is the multivariate equivalent to the univariate analysis of variance, where comparisons are made on a single response variable. The multivariate analysis of variance partitions the variances and covariances of the variables into the parts due to residual error and to treatment effects.

The one-way case is the multivariate equivalent to the single treatment factor completely randomized design. The groups would be formed by different treatments and each group would have a vector of mean values for the different response variables. Treatment structure can be accounted for by using multi-way designs. For example, when there are two treatment factors a two-way multivariate analysis of variance can test for interactions between the two factors and for the main effects of the two factors. If an interaction term in a multivariate analysis of variance is found to be statistically significant then it may not be possible to sensibly interpret any lower-order treatment effects involved in that interaction.

In a similar way to the univariate analysis of variance the multivariate analysis of variance performs an overall test to see if the group vector mean values are different. If this test is found to be significant then further tests are available which may indicate the major source of differences between the treatments (groups) or which variables are most affected by the treatments.

There are tests available to check the assumption of equal variance–covariance matrices. These tests, however, are sensitive to departures from normality so a significant result may indicate either that the data are not from a multivariate normal distribution or that the data are not homogeneous, or it could indicate both. Transformations can be used to correct non-normality in the data although, as Hand and Taylor (1987) say, the normality test is not usually applied to the multivariate normal distribution. Generally the normality testing takes the form of transforming the individual variables so each of them is normally distributed, with the hope that if each variable is from a normal distribution then the multivariate normality assumption is also satisfied.

Assumptions It is assumed that the response variables form independent observations from multivariate normal distributions. It is also assumed that the variance–covariance matrix for each group is equal.

A useful introduction to the concepts of multivariate analysis of variance is provided by Hand and Taylor (1987) who have approached the subject from a non-mathematical viewpoint. Model formulae for a one-way and two-way multivariate analysis of variance are presented in Morrison (1990) with worked examples.

7.1.3 Discriminant analysis

The idea behind the technique of discriminant analysis is that there are two known groups of observations and that these groups differ with respect to values from a set of independent variables. The variables can be used to produce a linear discrimination rule that can identify which group an individual belongs to based on the values of the observed variables for that individual. The discrimination rule can then be used to predict which group a future individual would belong to. As an illustration, consider the problem of wireworm infestation in grassland fields where a field can be either infested or not infested (the two groups). The wireworms may cause damage to a crop if an infested field were to be ploughed up and planted. If it could be shown that a set of site characteristics, such as sand content of the soil and amount of rainfall, differed between the two groups (infested and not infested) then it may be possible to look at a new field and from the sand content of the soil and the amount of rainfall that usually falls on that field, predict whether or not it is likely to be infested with wireworms.

The actual method used is to form a linear combination of the independent variables that discriminates the most between the two groups. This linear combination is called the discriminate function. A score can then be calculated from the linear combination for each individual that is to be classified into one of the groups and the value of the score will determine which group the individual is allocated to. Continuing with the example above, the linear combination of rainfall and sand content could result in a linear discrimination rule that says if the score is greater than or equal to a certain value, x, the field will be infested, and that if the score is less than x it is not infested. This score can then be calculated for future fields to predict the presence or absence of infestation. A Hotellings T^2 test should be used before calculating the linear combination to determine whether the variable values for the two groups do actually differ. If they do not, they will offer no real hope of being able to distinguish between the two groups.

There are different discrimination rules that can be used but the basic principles are the same. In any event there will always be the possibility of misclassifying an individual and a discrimination rule will be most effective if it has a low probability of misclassifying an individual. It is possible to calculate misclassification rates for a rule. These misclassification rates can be used to indicate the success or performance of a discrimination rule or to compare the relative merits of different rules that are developed.

Assumptions The linear discriminate function is used with continuous data and it is assumed that the variables are independent and randomly sampled and that they are from a multivariate normal distribution. It is further assumed that the samples are homoscedastic, i.e. they have equal variance covariance matrices. If these assumptions are not satisfied the other forms of discrimination rules, such as logistic discrimination rules or quadratic discriminant rules, can be developed. However, Fisher's linear discrimination rule is fairly robust against violations of the multivariate normality assumption.

Morrison (1990) provides the formulae for Fisher's discriminant rule and for calculating the misclassification probabilities within a readable text. A useful discussion of the assumptions and alternatives to Fisher's discriminant rule can be found in Everitt and Dunn (1991).

7.1.4 Principal component analysis

The main idea behind principal component analysis is to use a set of uncorrelated variables, derived from the original data, to describe or explain the total variation in a multivariate data set. The hope is that the first few uncorrelated variables will describe most, if not all, of the variation in the original data. This would result in the data dimensions being reduced from the original number of variables, with little information held in the original data being lost. These new independent variables may then be used in further analyses, such as multiple regression. Although the resultant independent variables can sometimes be very difficult to interpret, they can also help to promote an understanding of the data.

The original data set consists of a single response variable and a series of correlated variables that are thought to explain the variation in the response variable. The correlated variables are transformed by taking linear combinations of them. These linear combinations are formed so that the first transformed variable explains more of the variation in the original response variable than any of the other transformed variables. The second transformed variable then explains more of the remaining variability in the response variable than the remaining transformed variables. And so it continues until all the variability in the response variable has been explained by the new transformed variables. These transformed variables or linear combinations are called principal components. If the principal component analysis has been successful then the majority of the variability in the response variable is explained by the first few (two or three) principal components and the original dimensions of the data set have then been successfully reduced to two or three.

One of the main problems with this technique is that it may be very difficult to place any sort of meaningful interpretation on the principal components. With luck it will be possible to 'label' the principal component with an identifier linking it to one or more of the original variables. For example, if you think of an apple tree, there are many variables that could be measured: some variables would indicate general 'size' features such as tree height, or circumference of the trunk; other variables could concern the fruit crop, such as weight of the fruit, or disease levels in the fruit. Principal component analysis may reduce these four variables down to two, where each component is a linear combination of all four original variables but the coefficients in the components are such that one principal component is mainly concerned with the size variables and the other is mainly influenced by the fruit variables. If the original variables are highly correlated then principal component analysis should be successful in reducing the dimensions of the data. In fact, there is little to be gained in carrying out a principal component analysis on uncorrelated multivariate data.

Another problem can be in actually determining the number of principal components to use. It may not always be clear-cut, particularly if the first principal component does not explain a particularly large proportion of the variability in the response variable or if the original variables were not very highly correlated. There are tests that can be performed to determine the appropriate number of components to use, but these assume the original data are from a multivariate normal distribution. One general rule of thumb that is frequently applied is to use the number of components that explain a given percentage (e.g. 90%) of the total variation in the data (Everitt and Dunn, 1991).

Assumptions   Measurement of the variables needs to be on the interval or ratio scale. It is not necessary to assume the original variables are from a multivariate normal distribution although it can make the interpretation easier if this assumption holds.

Chatfield and Collins (1980) provide the formulae required for calculating principal components with examples and discuss the drawbacks of principal component analysis and its uses in further analyses. Some examples of interpreting principal components can be found in Morrison (1990).

7.1.5   Factor analysis

Factor analysis is used to explain multivariate data sets in terms of a small number of underlying factors. This is achieved by fitting a model to the data so any conclusions drawn from this technique will only be valid if the model assumptions are correct. In contrast to principal component analysis, factor analysis attempts to explain the covariance structure of the variables by fitting a model to the data. The model attempts to explain the observed variables in terms of a set of unobserved variables and a residual term. The unobserved variables are the model factors and the residual term is due to random error, and the aim is to identify a set of model factors that are fewer in number than the original variables without losing information contained in the original data.

Once the underlying model factors have been calculated it is possible to rotate them in an attempt to gain more information. The aim of the rotation is to try and identify the underlying structure in the model and to simplify the factor structure. For example, the original factors may not have much 'meaning' but when rotated they could lead to identifiable 'labels' for a factor in a manner similar to principal components. The factors can also be ranked in order of importance according to the size of their contribution towards the overall variance in the original variables. As with principal component analysis, there is little to be gained in using factor analysis on variables that are uncorrelated.

Assumptions   The original variables must have a multivariate normal distribution. The model factors should be independent although this assumption can be dropped if it is more meaningful to do so in order to gain a better insight into what is going on. The model residuals should be uncorrelated with each other and with the other model factors. The model factors should come from a multivariate $N(0, 1)$ distribution and the residuals should come from a multivariate normal distribution with a zero mean.

The coverage of factor analysis by Chatfield and Collins (1980) forms a very brief but informative introduction. More detailed coverage can be found in Everitt and Dunn (1991).

7.1.6   Cluster analysis

Cluster analysis is mainly a data exploration technique where a number of variables have been measured for a series of individuals or objects. The basic aims are to determine if there is any inherent grouping amongst the individuals or objects in the data set, and to gain an understanding of the structure of the data. There is no prior knowledge of distinct groups so this technique contrasts with discriminant analysis where the groups are known beforehand. Classifying the data into a set of groups containing similar individuals or objects can be a means of simplifying the data by reducing it without a drastic loss of information. Clustering can help summarize and describe data and may assist in formulating hypotheses about the data based on the information presented by the clusters. There are various clustering techniques that can be used with qualitative or quantitative data that are discrete or continuous.

Clusters can be formed by using an optimizing process or by using a hierarchical technique. With the former technique some form of optimizing criteria are defined and individuals are added to and taken away from the clusters until the optimizing criteria are satisfied. The most frequently used optimizing criteria are based on minimizing within-cluster dispersion or maximizing between-cluster dispersion. The number of clusters in the solution has to be predefined by the experimenter and there are techniques available to help determine the appropriate number of clusters in the data set.

A hierarchical technique can be used to cluster individuals and variables, and once an individual or variable is allocated to a cluster it remains in that cluster. There are a number of schemes that use hierarchical techniques but essentially they all operate in one of two ways. Each observation can be considered as a cluster on its own and clusters can be joined together in stages until all observations are in a single big cluster. Alternatively, all observations can start in a single cluster which is then split in stages until each observation forms its own cluster. The criteria for joining or splitting are based on various similarity or distance measures that can be used with different forms of data. An algorithm is then used to define the joining or splitting required to produce the clusters. Consequently these methods do not indicate a 'final' optimum number of clusters in the data and as different similarity or distance measures and algorithms can produce very different hierarchical structures, the 'correct' choice of the number of clusters can be problematic. The entire process of producing the hierarchical clusters is best represented pictorially as a dendogram which is a form of hierarchical tree. This shows the clusters formed at each stage of fusing and can be used to identify the number of clusters present in the data set.

There are other graphical ways of presenting the cluster information. Examples are Chernoff's faces, Andrew's plots and stars. In Chernoff's faces, each cluster is represented by a face, and the facial features represent the variables, with the 'size' of the feature (round eyes, slit eyes, big or little ears) indicating the variable values. In Andrew's plots, each individual is represented by a line and clustering is indicated by the closeness of the lines on the graph. Stars can also be used, where each individual or object is represented by a star. There are as many points to the star as there are variables, and the distance of each point from the centre is determined by the variable value.

Everitt (1993) provides a good introduction to cluster analysis, with plenty of examples and discussion of the pros and cons of the techniques covered.

7.1.7   Canonical correlation

Canonical correlation looks for a relationship between two groups of variables. It does this by calculating a series of linear combinations from the first group of variables and a series of linear combinations from the second group of variables. These linear combinations are called canonical variables and they are correlated within each group. The aim is to find the linear combination of the first group and the linear combination of the second group that has the maximum correlation.

The only canonical variables that are correlated across the groups are the direct pairs, such as the second canonical variables from each group, and these correlations are called the canonical correlations. The first pair of canonical variables will be the ones with the greatest correlation between them, the second pair will have the next highest correlation, and so on until the last pair of canonical variables, which will have the lowest correlation between them.

It is possible to test for any evidence of a relationship between the two groups by testing the overall significance of all the canonical correlations. If this proves to be statistically significant, indicating there is a relationship between the two groups, then whether the relationship is entirely accounted for by the first or first few canonical variables can be determined. This is done by dropping canonical correlations, starting with the first, in succession from the test and conducting the test at each stage on the remaining correlations. When a non-significant test result is found then all the canonical correlations contributing towards the overall significant relationship have been found.

Morrison (1990) and Chatfield (1989) provide the necessary formulae to conduct this analysis and provide a brief description of the technique illustrated by example.

Assumptions It can be assumed that the original variables have zero mean and unit variance but this is not essential.

7.2 TIME SERIES ANALYSIS

This is a very brief introduction to time series analysis which introduces some of the terminology used in this field. For a readable introduction to time series analysis, Chatfield (1989) is an informative text that emphasizes a practical approach to time series analysis and provides some useful references for further reading.

Time series analysis is a field of statistics that deals with a series of measurements over time and 'a time series is a collection of observations made sequentially in time' (Chatfield, 1989). Unlike most statistical situations it is usually only possible with time series to take a single observation at any point in time so there is usually no replication of the series. The values of the time series are usually correlated since what has already happened will usually influence what happens next; this form of correlation is called autocorrelation. The autocorrelation function, which measures the correlation between observations at different distances apart, is a major diagnostic tool used in time series analysis.

Time series are used to describe, explain or compare historical sequences because they can describe the main features of the data and their development over time. A series is best represented by a plot of the observations through time. This plot can indicate any particular points of interest such as regular repeating patterns, the overall trend of the points in the series (upwards, downwards, horizontal), outliers, or turning points where the trend of the series changes direction. They can also be used to predict or forecast future values of a time series once they have been modelled. Further uses can be found in process control where a variable of the process needs to be on target. If the variable is measured through time the process can be monitored and the time series would indicate when the process was moving off target, indicating the need for corrective action.

A time series is usually discrete because measurements are made at discrete time intervals even when the variable being measured is continuous (e.g. temperature may only be measured twice daily), or because the measurements only exist at discrete points in time (e.g. a cow's milk yield at milking). The time intervals between measurements are usually equal in length, such as hourly, daily or weekly, and measurements of daily rainfall throughout a period of a few weeks or more would form a time series of rainfall. However, the intervals do not have to be equal.

7.2.1 Components of a time series

There are four main components of variation in time series: short-term and long-term fluctuations, and cyclical and seasonal variation. Describing a series initially involves identifying which of the components are present. Fluctuations in a series can be either short term representing residual variation or long term representing changes in the level of the series. Seasonal variation is indicated by a repeating pattern of variation. For example, variations in some series over the summer, autumn, winter and spring months may well be repeated from one year to the next. Cyclical variation is indicated by wave-like oscillations in the level of the series but the pattern is not regular and repeating. All time series analysis starts with the production of a time series plot which indicates the presence of some or all of the components just mentioned. This gives an indication of the sorts of models that may be appropriate for modelling the series. For example, long-term fluctuations can be modelled with trend curves if the fluctuations are smooth, and with moving average models otherwise.

7.2.2 Modelling

A time series model can describe the sequence of observations as being dependent on time or dependent on the series past behaviour. It could be that a number of different models can be formulated that adequately describe the existing time series or adequately reproduce it. Each model will need to be validated and a requirement of an adequate model is that the model residuals are small and purely random. The autocorrelation function is important in helping to determine appropriate models for the data and in validating the models.

Stochastic models are frequently used to model the time series once sources of systematic variation such as trend or seasonal effects have been removed from the series. Differencing is a technique used to remove the trend from a series so that modelling the short-term fluctuations can be attempted. There are various stochastic models suitable for this when the fluctuations are not completely random but in any series there will be a random element of short-term fluctuations that cannot be modelled. Moving average models (MA) express the current state of the response in terms of past random behaviour. Autoregressive models (AR) assume the time series is entirely dependent on its own past behaviour. Models can be a mixture of the two, i.e. autoregressive moving average (ARMA) models. Autoregressive integrated moving average (ARIMA) models are combined AR and MA models fitted to a differenced time series. Box-Jenkins models can also incorporate seasonal effects into a model based on ARIMA models, called SARIMA models. All of these different sorts of models may provide an adequate description of the time series under investigation but the favoured model would be one that is relatively simple and has the smallest residuals. The principle of parsimony is a useful rule to operate by.

In other instances it is the actual trends themselves that are of more interest and which are modelled with the response variable being dependent on time. The trend models indicate overall changes in the level of a series that has small short-term fluctuations. Such models can be simple polynomial models or more complex models such as logistic models. A disadvantage is that although they may describe the shape of the series they will not necessarily provide a sensible explanation for the series. Smoothing techniques such as calculating a moving average can help identify trends and cycles as they smoothe out local fluctuations in the level of the series.

7.2.3 Forecasting

Forecasting with time series is an attempt to anticipate the future behaviour of a series and as such it is a form of extrapolation. Consequently an implicit assumption in forecasting models is that the basic structure of the model does not change throughout the entire time series, the exception to this assumption being when anticipated changes can be built in to the model. No model will be able to forecast future observations completely accurately so there will always be an element of error in the forecasts. The better models will be those which maintain a tight control over the errors and for an adequate model the errors should be random. These errors should be monitored to ensure the forecasting model stays on course. It is always possible to change the forecasting model in the light of new information obtained.

REFERENCES

Chatfield C (1989) *The Analysis of Time Series: An Introduction*, fourth edition. Chapman and Hall, London, Chapter 1.

Chatfield C and Collins AJ (1980) *Introduction to Multivariate Analysis*. Chapman and Hall, London, Chapters 1, 4, 5, 7, 9 and 11.

Everitt BS (1993) *Cluster Analysis*, third edition. Edward Arnold, London, Chapters 2–5.

Everitt BS and Dunn G (1991) *Applied Multivariate Data Analysis*. Edward Arnold, London, Chapters 4, 12 and 13.

Hand DJ and Taylor CC (1987) *Multivariate Analysis of Variance and Repeated Measures*. Chapman and Hall, London, Chapters 3 and 4.

Morrison DF (1990) *Multivariate Statistical Methods*, third edition. McGraw-Hill, New York, Chapters 5–9.

CHAPTER 8

Aspects of Computing

In the majority of experiments data will be stored, manipulated and analysed electronically. Consideration must therefore be given to the computing facilities that are required for the task. This section covers some points that should be considered in the selection and use of computer packages. In recent years the authors have mainly used Minitab and Genstat for statistical analyses, with occasional forays into other statistical software. Some pros and cons of these two packages are mentioned below. It would be very rare for computer hardware and software to be obtained for only one experiment so the requirements of all the types of work to be undertaken must be met. The choice of software will, in some cases, be limited by the hardware available. Modern software packages require large amounts of computer memory (RAM) and hard disk storage area.

Computer packages will perform many tasks, although all will have weaknesses and strengths. To make the best use of a package its full potential should be realized. It is important to find out the capabilities of any package so that it can be used to its full advantage. It is also important to know how a package handles various data types. How are missing data expressed and handled? Can it cope with text strings in data files? Is information provided on how the packages perform various functions and calculations, and are they consistent with other packages that the user may be using? It is common for a user to have access to several different packages. In such situations the strengths of each package should be taken advantage of and the weaknesses avoided. For example, if it is relatively easy to enter data in one package but another has more powerful analytical techniques then both packages should be used. There is little to be gained from struggling to enter data in one package just because it can perform all of the required analyses easily when it can be entered much more simply in another package and transferred across.

To make use of several packages, each must be able to read and produce data files and output files from the other packages. In most instances the files that a package uses are specific to that package but there should be the facility to import and export files from other packages. The means of transferring files from one package to the next involves using 'flat' files, which on a PC are known as ASCII (American Standard Code for Information Interchange) files. Such files can be imported and exported to and from the various pieces of software. These files contain only numbers and characters and have no package-specific characters, such as formatting characters. The data items are usually separated by spaces but can be separated by commas (CSV files). In addition, some packages can handle files produced by others, for example some spreadsheet packages can convert files produced by other spreadsheet packages and it is possible to copy and paste data between packages that are fully Windows compatible. When considering software requirements, account should be taken of all stages of work in which computer files may be used. Data may be entered and checked in one package, the analysis performed in another one or more, and further packages used to produce a final report and graphs and figures.

Whatever type of package is being considered, ease of use must be one of the first considerations, and can depend very much on the individual who will be using the software. A person who is highly computer literate may find one package very easy to use but to someone who only uses a computer occasionally the same package may seem totally incomprehensible. The type and style of package that the user is familiar with will also have a bearing on choice. A user who is familiar only with spreadsheet packages may find statistical packages very confusing. Menu-driven interactive packages may be very simple and lead the user step by step through a procedure, but can be very frustrating to someone who is used to programming and cannot find the correct menu, or simply finds the menu slow and tedious. If a task needs to be repeated several times an interactive system should have a macro facility which allows a sequence of steps to be stored and re-used. Programming languages require the user to develop the program code, which can then be stored in a program and run as many times as required.

There are different ways of presenting data, such as in databases or spreadsheets or flat files. Databases are usually only used to enter, store and display data in a variety of formats and would not usually be used for any type of statistical analysis other than as an input data source to a different package. Spreadsheets can be used to enter data and have some very useful data manipulation functions. With these packages data are stored as rows and columns which can easily be related to experiment units and variables. Most spreadsheet packages will perform some statistical analyses. Statistical packages are written specifically for data handling and statistical analysis, and can use 'flat' files for data input or their own package-specific files. Statistics packages range from very simple menu-driven packages which only perform the most basic types of analysis, to very complex packages which need to be programmed in their own language. Some of these packages include facilities to enter and manipulate data whilst others need data in a format ready to analyse. Some give a comprehensive coverage of the most commonly used statistical techniques whilst others specialize in a particular aspect of statistics. The final choice of software will depend on personal preference, requirements and resources, which includes financial resources as well as the computer hardware available. Whatever the final decision, if the software is to be used for data entry it should be 'user friendly', particularly if data entry is only an occasional task for the user. The facility to label columns of data so that they can be easily identified at a future date should be available. It should allow data items and columns to be inserted, deleted or moved easily but not so easily that data are accidentally moved or deleted. It should be able to handle numerical and character data and be able to produce and read flat files.

The two statistics packages Minitab and Genstat, although not quite at the extremes of basic but simple and powerful but complex, do offer very different approaches to statistical analyses. At the time of writing the versions of these packages which are available are Minitab version 10 for Windows (Minitab Inc., 1994) and Genstat 5 release 3.1 (Genstat 5 Committee, 1993), which runs under Windows but is not a Windows package. Minitab is used interactively and, if used in Windows, offers a window for entering commands (the session window), one for entering data (the data window) and additional windows for other functions such as displaying graphics and summarizing the worksheet contents. It can be programmed either by typing in commands at the session window or by using a 'windows style' menu system. Genstat offers the options of either interactive or batch mode use. In batch mode a program is written in flat file format and then run in Genstat and an output file produced. It does have a simple menu system but it is not as 'user friendly' as Minitab in that respect. It does, however, have the advantage that it is more powerful than Minitab and can handle more complex analyses. The next version of Genstat will be a fully Windows-compatible version.

The data window in Minitab allows data to be entered directly and easily into worksheets. This data window has the same appearance as a spreadsheet but should never be thought of as a spreadsheet as it handles data very differently. Formulae cannot be entered into cells of a Minitab worksheet as they can into the cells of a spreadsheet; the cell must only contain numerical or character data. If a data item in a cell is changed and calculations have been performed on that data item to produce new columns of data then the calculations will need to be repeated on the revised data. Minitab saves data as worksheets, but these can only be read by Minitab and are not compatible across different versions of Minitab or across different platforms such as mainframe or PC. However, it can produce portable worksheets which are compatible. It can also create data files that are in flat file format. However, Minitab uses the ampersand sign (&) as an indication that data are continued on the next line, and most other packages will not recognize this so the file produced needs to be edited before it is read into another package unless it is produced with a user-specified format. Whilst Genstat 5 release 3.1 does have a facility to enter data directly through the menu system, data can be entered as part of the program or entered directly when in interactive mode, or data can be entered into a separate ASCII data file which the Genstat program then reads when it is run. Genstat does not use files that are specific to Genstat; it reads and outputs ASCII files.

When going through the data-checking process, computers can save valuable time and effort both in data verification and data validation. When verifying data, double data entry can be used to ensure that the data entered are those which were collected. The process of double data entry involves entering the data in the chosen format and then re-entering the data. The second entry can either be done in a new file or as additional columns or variables in the original files. The two columns can then be compared and any discrepancies checked against the original data. If the second entry were done as part of the original file, discrepancies could easily be found by subtracting the two data sets; the values in which there were discrepancies would be non-zero.

Data validation is also a straightforward process in most packages. One of the easiest ways to validate data is to produce summary statistics. In Minitab this is easily done using the describe command which produces a set of summary statistics for the requested variables. Genstat will produce summary statistics for each of the variables it reads. Most other statistics packages and spreadsheets will also produce these statistics (or at least the minimum and maximum values) relatively simply. All of the descriptive statistics mentioned in this text can be readily produced in either Genstat or Minitab.

Minitab will produce either high-quality graphics or line printer graphics either from the menus or directly by typing in the relevant commands. In this way it is relatively quick and easy to produce most types of graphical output. The graphs can then be saved or 'cut and pasted' into word processor packages. The process in Genstat is slightly more complex, particularly for high-quality graphics.

The advancements in computing technology have changed the way in which statistical analysis can be approached. Before computer facilities were so widely available analysis would be undertaken by hand. In the past, even a simple analysis of variance would take several hours to analyse; now it may only be a matter of minutes. These changes have allowed more analysis to be undertaken in a very short space of time and also make statistical techniques available to anyone who wishes to use them. This is generally thought to be a great advantage; however, the disadvantage is that a computer package will usually give an answer but will not give any indication of whether the correct analysis has been performed. Just because the computer produces an analysis it does not mean it is correct. However, some packages will give warnings if assumptions of the analysis are violated. For the packages that do give a warning that assumptions are violated, it is important to know how the package checks these assumptions as some checks are more rigorous than others.

The majority of computer packages will handle the most common parametric techniques but they will differ in the extent to which they check the various test assumptions. For example, Minitab will perform the t-tests discussed in Section 6.1 either from the menus or by a one line command without checking any of the assumptions, but Genstat will simultaneously check the assumption of equal variance and print a warning if this assumption is violated for the two-sample test. Both packages also have the facility to give the p-values associated with statistics from particular distributions (such as the t distribution and the normal distribution).

Minitab will perform analysis of variance for all of the models discussed in this book, the major drawback being that Minitab cannot cope with missing values and will not perform the analysis of variance on data sets with missing observations. It is quite easy to use for the simple designs such as completely randomized and randomized blocks, but for the more complex designs it is less easy as the model formula needs to be fully understood to be able to describe the model. Models with more than one strata (such as the split-plot design) need more care as all the information required for a split-plot analysis of variance table cannot be presented in a single table in Minitab. The various sums of squares, degrees of freedom and variance ratios are presented in more than one table and the user has to extract the relevant pieces of information. Minitab does have the facility to produce residual plots and test the assumptions of the analysis although none of this is performed by default. It does have the facility to handle unbalanced data but this does need some thought. Genstat is a very powerful package in terms of the analysis of variance technique. It will estimate values for missing observations, although this facility should be treated with caution since if the majority of values are missing it will estimate them all and also if a whole level of a treatment is missing it will also estimate that level. It is relatively straightforward to perform analysis of variance in Genstat. The model is described in terms of block structure and treatment structures and needs only to be specified once for a set of variables with the same structure. Some of the assumptions of the analysis are automatically tested and others can be easily tested using options or addition commands. The plot of the residuals is not produced by default but can easily be output. It is possible to include orthogonal contrasts in the analysis of variance and it also has good facilities for handling unbalanced data, analysis of covariance, and more complex designs.

Whatever package is used to produce an analysis of variance, the user should have some idea of the analysis expected, including which terms should be produced in the analysis of variance table and the appropriate degrees of freedom for each of them. It is only with this knowledge that the user will be sure that the package has produced the analysis that was expected. This is a useful exercise to perform before the experiment is run to ensure that the design is analysable. In Genstat it is possible to enter the treatment structure and produce an analysis without any data, usually referred to as a dummy analysis. This facility is not available in Minitab but random data can be generated which can be used to check that the model is analysable.

Simple linear regression is straightforward in Minitab and will produce the results with high-resolution graphics using either the menus or the commands. Additional columns can be calculated to be used to adjust the model or fit other models. For example, a quadratic model can be fitted by calculating an x^2-term and fitting this and the x-term using the multiple linear regression option. Many of the goodness-of-fit criteria are available in Minitab; some are produced by default and others need to be requested. There is no facility to fit non-linear curves or to compare curves. Genstat offers a much more comprehensive coverage of regression. It will fit models ranging from simple linear models to non-linear models with non-standard formulae and has the facilities to produce the goodness-of-fit criteria mentioned in Section 6.1.7.6 and to compare regression models.

Non-parametric techniques are available in both Minitab and Genstat, each offering a different selection of tests. Unless a package is chosen that specializes in non-parametric techniques, then a full range of tests is unlikely to be available.

The final choice of package will depend on requirements and resources but it may be that the best option is to have more than one package so that the strengths of several can be exploited.

REFERENCES

Genstat 5 Committee (1993) *Genstat 5 Release 3 Reference Manual*. Oxford University Press, Oxford.
Minitab Inc. (1994) *Minitab Reference Manual Release 10 for Windows*. Minitab Inc.

APPENDIX I

Glossary of Statistical Terms

This glossary is not intended to give strict formal definitions of the terms included. It is the intention that the 'definitions' will promote the idea behind the terms without getting too involved in some of the detail that can accompany many more formal definitions.

Acceptance Region The group of values that would lead to the acceptance of the null hypothesis.

Accuracy 'The closeness of computations or estimates to the exact or true values' (Marriott, 1990). It refers to the average value of the measurements obtained.

Aliasing See *Confounding*.

Alternative Hypothesis A statement indicating how the value of the statistic of interest deviates from the value specified in the null hypothesis.

Arithmetic Mean A measure of location that is calculated as 'the sum of the observations divided by the number of observations' (Clarke and Cook, 1992).

Average Value A single value that is representative of the location of all individual values. There are several measures of average value.

Bias The continual handicapping or favouring of a particular treatment in successive replications.

Binary Data Data that have just two possible outcomes, e.g. presence/absence data.

Block A group of homogeneous experiment units.

Classification Factor A factor that groups similar units of a variate together.

Coefficient A constant multiplier that measures some property of a variable or function of a variable.

Coefficient of Variation (%CV) A measure of sample variability in relation to its location.

Confidence Interval The difference between the upper and lower confidence limits. The interval has a given probability of covering the true population mean value.

Confidence Limits These define the upper and lower bounds of a confidence interval which has a given probability of covering the true population mean value.

Confounding Confounding occurs 'when certain effects can be estimated only for treatments in combination and not for separate treatments' (Marriott, 1990).

Contingency Tables A table of frequency counts for mutually exclusive categories.

Continuous Data Data that 'may take in all values within a given range' of values (Clarke and Cook, 1992).

Contrasts Planned comparisons.

Control Treatment A null or standard treatment.

Correlation A measure of the linear association between two variables.

Covariate A variate that is believed to be related in some way to the response of interest. It is not affected by the treatments but provides information about the way in which experiment units will respond to the treatment.

Critical Region   The values of the test statistic that lead to the rejection of the null hypothesis.

Critical Value   The value that defines the acceptance and rejection regions.

Data Validation   The process used to ensure that the data make sense.

Data Verification   The process of checking that the data recorded are those that were collected.

Degrees of Freedom   The degrees of freedom of a system is the total number of observations in that system minus the number of mathematical constraints imposed within that system; these vary according to the complexity of the system. Alternatively, the degrees of freedom associated with any component are the number of independent parameters required to describe that component in the model.

Dependent Variable   In regression the dependent variable, usually y, is expressed as a function of another variable or variables.

Descriptive Statistics   A set of summary statistics that describe a set of data. For example, mean values, standard deviations, minimum and maximum values.

Dichotomous Data   See *Binary data*.

Discrete Data   Data that have 'a finite number of possible values' from a range of values (Clarke and Cook, 1992).

Error Term   See *Residual value*.

Error Variance   See *Residual variance*.

Experiment Unit   The smallest group of experiment material to receive a single treatment or treatment combination. Sometimes referred to as a plot.

Factor   See *Treatment factor*.

Fitted Value   An estimate of the 'true' value calculated from fitting a model to data.

Heterogeneous Units   Experiment units that respond in a diverse way to a treatment.

Homogeneous Units   Experiment units that respond in the same way to a treatment.

Hypothesis   The statement of an objective in statistical terms.

Hypothesis Test   A statistical test of the objective stated in the hypothesis.

Independent Variable   In regression the dependent variable, usually y, is expressed in terms of another variable or variables, usually x. The x's are the independent variables.

Interaction   Interactions occur when the differences between the response to the levels of a factor themselves differ at each level of another factor.

Inter-Quartile Range   'The difference between the upper and lower quartiles' (Clarke and Cook, 1992).

Interval Data   Continuous data 'where there are equal differences between successive integers but where the zero point is arbitrary' (Chatfield, 1995).

Level   See *Treatment factor level*.

Leverage   Influential data points are ones which, when removed from the regression analysis, cause a different model to be fitted. Such points are said to have high leverage.

Maximum Value   The highest numerical value in a group of numbers.

Mean Value   See *Arithmetic mean*.

Median Value   The middle value of a group of observations when they are arranged in order of magnitude.

Minimum Value   The lowest numerical value in a group of numbers.

Missing Value   A value that was not recorded or was lost either by error or design.

Mode   The value of the most frequently occurring observation.

Nominal Data   Data that are classified and cannot be ordered in any way.

Null Hypothesis   A statement that the statistic of interest takes a specified value.

Objective   A clear concise statement of the need for the experiment.

Ordinal Data   Data that are classified and can be placed in a fixed order.

Orthogonal   Statistically independent.

Orthogonal Contrasts   Mutually independent planned comparisons.

Outlier   'A 'wild' or extreme observation which does not appear to be consistent with the rest of the data' (Chatfield, 1995).

Parameter   A 'quantity which may vary over a certain set of values' but is constant for a given population (Marriott, 1990). The quantity for the population is usually unknown and has to be estimated from a sample.

Pilot Study   A small-scale run of the experiment to gain information about variability and to test any novel application methods prior to the running of the main experiment.

Plot   See *Experiment unit.*

Population   The total number of individuals from which a sample is assumed to have been taken.

Population Mean   The arithmetic mean of the entire population.

Population Standard Deviation   The standard deviation of the entire population.

Population Variance   The square of the population standard deviation.

Power of a Test   This indicates the certainty with which a difference will be detected by a test given that it is a real difference.

Precision   This refers to the variability of the measurements and is sometimes called repeatability.

Probability   The quantitative measure assigned to situations where uncertainty occurs.

Qualitative Data   Data 'whose values cannot be put in any numerical order' (Clarke and Cook, 1992).

Quantitative Data   Data 'whose values can be put in numerical order' (Clarke and Cook, 1992).

Quartiles   The values below which 25%, 50% and 75% of the observations in a sample fall when the observations are arranged in rank order: 25% of the observations lie below the first quartile, 50% below the median (or second quartile) and 75% below the third quartile.

Randomization   The allocation of treatments to experiment units in an unbiased manner.

Range   'The difference in value between the largest and the smallest observations in a set' (Clarke and Cook, 1992).

Ratio Data   Continuous data where it is possible 'to compare the relative magnitude of the values as well as the differences' (Chatfield, 1995).

Rejection Region   The group of values that would lead to the rejection of the null hypothesis.

Replication   The application of a treatment of interest to one or more experiment units.

Residual Value   'The difference between an occurring value and its "true" or "expected" value' (Marriott, 1990).

Residual Variance   The inherent variability between samples regardless of the treatment applied. 'It measures the variability due to unexplained causes or experimental error' (Marriott, 1990).

Response Variable A variable that is measured and affected by the treatments applied in an experiment.

Robust 'A statistical procedure is described as robust if it is not very sensitive to departure from the assumptions on which it depends' (Marriott, 1990).

Sample 'A part of a population' (Marriott, 1990).

Sample Mean The arithmetic mean of a sample drawn from a population.

Sample Standard Deviation The standard deviation of a sample drawn from a population.

Sample Variance The square of the sample standard deviation.

Significance Level of Test The probability of rejecting the null hypothesis when it is correct.

Standard Deviation The standard deviation of a group of observations is the positive square root of the mean of the squared deviations from the mean value of the group of observations, and is a measure of how the individual observations lie around their arithmetic mean value.

Standard Error The standard deviation of a 'sampling distribution of a statistic' or observation (Marriott, 1990).

Standard Error of Difference The standard deviation of a distribution of differences between two mean values for samples of a given size.

Standard Error of Mean The standard deviation of a distribution of means for samples of a given size.

Statistic 'A summary value calculated from the observations in a sample' (Marriott, 1990).

Test Statistic 'A function of a sample of observations which provides the basis for testing a statistical hypothesis' (Marriott, 1990).

Treatment 'Anything which is capable of controlled application according to the requirements of the experiment' (Marriott, 1990). It is any combination of treatment factor and treatment level.

Treatment Factor The method or type of application.

Treatment Factor Level The rate of application or dose.

Truncated Data Data which ignore all values above or below a pre-specified value.

Type I Error Rejecting the null hypothesis when it is correct.

Type II Error Accepting the null hypothesis when it is false.

Variable A variable is 'a quantity which may take one of a specified set of values' (Marriott, 1990).

Variance The square of the standard deviation.

REFERENCES

Chatfield C (1995) *Problem Solving: A Statistician's Guide*, second edition. Chapman and Hall, London, Chapter 6.

Clarke GM and Cooke D (1992) *A Basic Course in Statistics*, third edition. Edward Arnold, London, Chapters 1, 2 and 4.

Marriott FHC (1990) *A Dictionary of Statistical Terms*, fifth edition. Addison-Wesley Longman.

Analysis of Variance Formulae

The calculations involved in producing the analysis of variance table results are detailed below. The three fundamental designs, single-treatment factor completely randomized, randomized blocks and Latin square, are detailed below and illustrated with the examples covered in the main text. For the designs with a treatment structure the calculations have only been shown using the randomized blocks designs. However, the treatment structure in these designs affects only the calculations for the treatment effects so the calculations for the blocking structure (completely randomized, randomized blocks and Latin square) are as illustrated in the appropriate single-treatment factor design. The two-treatment factor factorial and split-plot designs are covered in detail. The multi-treatment factor designs are a straightforward extension of the two-treatment factor designs and the basic structure behind the various components of the formulae is the same irrespective of the design used. These formulae can be found in a number of statistical texts, albeit presented in slightly different formats. Montgomery (1991) gives a clear presentation of the formulae for the single-treatment factor designs and how they relate to the analysis of variance table. Cochran and Cox (1957) present the formulae as part of complete worked examples. Steel and Torrie (1980) take a similar step-by-step approach, presenting the final analysis of variance table. All three texts explain the derivation of the formulae. Montgomery (1991) presents formulae for a randomized blocks two-way factorial. The two-factor split-plot design is covered by Cochran and Cox (1957) and Steel and Torrie (1980), the latter being the more thorough presentation of the two.

The treatments to be used in an experiment can be selected in different ways, which will influence the inferences that can be drawn from any hypothesis testing. If the experimenter selects a specific set of treatments (unstructured or structured) the model is termed a 'fixed effects' model. Any inferences drawn from the resultant analysis will apply only to that set of treatments. If the experimenter randomly selects treatments to represent a wider population of treatments then the model is termed a 'random effects' model. Any inferences drawn from such a model apply to the wider treatment population from which the treatments actually used were drawn (Montgomery, 1991). It is also possible to have combined models with both fixed and random effects. This text considers only fixed effects models so formulae and conclusions presented only apply to these models.

COMPLETELY RANDOMIZED DESIGN

The analysis of variance table is used to decide which of the effects being tested gives a significant result; it is in effect the decision rule for the tests. For a completely randomized design with t treatments and at least two replicates of each treatment it takes the following form:

Source	df	ss	ms	vr
Treatment (T_i)	df_T	ss_T	ms_T	F_T
Residual (e_{ij})	df_E	ss_E	s^2	
Total	df_{TOT}	ss_{TOT}		

The following equations show how each component is calculated.

Terms used

t = number of treatments
r_i = number of replicates of the ith treatment

N = total number of experiment units $\left(\sum_{i=1}^{t} r_i\right)$

T_i = the sum of the observations receiving the ith treatment
y_{ij} = the observation from the unit with the jth replicate of ith treatment

T = the sum of all the observations $\left(\sum_{i=1}^{t}\sum_{j}^{r_i} y_{ij}\right)$

Degrees of freedom (df)

$$df_T = t - 1$$
$$df_E = N - t$$
$$\quad\; = df_{TOT} - df_T$$
$$df_{TOT} = N - 1$$

Sums of squares (ss)

$$ss_{TOT} = \sum_{i=1}^{t}\sum_{j=1}^{r_i} y_{ij}^2 - \frac{T^2}{N}$$

$$ss_T = \sum_{i=1}^{t} \frac{T_i^2}{r_i} - \frac{T^2}{N}$$

$$ss_E = \sum_{i=1}^{t}\sum_{j=1}^{r_i} y_{ij}^2 - \sum_{i=1}^{t} \frac{T_i^2}{r_i}$$

$$\quad\; = ss_{TOT} - ss_T$$

Mean square (ms)

$$ms_T = ss_T/df_T$$
$$s^2 = ss_E/df_E$$

Variance ratio (vr)

$$F_T = ms_T/s^2$$

Hypothesis test

The null hypothesis of no difference between the t treatments is rejected at the $\alpha\%$ level of significance if the calculated variance ratio (F_T) is greater than the tabulated value with $(t-1)$, $(N-t)$ degrees of freedom at significance level α.

$$F_T \geq F_{(t-1),\,(N-t),\,\alpha}$$

Example

The experiment used to illustrate these calculations is that which is designed in Section 4.2.1.6 and analysed and interpreted in Section 6.1.5.1. The experiment is an investigation into the effect on hen total egg weight of using different stocking densities. There are four stocking densities with two densities being replicated twice and the other two being replicated three times. The 10 treatments are freely randomized. The analysis of variance table is shown below:

Source	df	ss	ms	vr
Stocking density (T_i)	3	2.013	0.671	0.10
Residual (e_{ij})	6	38.423	6.404	
Total	9	40.436		

Summations and terms used

t = number of treatments = 4

N = total number of experiment units $\left(\sum_{i=1}^{4}\right) r_i = 3 + 2 + 2 + 3 = 10$

T_1 = sum of the observations from the 1st treatment = 141.4
T_2 = sum of the observations from the 2nd treatment = 94.7
T_3 = sum of the observations from the 3rd treatment = 92.1
T_4 = sum of the observations from the 4th treatment = 140.6

$\sum_{i=1}^{4} T_i^2$ = the sum of the squares of the treatment totals = 57212.82

$\sum_{i=1}^{4} \sum_{j=1}^{r_i} y_{ij}$ = sum of all the observations = 468.8

$\sum_{i=1}^{4} \sum_{j=1}^{r_i} y_{ij}^2$ = the sum of the squares of each observation = 22017.78

Degrees of freedom (df)

$$df_{TOT} = N - 1 \quad\quad = 10 - 1 \quad\quad = 9$$
$$df_T = t - 1 \quad\quad = 4 - 1 \quad\quad = 3$$
$$df_E = N - df_T \quad = 10 - 4 \quad = 6$$
$$\quad\quad = df_{TOT} - df_T \quad = 9 - 3 \quad\quad = 6$$

Sums of squares (ss)

$$SS_{TOT} = \sum_{i=1}^{4} \sum_{j=1}^{r_i} y_{ij}^2 - \frac{T^2}{N} = 22017.78 - \left(\frac{468.8^2}{10}\right)$$
$$= 22017.78 - 21977.344 = 40.436$$

$$SS_T = \sum_{i=1}^{4} \frac{T_i^2}{r_i} - \frac{T^2}{N} = \frac{1}{3} \times 141.4^2 + \frac{1}{2} \times 94.7^2 + \frac{1}{2} \times 92.1^2 + \frac{1}{3} \times 140.6^2 - \left(\frac{468.8^2}{10}\right)$$
$$= 6664.653333 + 4484.045 + 4241.205 + 6589.453333 - 21977.344 = 2.013$$

$$SS_E = \sum_{i=1}^{4} \sum_{j=1}^{r_i} y_{ij}^2 - \sum_{i=1}^{4} \frac{T_i^2}{r_i}$$

$$= 22\,017.78 - \left(\frac{1}{3} \times 141.4^2 + \frac{1}{2} \times 94.7^2 + \frac{1}{2} \times 92.1^2 + \frac{1}{3} \times 140.6^2\right)$$

$$= 22\,017.78 - 6664.653333 - 4484.045 - 4241.205 - 6589.453333 = 38.423$$
$$= ss_{TOT} - ss_T$$
$$= 40.436 - 2.013 = 38.423$$

Mean square (ms)

$ms_T = ss_T/df_T = 2.013/3 = 0.671$
$s^2 \quad = ss_E/df_E = 38.423/6 = 6.404$

Variance ratio (vr)

$F_T = ms_T/s^2 = 0.671/6.404 = 0.10$

Hypothesis test result

$F_T = 0.10 \leq F_{(3,6,0.05)} = 4.76$

At the 0.05 level of significance the null hypothesis of no difference between the four stocking density population mean values is accepted because the calculated variance ratio (F_T) is less than the tabulated F value with $(4-1)$ $(10-4)$ degrees of freedom.

RANDOMIZED BLOCKS DESIGN

The analysis of variance table is used to decide which of the effects being tested gives a significant result; it is in effect the decision rule for the tests. For a randomized blocks design in which there are b blocks and t treatments with one complete replicate of all treatments in each block, it takes the following form:

Source	df	ss	ms	vr
Block (Bl_h)	df_{Bl}	ss_{Bl}	ms_{Bl}	
Treatment (T_i)	df_T	ss_T	ms_T	F_T
Residual (e_{ij})	df_E	ss_E	s^2	
Total	df_{TOT}	ss_{TOT}		

The following equations show how each component is calculated.

Terms used

b	= number of blocks	
t	= number of treatments	
N	= total number of experiment units (bt)	
Bl_h	= the sum of the observations in the jth block	
T_i	= the sum of the observations from the ith treatment	
y_{hi}	= the observation from the unit receiving the ith treatment in the hth block	

$T \quad$ = the sum of all the observations $\left(\sum\limits_{h=1}^{b}\sum\limits_{i=1}^{t} y_{hi}\right)$

Degrees of freedom (df)

$df_{TOT} = bt - 1$
$df_{Bl} \quad = b - 1$

$$df_T = t - 1$$
$$df_E = (b-1)(t-1)$$
$$= df_{TOT} - df_{Bl} - df_t$$

Sums of squares (ss)

$$ss_{TOT} = \sum_{h=1}^{b} \sum_{i=1}^{t} y_{hi}^2 - \frac{T^2}{N}$$

$$ss_{Bl} = \frac{1}{t} \sum_{h=1}^{b} B_h^2 - \frac{T^2}{N}$$

$$ss_T = \frac{1}{b} \sum_{i=1}^{t} T_i^2 - \frac{T^2}{N}$$

$$ss_E = \sum_{h=1}^{b} \sum_{i=1}^{t} y_{hi}^2 - \frac{1}{t} \sum_{h=1}^{b} B_h^2 - \frac{1}{b} \sum_{i=1}^{t} T_i^2 + \frac{T^2}{N}$$

$$= ss_{TOT} - ss_{BL} - ss_T$$

Mean square (ms)

$$ms_{Bl} = ss_{Bl}/df_{Bl}$$
$$ms_T = ss_T/df_T$$
$$s^2 = ss_E/df_E$$

Variance ratio (vr)

$$F_T = ms_T/s^2$$

Hypothesis test

The null hypothesis of no difference between the t treatments is rejected at the $\alpha\%$ level of significance if the calculated variance ratio (F_T) is greater than the tabulated values with $(t-1)$, $(t-1)(b-1)$ degrees of freedom at significance level α.

$$F_T \geq F_{(t-1), (t-1)(b-1), \alpha}$$

Example

The experiment used to illustrate these calculations is that which is designed in Section 4.2.2.6 and analysed and interpreted in Section 6.1.5.2. The experiment is an investigation into the effect of nine fungicides and an untreated control on grain size of winter wheat. There are four blocks of 10 experiment units and the 10 treatments are freely randomized within each block. The analysis of variance table is shown below:

Source	df	ss	ms	vr
Block (Bl_h)	3	6.093	2.031	
Fungicide (T_i)	9	190.504	21.167	11.11
Residual (e_{hi})	27	51.422	1.905	
Total	39	248.019		

Summations and terms used

b = number of blocks = 4
t = number of treatments = 10
N = total number of experiment units $(bt) = 4 \times 10 = 40$

Bl_1 = sum of the observations in the 1st block = 480.9
Bl_2 = sum of the observations in the 2nd block = 481.5
Bl_3 = sum of the observations in the 3rd block = 490.2
Bl_4 = sum of the observations in the 4th block = 487.2

$$\sum_{h=1}^{4} Bl_h^2 = \text{sum of the squares of the block totals} = 940\,766.940$$

T_1 = sum of the observations from the 1st treatment = 173.2
T_2 = sum of the observations from the 2nd treatment = 189.9
T_3 = sum of the observations from the 3rd treatment = 205.1
T_4 = sum of the observations from the 4th treatment = 190.4
T_5 = sum of the observations from the 5th treatment = 205.1
T_6 = sum of the observations from the 6th treatment = 193.4
T_7 = sum of the observations from the 7th treatment = 195.3
T_8 = sum of the observations from the 8th treatment = 190.8
T_9 = sum of the observations from the 9th treatment = 196.9
T_{10} = sum of the observations from the 10th treatment = 199.7

$$\sum_{i=1}^{10} T_i^2 = \text{sum of the squares of the treatment totals} = 377\,044.42$$

$$\sum_{h=1}^{4} \sum_{i=1}^{10} y_{hi} = \text{the sum of all the observations} = 1939.8$$

$$\sum_{h=1}^{4} \sum_{i=1}^{10} y_{hi}^2 = \text{the sum of the squares of each observation} = 94\,318.62$$

Degrees of freedom (df)

$$
\begin{aligned}
df_{TOT} &= bt - 1 & &= (4 \times 10) - 1 & &= 39 \\
df_{Bl} &= b - 1 & &= 4 - 1 & &= 3 \\
df_T &= t - 1 & &= 10 - 1 & &= 9 \\
df_E &= (b-1)\,(t-1) & &= (4-1)(10-1) & &= 27 \\
&= df_{TOT} - df_{Bl} - df_T & &= 39 - 3 - 9 & &= 27
\end{aligned}
$$

Sums of squares (ss)

$$ss_{TOT} = \sum_{h=1}^{4} \sum_{i=1}^{10} y_{hi}^2 - \frac{T^2}{N} = 94\,318.62 - \frac{1939.8^2}{40}$$

$$= 94\,318.62 - 94\,070.601 = 248.019$$

$$ss_{Bl} = \frac{1}{t} \sum_{h=1}^{4} B_h^2 - \frac{T^2}{N} = \frac{1}{10} \times 940\,766.940 - \frac{1939.8^2}{40}$$

$$= 94\,076.6940 - 94\,070.601 = 6.903$$

$$ss_T = \frac{1}{b} \sum_{i=1}^{10} T_i^2 - \frac{T^2}{N} = \frac{1}{4} \times 377\,044.42 - \frac{1939.8^2}{40}$$

$$= 94\,261.105 - 94\,070.601 = 190.504$$

$$SS_E = \sum_{h=1}^{4}\sum_{i=1}^{10} y_{hi}^2 - \frac{1}{t}\sum_{h=1}^{4} B_h^2 - \frac{1}{t}\sum_{i=1}^{10} T_i^2 + \frac{T^2}{N}$$

$$= 94\,318.62 - \frac{1}{10} \times 940\,766.940 - \frac{1}{4} \times 377\,044.42 + \frac{1939.8^2}{40}$$

$$= 94\,318.62 - 94\,076.6940 - 94\,261.105 + 94\,070.601 = 51.422$$

$$= SS_{TOT} - SS_{BI} - SS_T$$

$$= 248.019 - 6.093 - 190.504 = 51.422$$

Mean square (ms)

ms_{BI}	$= ss_{BI}/df_{BI}$	$= 6.903/3$	$= 2.031$
ms_T	$= ss_T/df_T$	$= 190.504/9$	$= 21.167$
s^2	$= ss_E/df_E$	$= 51.422/27$	$= 1.905$

Variance ratio (vr)

$$F_T \quad = ms_T/s^2 \qquad = 21.167/1.905 \ = 11.11$$

Hypothesis test result

$$F_T = 11.11 \geq F_{9,27,0.005} = 2.25$$

At the 0.05 level of significance the null hypothesis of no difference between the 10 fungicide population mean thousand grain weight values is rejected because the calculated variance ratio (F_T) is greater than the tabulated F value with $(10-1),(10-1)(4-1)$ degrees of freedom.

LATIN SQUARE DESIGN

The analysis of variance table is used to decide which of the effects being tested gives a significant result; it is in effect the decision rule for the tests. For a single Latin square design with r rows, c columns and t treatments, where each treatment occurs once in each row and column, it takes the following form:

Source	df	ss	ms	vr
Row (R_i)	df_R	ss_R	ms_R	
Column (C_j)	df_C	ss_C	ms_C	
Treatment (T_k)	df_T	ss_T	ms_T	F_T
Residual (e_{ijk})	df_E	ss_E	s^2	
Total	df_{TOT}	ss_{TOT}		

The following equations show how each component is calculated.

Terms used

r	= number of rows
c	= number of columns
t	= number of treatments
N	= total number of experiment units (t^2)
R_i	= the sum of the observations in the ith row

C_j = the sum of the observations in the jth column
T_k = the sum of the observations receiving the kth treatment
y_{ijk} = the observation from the unit receiving the kth treatment in the ith row and jth column

T = the sum of all the observations $\left(\sum\limits_{i=1}^{r} \sum\limits_{j=1}^{c} \sum\limits_{k=1}^{k} y_{ijk} \right)$

Degrees of freedom (df)

$$df_{TOT} = t^2 - 1$$
$$df_R = t - 1$$
$$df_C = t - 1$$
$$df_T = t - 1$$
$$df_E = t^2 - 3t + 2$$
$$= df_{TOT} - df_R - df_C - df_T$$

Sums of squares (ss)

$$ss_{TOT} = \sum_{i=1}^{t}\sum_{j=1}^{t}\sum_{k=1}^{t} y_{ijk}^2 - \frac{T^2}{N}$$

$$ss_R = \frac{1}{t}\sum_{i=1}^{t} R_i^2 - \frac{T^2}{N}$$

$$ss_C = \frac{1}{t}\sum_{j=1}^{t} C_j^2 - \frac{T^2}{N}$$

$$ss_T = \frac{1}{t}\sum_{k=1}^{t} T_k^2 - \frac{T^2}{N}$$

$$ss_E = \sum_{i=1}^{t}\sum_{j=1}^{t}\sum_{k=1}^{t} y_{ijk}^2 - \frac{1}{t}\sum_{i=1}^{t} R_i^2 - \frac{1}{t}\sum_{j=1}^{t} C_j^2 - \frac{1}{t}\sum_{k=1}^{t} T_k^2 + \frac{T^2}{N}$$

$$= ss_{TOT} - ss_R - ss_C - ss_T$$

Mean square (ms)

$$ms_R = ss_R/df_R$$
$$ms_C = s_C/df_C$$
$$ms_T = ss_T/df_T$$
$$s^2 = ss_E/df_E$$

Variance ratio (vr)

$$F_T = ms_T/s^2$$

Hypothesis test

The null hypothesis of no difference between the t treatments is rejected at the $\alpha\%$ level of significance if the calculated variance ratio (F_T) is greater than the tabulated values with $(t-1),(t^2 - 3t + 2)$ degrees of freedom at significance level α.

$$F_T \geq F_{(t-1),(t^2-3t+2),\alpha}$$

EXAMPLE

The experiment used to illustrate these calculations is that which is designed in Section 4.2.3.6 and analysed and interpreted in Section 6.1.5.3. The experiment is an investigation into the effect of four diets on dry matter disappearance. The design used repeated 4×4 Latin squares but for illustrative purposes the analysis shown below is conducted on a single square. In practice, single Latin squares of less than 5×5 are rarely used. The analysis of variance table is shown below.

Source	df	ss	ms	vr
Row (R_i)	3	12.850	4.283	
Column (C_j)	3	12.673	4.224	
Treatment (T_k)	3	349.548	116.516	32.12
Residual (e_{ijk})	6	21.767	3.628	
Total	15	396.839		

The following equations show how each component is calculated.

Summations and terms used

r　　　= number of rows = 4
c　　　= number of columns = 4
t　　　= number of treatments = 4
N　　　= total number of experiment units (t^2) = $4 \times 4 = 16$

R_1　　= sum of the observations in the 1st row　= 247.3799
R_2　　= sum of the observations in the 2nd row　= 238.0278
R_3　　= sum of the observations in the 3rd row　= 239.8076
R_4　　= sum of the observations in the 4th row　= 243.4025

$$\sum_{i=1}^{4} R_i^2 = \text{sum of the squares of the row totals} = 234\,606.5105$$

C_1　　= sum of the observations in the 1st column　= 236.9952
C_2　　= sum of the observations in the 2nd column　= 246.8246
C_3　　= sum of the observations in the 3rd column　= 241.3633
C_4　　= sum of the observations in the 4th column　= 243.4347

$$\sum_{j=1}^{4} C_j^2 = \text{sum of the squares of the column totals} = 234\,605.8037$$

T_1　　= sum of the observations from the 1st treatment　= 222.1808
T_2　　= sum of the observations from the 2nd treatment　= 230.5456
T_3　　= sum of the observations from the 3rd treatment　= 244.4225
T_4　　= sum of the observations from the 4th treatment　= 271.4689

$$\sum_{k=1}^{4} T_k^2 = \text{sum of the squares of the treatment totals} = 235\,953.3037$$

$$\sum_{i=1}^{4}\sum_{j=1}^{4}\sum_{k=1}^{4} y_{ijk} = \text{sum of all the observations} = 968.6178$$

$$\sum_{i=1}^{4}\sum_{j=1}^{4}\sum_{k=1}^{4} y_{ijk}^2 = \text{sum of the squares of all observations} = 59\,035.61657$$

Degrees of freedom (df)

$$
\begin{aligned}
\mathrm{df_{TOT}} &= t^2-1 & &= 16-1 = 15 \\
\mathrm{df_R} &= t-1 & &= 4-1 = 3 \\
\mathrm{df_C} &= t-1 & &= 4-1 = 3 \\
\mathrm{df_T} &= t-1 & &= 4-1 = 3 \\
\mathrm{df_E} &= t^2-3t+2 & &= 16-(3\times4)+2=6 \\
&= \mathrm{df_{TOT}-df_R-df_C-df_T} = 15-3-3-3=6
\end{aligned}
$$

Sums of squares (ss)

$$
\mathrm{ss_{TOT}} = \sum_{i=1}^{4}\sum_{j=1}^{4}\sum_{k=1}^{4} y_{ijk}^2 - \frac{T^2}{N} = 59\,035.61657 - \frac{968.6178^2}{16}
$$

$$
= 59\,035.61657 - 58\,638.77765 = 396.839
$$

$$
\mathrm{ss_R} = \frac{1}{t}\sum_{i=1}^{4} R_i^2 - \frac{T^2}{N} = \frac{1}{4}\times 234\,606.5105 - \frac{968.6178^2}{16}
$$

$$
= 58\,651.62763 - 58\,638.77765 = 12.850
$$

$$
\mathrm{ss_C} = \frac{1}{t}\sum_{j=1}^{4} C_j^2 - \frac{T^2}{N} = \frac{1}{4}\times 234\,605.8037 - \frac{968.6178^2}{16}
$$

$$
= 58\,651.45093 - 58\,638.77765 = 12.673
$$

$$
\mathrm{ss_T} = \frac{1}{t}\sum_{k=1}^{4} T_k^2 - \frac{T^2}{N} = \frac{1}{4}\times 235\,953.3037 - \frac{968.6178^2}{16}
$$

$$
= 58\,988.32593 - 58\,638.77765 = 349.548
$$

$$
\mathrm{ss_E} = \sum_{i=1}^{4}\sum_{j=1}^{4}\sum_{k=1}^{4} y_{ijk}^2 - \frac{1}{t}\sum_{i=1}^{4} R_i^2 - \frac{1}{t}\sum_{j=1}^{4} C_j^2 - \frac{1}{t}\sum_{k=1}^{4} T_k^2 + 2\times \frac{T^2}{N}
$$

$$
= 59\,035.61657 - \frac{1}{4}\times 234\,606.5105 - \frac{1}{4}\times 234\,605.8037 - \frac{1}{4}\times 235\,953.3037 + 2\times \frac{968.6178^2}{16}
$$

$$
= 59\,035.61657 - 58\,651.62763 - 58\,651.45093 - 58\,988.32593 + 117\,277.5553 = 21.767
$$

$$
= \mathrm{ss_{TOT} - ss_R - ss_C - ss_T}
$$

$$
= 396.839 - 12.850 - 12.673 - 349.548 = 21.767
$$

Mean square (ms)

$$
\begin{aligned}
\mathrm{ms_R} &= \mathrm{ss_R/df_R} = 12.850/3 = 4.283 \\
\mathrm{ms_C} &= \mathrm{ss_C/df_C} = 12.673/3 = 4.224 \\
\mathrm{ms_T} &= \mathrm{ss_T/df_T} = 349.548/3 = 116.516 \\
s^2 &= \mathrm{ss_E/df_E} = 21.767/6 = 3.628
\end{aligned}
$$

Variance ratio (vr)

$$
F_T = \mathrm{ms_T}/s^2 = 116.516/3.628 = 32.12
$$

Hypothesis test result

$\quad$ F_T $\quad = 32.12 \geq F_{3,6,0.05} = 4.76$

At the 0.05 level of significance the null hypothesis of no difference between the four population mean DMD values is rejected because the calculated variance ratio (F_T) is greater than the tabulated F value with $(4-1)$, $(4^2-(3\times4)+2)$ degrees of freedom.

TWO-TREATMENT FACTOR FACTORIAL DESIGNS IN RANDOMIZED BLOCKS

The analysis of variance table is used to decide which of the effects being tested gives a significant result; it is in effect the decision rule for the tests. For an experiment in which there are b blocks and two treatment factors, factor 1 with x levels and factor 2 with y levels, it takes the following form when there is one complete replicate of all treatments, xy, in each block:

Source	df	ss	ms	vr	
Block (Bl_h)	df_{Bl}	ss_{Bl}	ms_{Bl}		
Factor X (x_i)	df_X	ss_X	ms_X	F_X	
Factor Y (y_j)	df_Y	ss_Y	ms_Y	F_Y	
XY interaction (xy_{ij})		df_{XY}	ss_{XY}	ms_{XY}	F_{XY}
Residual (e_{hij})	df_E	ss_E	s^2		
Total	df_{TOT}	ss_{TOT}			

The following equations show how each component is calculated.

Terms used

$\quad b \quad$ = number of blocks
$\quad x \quad$ = number of factor 1 treatment levels
$\quad y \quad$ = number of factor 2 treatment levels
$\quad N \quad$ = total number of experiment units (bxy)
$\quad Bl_h \quad$ = the sum of the observations in the hth block
$\quad x_i \quad$ = the sum of the observations from the ith factor 1 treatment
$\quad y_j \quad$ = the sum of the observations from the jth factor 2 treatment
$\quad x_iy_j \quad$ = the sum of the observations from the treatment combination corresponding to the ith factor 1 treatment and the jth factor 2 treatment
$\quad y_{hij} \quad$ = the observation from the unit receiving the ith factor 1 treatment and the jth factor 2 treatment in the hth block

$\quad T \quad$ = the sum of all the observations $\left(\sum\limits_{h=1}^{b} \sum\limits_{i=1}^{x} \sum\limits_{j=1}^{y} y_{hij} \right)$

Degrees of freedom (df)

$\quad df_{Bl} \quad = b-1$
$\quad df_X \quad = x-1$
$\quad df_Y \quad = y-1$
$\quad df_{XY} = (x-1)(y-1)$
$\quad df_E \quad = (b-1)(xy-1)$
$\quad\quad\quad = df_{TOT}-df_{Bl}-df_X-df_Y-df_{XY}$
$\quad df_{TOT} = bxy-1$

Sums of squares (ss)

$$SS_{TOT} = \sum_{h=1}^{b}\sum_{i=1}^{x}\sum_{j=1}^{y} y_{hij}^2 - \frac{T^2}{N}$$

$$SS_{BI} = \frac{1}{xy}\sum_{h=1}^{b} B_h^2 - \frac{T^2}{N}$$

$$SS_X = \frac{1}{by}\sum_{i=1}^{x}(x_i)^2 - \frac{T^2}{N}$$

$$SS_Y = \frac{1}{bx}\sum_{j=1}^{y}(y_j)^2 - \frac{T^2}{N}$$

$$SS_{X.Y} = \frac{1}{b}\sum_{i=1}^{x}\sum_{j=1}^{y}(x_i y_j)^2 - \frac{T^2}{N}$$

$$SS_{X.Y} = SS_{XY} - SS_X - SS_Y$$

$$SS_E = SS_{TOT} - SS_{BI} - SS_X - SS_Y - SS_{XY}$$

Mean square (ms)

$$
\begin{aligned}
ms_{BI} &= ss_{BI}/df_{BI}\\
ms_X &= ss_X/df_X\\
ms_Y &= ss_Y/df_Y\\
ms_{XY} &= ss_{XY}/df_{XY}\\
s^2 &= ss_E/df_E
\end{aligned}
$$

Variance ratio (vr)

$$
\begin{aligned}
F_X &= ms_X/s^2\\
F_Y &= ms_Y/s^2\\
F_{XY} &= ms_{XY}/s^2
\end{aligned}
$$

Hypotheses tests

The null hypothesis of no difference between the treatment combinations corresponding to the x factor 1 treatments and the y factor 2 treatments is rejected if the calculated variance ratio (F_{XY}) is greater than the tabulated values with $(x-1)(y-1)$, $(b-1)(xy-1)$ degrees of freedom at significance level α.

$$F_{XY} \geq F_{(x-1)(y-1),(b-1)(xy-1),\alpha}$$

The null hypothesis of no difference between the x factor 1 treatments is rejected if the calculated variance ratio (F_X) is greater than the tabulated values with $(x-1)$, $(b-1)(xy-1)$ degrees of freedom at significance level α.

$$F_X \geq F_{(x-1),(b-1)(xy-1),\alpha}$$

The null hypothesis of no difference between the y factor 2 treatments is rejected if the calculated variance ratio (F_Y) is greater than the tabulated values with $(y-1)$, $(b-1)(xy-1)$ degrees of freedom at significance level α.

$$F_Y \geq F_{(y-1),(b-1)(xy-1),\alpha}$$

Example

The experiment used to illustrate these calculations is that which was designed in Section 4.3.1.6 and analysed and interpreted in Section 6.1.5.4. This experiment investigates the effect on protein of heifers' milk when fed with two diets (D1 and D2) and housed in two different environments (H1 and H2). The experiment uses nine randomized blocks with one complete replicate of the four treatments in each block. The animals are milked throughout their lactation period and the actual observations used are the mean milk protein values where the protein is expressed as a percentage of the total milk yield. The analysis of variance table obtained is presented below.

Source	df	ss	ms	vr
Block (Bl_h)	8	0.17671	0.02209	
Diet (D_i)	1	0.68890	0.68890	21.40
Housing (H_j)	1	0.06760	0.06760	2.10
DH interaction (DH_{ij})	1	0.00538	0.00538	0.17
Residual (e_{hij})	24	0.77247	0.03219	
Total	35	1.71106		

The following equations show how each component in the analysis of variance table is calculated.

Summations and terms used

b = number of blocks $= 9$
x = number of diets $= 2$
y = number of housing environments $= 2$
N = total number of experiment units $(bxy) = 36$

Bl_1 = sum of observations in the 1st block $= 12.00$
Bl_2 = sum of observations in the 2nd block $= 12.82$
Bl_3 = sum of observations in the 3rd block $= 12.95$
Bl_4 = sum of observations in the 4th block $= 12.63$
Bl_5 = sum of observations in the 5th block $= 12.28$
Bl_6 = sum of observations in the 6th block $= 12.79$
Bl_7 = sum of observations in the 7th block $= 12.51$
Bl_8 = sum of observations in the 8th block $= 12.47$
Bl_9 = sum of observations in the 9th block $= 12.73$

$$\sum_{h=1}^{9} Bl_h^2 = \text{sum of squares of the block totals} = 1424.0082$$

The sum of the observations for each treatment and treatment combination are given below. The table margins are the main treatment totals and the table cells are the treatment combination totals, $x_i y_j$.

		Housing		
		H1	H2	x_i
Diet	D1	30.04	29.04	59.08
	D2	27.33	26.77	54.1
	y_j	57.37	55.81	113.18

$\sum_{i=1}^{2}(x_i)^2$ = the sum of the squares of the diet treatment totals

$$= (59.08^2 + 54.1^2) = 6417.2564$$

$\sum_{j=1}^{2}(y_j)^2$ = the sum of the squares of the housing treatment totals

$$= (57.37^2 + 55.81^2) = 6406.073$$

$\sum_{i=1}^{2}\sum_{j=1}^{2}(x_iy_j)^2$ = the sum of the squares of the treatment combination totals corresponding to the ith diet treatment and the jth housing treatment

$$= (30.04^2 + 29.04^2 + 27.33^2 + 26.77^2) = 3209.285$$

$\sum_{h=1}^{9}\sum_{i=1}^{2}\sum_{j=1}^{2}y_{hij}$ = the sum of all the observations = 113.18

$\sum_{h=1}^{9}\sum_{i=1}^{2}\sum_{j=1}^{2}y_{hij}^2$ = the sum of the squares of all the observations = 357.5364

Degrees of freedom (df)

$$\begin{aligned}
df_{Bl} &= b - 1 = 9 - 1 = 8\\
df_X &= x - 1 = (2 - 1) = 1\\
df_Y &= y - 1 = (2 - 1) = 1\\
df_{XY} &= (x - 1)(y - 1) = (2 - 1)(2 - 1) = 1\\
df_E &= df_{TOT} - df_{Bl} - df_X - df_Y - df_{XY}\\
&= (b - 1)(xy - 1) = (9 - 1)(2 \times 2 - 1) = 24\\
df_{TOT} &= bxy - 1 = (9 \times 2 \times 2) - 1 = 35
\end{aligned}$$

Sums of squares (ss)

$$SS_{TOT} = \sum_{h=1}^{9}\sum_{i=1}^{2}\sum_{j=1}^{2}y_{hij}^2 - \frac{T^2}{N} = 357.5364 - \frac{113.18^2}{36} = 1.71106$$

$$SS_{Bl} = \frac{1}{xy}\sum_{h=1}^{9}B_h^2 - \frac{T^2}{N} = \frac{1}{2\times 2}\times 1424.0082 - \frac{113.18^2}{36} = 0.17671$$

$$SS_X = \frac{1}{by}\sum_{i=1}^{2}(x_i)^2 - \frac{T^2}{N} = \frac{1}{9\times 2}\times 6417.2564 - \frac{113.18^2}{36} = 0.68890$$

$$SS_Y = \frac{1}{bx}\sum_{j=1}^{2}(y_j)^2 - \frac{T^2}{N} = \frac{1}{9\times 2}\times 6406.073 - \frac{113.18^2}{36} = 0.6760$$

$$SS_{X.Y} = \frac{1}{b}\sum_{i=1}^{2}\sum_{j=1}^{2}(x_iy_j)^2 - \frac{T^2}{N} = \frac{1}{9}\times 3209.285 - \frac{113.18^2}{36} = 0.76188$$

$$SS_{XY} = SS_{X.Y} - SS_X - SS_Y$$

$$= 0.76188 - 0.68890 - 0.06760 = 0.00538$$

$$SS_E = SS_{TOT} - SS_{Bl} - SS_X - SS_Y - SS_{XY}$$

$$= 1.71106 - 0.17671 - 0.68890 - 0.06760 - 0.00538 = 0.77247$$

Mean square (ms)

$$
\begin{aligned}
ms_{Bl} &= ss_{Bl}/df_{Bl} &&= 0.17671/8 &&= 0.02209 \\
ms_X &= ss_X/df_X &&= 0.68890/1 &&= 0.68890 \\
ms_Y &= ss_Y/df_Y &&= 0.06760/1 &&= 0.06760 \\
ms_{XY} &= ss_{XY}/df_{XY} &&= 0.00538/1 &&= 0.00538 \\
s^2 &= ss_E/df_E &&= 0.77247/24 &&= 0.03219
\end{aligned}
$$

Variance ratio (vr)

$$
\begin{aligned}
F_X &= ms_X/s^2 &&= 0.68890/0.03219 = 21.40 \\
F_Y &= ms_Y/s^2 &&= 0.06760/0.03219 = 2.10 \\
F_{XY} &= ms_{XY}/s^2 &&= 0.00538/0.03219 = 0.17
\end{aligned}
$$

Hypotheses test results

$$F_{XY} = 0.17 < F_{1,24,0.05} = 4.26$$

At the 0.05 level of significance the null hypothesis of no difference between the four interaction population mean values is accepted because the calculated variance ratio (F_{XY}) is less than the tabulated F value with $(2-1)(2-1),(9-1)(4-1)$ degrees of freedom.

$$F_X = 21.4 > F_{1,24,0.05} = 4.26$$

At the 0.05 level of significance the null hypothesis of no difference between the two diet treatment population mean values is rejected because the calculated variance ratio (F_X) is greater than the tabulated F value with $(2-1),(9-1)(4-1)$ degrees of freedom.

$$F_Y = 2.10 < F_{1,24,0.05} = 4.26$$

At the 0.05 level of significance the null hypothesis of no difference between the two housing environment treatment population mean values is accepted because the calculated variance ratio (F_Y) is less than the tabulated F value with $(2-1),(9-1)(4-1)$ degrees of freedom.

TWO-TREATMENT FACTOR SPLIT-PLOT DESIGNS IN RANDOMIZED BLOCKS

The analysis of variance table is used to decide which of the effects being tested gives a significant result; it is in effect the decision rule for the tests. For the two-factor split-plot design with a single replicate of each treatment in each of *b* blocks it takes the following form. There are two treatment factors and the first has *x* levels and is randomized on the *x* main-plots in each block. The second treatment factor has *y* levels and is randomized on the *y* sub-plots in each main-plot. There are a total of $m (= bx)$ main-plots.

Source	df	ss	ms	vr
Block (Bl_h)	df_{Bl}	ss_{Bl}	ms_{Bl}	
Factor X (x_i)	df_X	ss_X	ms_X	F_X
Main-plot residual ($em_{h(i)}$)	df_{EM}	ss_{EM}	ms_{EM}	
Main-plot total	df_{MTOT}	ss_{MTOT}		
Factor Y (y_j)	df_Y	ss_Y	ms_Y	F_Y
XY interaction (xy_{ij})	df_{XY}	ss_{XY}	ms_{XY}	F_{XY}
Sub-plot residual (e_{hij})	df_E	ss_E	s^2	
Total	df_{TOT}	ss_{TOT}		

The following equations show how each component is calculated.

Terms used

b	= number of blocks
x	= number of factor 1 treatment levels
y	= number of factor 2 treatment levels
N	= total number of experiment units (bxy)
m	= the total number of main-plots per block (bx)
Bl_h	= the sum of the observations in the hth block
x_i	= the sum of the observations from the ith factor 1 treatment
y_j	= the sum of the observations from the jth factor 2 treatment
x_iy_j	= the sum of the observations from the treatment combination corresponding to the ith factor 1 treatment and the jth factor 2 treatment
m_k	= the sum of the observations in the kth main-plot
y_{hij}	= the observation from the unit receiving the ith factor 1 treatment and the jth factor 2 treatment in the kth main-plot in the hth block
T	= the sum of all the observations $\left(\sum\limits_{h=1}^{b} \sum\limits_{i=1}^{x} \sum\limits_{j=1}^{y} y_{hij} \right)$

Degrees of freedom (df)

$$\begin{aligned}
df_{Bl} &= b - 1 \\
df_X &= x - 1 \\
df_Y &= y - 1 \\
df_{XY} &= (x-1)(y-1) \\
df_M &= m - 1 \\
&= (bx-1) \\
df_{EM} &= (b-1)(x-1) \\
&= df_M - df_X - df_{Bl} \\
df_E &= x(b-1)(y-1) \\
&= df_{TOT} - df_M - df_Y - df_{XY} \\
df_{TOT} &= bxy - 1
\end{aligned}$$

Sums of squares (ss)

Main-plot stratum

$$SS_M = \frac{1}{y} \sum_{k=1}^{m} (m_k)^2 - \frac{T^2}{N}$$

$$SS_{Bl} = \frac{1}{xy} \sum_{h=1}^{b} B_h^2 - \frac{T^2}{N}$$

$$SS_X = \frac{1}{by} \sum_{i=1}^{x} (x_i)^2 - \frac{T^2}{N}$$

$$SS_{EM} = SS_M - SS_X - SS_{Bl}$$

Sub-plot stratum

$$SS_{TOT} = \sum_{h=1}^{b} \sum_{i=1}^{x} \sum_{j=1}^{y} \sum_{k=1}^{m} y_{hijk}^2 - \frac{T^2}{N}$$

$$ss_Y = \frac{1}{bx}\sum_{j=1}^{y}(y_j)^2 - \frac{T^2}{N}$$

$$ss_{X.Y} = \frac{1}{b}\sum_{i=1}^{x}\sum_{j=1}^{y}(x_i y_j)^2 - \frac{T^2}{N}$$

$$ss_{XY} = ss_{X.Y} - ss_X - ss_Y$$
$$ss_E = ss_{TOT} - ss_M - ss_Y - ss_{XY}$$
$$ss_{TOT} = ss_{Bl} - ss_X - ss_{EM} - ss_Y - ss_{XY}$$

Mean square (ms)

$$
\begin{aligned}
ms_{Bl} &= ss_{Bl}/df_{Bl}\\
ms_X &= ss_X/df_X\\
ms_{EM} &= ss_{EM}/df_{EM}\\
ms_Y &= ss_Y/df_Y\\
ms_{XY} &= ss_{XY}/df_{XY}\\
s^2 &= ss_E/df_E
\end{aligned}
$$

Variance ratio (vr)

$$
\begin{aligned}
F_X &= ms_X/ms_{EM}\\
F_Y &= ms_Y/s^2\\
F_{XY} &= ms_{XY}/s^2
\end{aligned}
$$

Hypotheses tests

The null hypothesis of no difference between the treatment combinations corresponding to the x factor 1 treatments and the y factor 2 treatments is rejected if the calculated variance ratio (F_{XY}) is greater than the tabulated values with $(x-1)(y-1)$, $x(b-1)(y-1)$ degrees of freedom at significance level α.

$$F_{XT} \geq F_{(x-1)(y-1),x(b-1)(y-1),\alpha}$$

The null hypothesis of no difference between the x factor 1 treatments is rejected if the calculated variance ratio (F_X) is greater than the tabulated values with $(x-1)$, $(b-1)(x-1)$ degrees of freedom at significance level α.

$$F_X \geq F_{(x-1),(b-1)(x-1),\alpha}$$

The null hypothesis of no difference between the y factor 2 treatments is rejected if the calculated variance ratio (F_Y) is greater than the tabulated values with $(y-1)$, $x(b-1)(y-1)$ degrees of freedom at significance level α.

$$F_Y \geq F_{(y-1),x(b-1)(y-1),\alpha}$$

Example

The experiment used to illustrate these calculations is that which was designed in Section 4.3.3.6 and analysed and interpreted in Section 6.1.5.6. The experiment is an investigation into the effects of seven nitrogen rates and four fallow treatments on the specific weight of a winter wheat. The experiment is a split-plot design in four randomized blocks with all 28 treatment combinations occurring once in each block. The four fallow treatments are randomized on the main-plots and the seven nitrogen treatments are randomized on the sub-plots. The analysis of variance table obtained is presented below:

Source	df	SS	ms	vr
Block (Bl_h)	3	14.2054	4.7351	
Fallow (x)	3	44.4883	14.8294	10.64
Main-plot residual ($em_{h(i)}$)	9	12.5411	1.3935	
Main-plot total	15	71.2348		
Nitrogen (y_j)	6	157.1057	26.1843	76.64
Fallow.nitrogen interaction (xy_{ij})	18	12.6253	0.7014	2.05
Sub-plot residual (e_{hij})	72	24.5992	0.3417	
Total	111	265.5650		

The following equations show how each component in the analysis of variance table is calculated.

Summations and terms used

b = number of blocks = 4
x = number of fallow treatments = 4
y = number of nitrogen rates = 7
N = total number of experiment units (bxy) = 112
m = total number of main-plots (bx) = 16

Bl_1 = the sum of the observations in the 1st block = 2128.49
Bl_2 = the sum of the observations in the 2nd block = 2121.93
Bl_3 = the sum of the observations in the 3rd block = 2130.60
Bl_4 = the sum of the observations in the 4th block = 2148.81

$$\sum_{h=1}^{4} Bl_h^2 = \text{the sum of the squares of the block totals} = 18\,189\,940.00$$

m_1 = the sum of the observations in the 1st main-plot = 526.46
m_2 = the sum of the observations in the 2nd main-plot = 539.55
m_3 = the sum of the observations in the 3rd main-plot = 526.46
m_4 = the sum of the observations in the 4th main-plot = 536.02
m_5 = the sum of the observations in the 5th main-plot = 533.32
m_6 = the sum of the observations in the 6th main-plot = 536.11
m_7 = the sum of the observations in the 7th main-plot = 530.21
m_8 = the sum of the observations in the 8th main-plot = 522.29
m_9 = the sum of the observations in the 9th main-plot = 538.32
m_{10} = the sum of the observations in the 10th main-plot = 525.05
m_{11} = the sum of the observations in the 11th main-plot = 528.87
m_{12} = the sum of the observations in the 12th main-plot = 538.36
m_{13} = the sum of the observations in the 13th main-plot = 535.91
m_{14} = the sum of the observations in the 14th main-plot = 536.31
m_{15} = the sum of the observations in the 15th main-plot = 541.14
m_{16} = the sum of the observations in the 16th main-plot = 535.46

$$\sum_{k=1}^{16} (m_k)^2 = \text{the sum of squares of the main-plot totals} = 4\,547\,884.50$$

The sum of the observations for each treatment and treatment combination is given below. The table margins are the main treatment totals and the table cells are the treatment combination totals, $x_i y_j$.

		Nitrogen							
		1	2	3	4	5	6	7	x_i
Fallow	1	303.80	300.28	304.06	308.37	310.03	311.80	311.95	2150.29
	2	295.23	293.81	298.54	306.46	308.05	307.99	311.37	2121.45
	3	294.90	295.99	300.29	303.58	305.13	304.63	304.74	2109.26
	4	300.36	302.10	302.95	310.44	309.90	311.11	311.98	2148.84
	y_j	1194.29	1192.18	1205.84	1228.85	1233.11	1235.53	1240.04	

$$\sum_{i=1}^{4}(x_i)^2 = \text{the sum of the squares of the fallow treatment totals}$$

$$= (2150.29^2 + \ldots + 2148.84^2) = 18\,190\,790.00$$

$$\sum_{j=1}^{7}(y_j)^2 = \text{the sum of the squares of the nitrogen treatment totals}$$

$$= (1194.29^2 + \ldots + 1240.04^2) = 10\,396\,538.00$$

$$\sum_{i=1}^{4}\sum_{j=1}^{7}(x_i y_j)^2 = \text{the sum of the squares of the treatment combination totals corresponding to the } i\text{th fallow treatment and the } j\text{th nitrogen rate}$$

$$= (303.80^2 + 300.28^2 + \ldots + 311.11^2 + 311.98^2) = 2\,599\,363.25$$

$$\sum_{h=1}^{4}\sum_{i=1}^{4}\sum_{j=1}^{7} y_{hij} = \text{the sum of all observations} = 8529.84$$

$$\sum_{h=1}^{4}\sum_{i=1}^{4}\sum_{j=1}^{7} y^2_{hij} = \text{the sum of the squares of all the observations} = 649\,892.06$$

Degrees of freedom (df)

$$df_{BI} = b-1 = 4-1 = 3$$
$$df_X = x-1 = 4-1 = 3$$
$$df_Y = y-1 = 7-1 = 6$$
$$df_{XY} = (x-1)(y-1) = (4-1)(7-1) = 18$$
$$df_M = m-1 = 16-1 = 15$$
$$= bx-1 = (4 \times 4)-1 = 15$$
$$df_{EM} = (b-1)(x-1) = (4-1)(4-1) = 9$$
$$= df_M - df_X - df_{BI} = 15-3-3 = 9$$
$$df_E = x(b-1)(y-1) = 4(4-1)(7-1) = 72$$
$$= df_{TOT} - df_M - df_Y - df_{XY} = 111-15-6-18 = 72$$
$$df_{TOT} = bxy-1 = (4 \times 4 \times 7)-1 = 111$$

Sums of squares (ss)

Main-plot stratum:

$$ss_M = \frac{1}{y}\sum_{k=1}^{16}(m_k)^2 - \frac{T^2}{N} = \frac{1}{7} \times 4\,547\,884.50 - \frac{8529.84^2}{112} = 71.00$$

$$ss_{BL} = \frac{1}{xy}\sum_{h=1}^{4}B_h^2 - \frac{T^2}{N} = \frac{1}{4 \times 7} \times 18\,189\,940.00 - \frac{8529.84^2}{112} = 14.20$$

$$ss_X = \frac{1}{by}\sum_{i=1}^{4}(x_i)^2 - \frac{T^2}{N} = \frac{1}{4 \times 7} \times 18\,190\,790.00 - \frac{8529.84^2}{112} = 44.49$$

$$SS_{EM} = SS_M - SS_X - SS_{BI}$$

$$= 71.00 - 44.49 - 14.20 = 12.54$$

Sub-plot stratum:

$$SS_{TOT} = \sum_{h=1}^{4}\sum_{i=1}^{4}\sum_{j=1}^{7} y_{hij}^2 - \frac{T^2}{N} = 649\,892.06 - \frac{8529.84^2}{112} = 265.57$$

$$SS_Y = \frac{1}{bx}\sum_{j=1}^{7}(y_j)^2 - \frac{T^2}{N} = \frac{1}{4 \times 4} \times 10\,396\,538.00 - \frac{8529.84^2}{112} = 157.11$$

$$SS_{X.Y} = \frac{1}{b}\sum_{i=1}^{4}\sum_{j=1}^{7}(x_iy_j)^2 - \frac{T^2}{N} = \frac{1}{4} \times 2\,599\,363.25 - \frac{8529.84^2}{112} = 214.00$$

$$SS_{XY} = SS_{X.Y} - SS_X - SS_Y$$

$$= 214.00 - 44.49 - 157.11 = 12.63$$

$$SS_E = SS_{TOT} = SS_M - SS_Y - SS_{XY}$$

$$= 265.57 - 71.00 - 157.11 - 12.63 = 24.60$$

$$= SS_{TOT} - SS_{BI} - SS_X - SS_{EM} - SS_Y - SS_{XY}$$

$$= 265.57 - 14.20 - 44.49 - 12.54 - 157.11 - 12.63 = 24.60$$

Mean square (ms)

$$
\begin{aligned}
ms_{BI} &= ss_{BI}/df_{BI} = 14.2054/3 = 4.7351\\
ms_X &= ss_X/df_X = 44.4883/3 = 14.8294\\
ms_{EM} &= ss_{EM}/df_{EM} = 12.5411/9 = 1.3935\\
ms_Y &= ss_Y/df_Y = 157.1057/6 = 26.1843\\
ms_{XY} &= ss_{XY}/df_{XY} = 12.6253/18 = 0.7014\\
s^2 &= ss_E/df_E = 24.5992/72 = 0.3417
\end{aligned}
$$

Variance ratio (vr)

$$
\begin{aligned}
F_X &= ms_X/ms_{EM} = 14.8294/1.3935 = 10.64\\
F_Y &= ms_Y/s^2 = 26.1843/0.3417 = 76.64\\
F_{XY} &= ms_{XY}/s^2 = 0.7014/0.3417 = 2.05
\end{aligned}
$$

Hypotheses test results

$$F_{XY} = 2.05 > F_{18,72,0.05} = 1.7489$$

At the 0.05 level of significance the null hypothesis of no difference between the 28 interaction population mean values is rejected because the calculated variance ratio (F_{XY}) is greater than the tabulated F-value with $(4-1)(7-1)$, $4(4-1)(7-1)$ degrees of freedom.

$$F_X = 10.64 > F_{3,9,0.05} = 3.8626$$

At the 0.05 level of significance the null hypothesis of no difference between the four fallow treatment population mean values is rejected because the calculated variance ratio (F_X) is greater than the tabulated F-value with $(4-1)$, $(4-1)(4-1)$ degrees of freedom.

$$F_Y = 76.64 > F_{6,72,0.05} = 2.2274$$

At the 0.05 level of significance the null hypothesis of no difference between the seven nitrogen rate population mean values is rejected because the calculated variance ratio (F_Y) is greater than the tabulated F-value with $(7-1)$, $4(4-1)(7-1)$ degrees of freedom.

REFERENCES

Cochran WG and Cox GN (1957) *Experimental Designs*, second edition. John Wiley, New York, Chapter 4.

Montgomery DG (1991) *Design and Analysis of Experiments*, third edition. John Wiley, New York, Chapters 3 and 5.

Steel RGD and Torrie JH (1980) *Principles and Procedures of Statistics: A Biometrical Approach*, second edition. McGraw-Hill, New York, Chapters 7, 9 and 15.

Index

Illustrated Classics
ROBINSON CRUSOE
Daniel Defoe

My Early Life

I was born in the year 1632, in the city of York, of a good family, though not of that country, my father being a foreigner of Bremen, who settled first at Hull: he got a good estate by merchandise, and, leaving off his trade, lived afterwards at York, from whence he had married my mother, whose relations were named Robinson, a very good family in that country, and from whom I was called Robinson Kreutznaer; but, by the usual corruption of words in England, we are now called, nay, we call ourselves, and write our name, Crusoe, and so my companions always called me.

I had two elder brothers, one of which was lieutenant-colonel to an English regiment of foot in Flanders, and was killed at the battle near Dunkirk against the Spaniards: what became of my second brother I never knew, any more than my father or mother did know what was become of me.

Being the third son of the family, and not bred to any trade, my head began to be filled very early with rambling thoughts. My father had given me a competent share of learning, and designed me for the law; but I would be satisfied with nothing but going to sea, and my inclination to this led me so strongly against the will, nay, the commands of my father, and against all the entreaties and persuasions of my mother and friends, that there seemed to be something fatal in that propension of nature tending directly to the life of misery which was to befall me.

My father, a wise and grave man, gave me serious and excellent counsel against what he foresaw was my design. I was sincerely affected with this discourse, but, alas! a few days wore it all off; and, in short, to prevent any of my father's further importunities, in a few weeks after I resolved to run quite away from him.

My First Voyage

Being one day at Hull, where I went casually, and without any purpose of making an elopement, and one of my friends being going to sea to London in his father's ship, and prompting me to go with them, I consulted neither father nor mother any more, and in an ill hour, on the 1st of September, 1651, I went on board a ship bound for London. Never any young adventurer's misfortunes, I believe, began sooner, or continued longer than mine. The ship was no sooner gotten out of the Humber but the wind began to blow, and the waves to rise in a most frightful manner. After a violent storm, our ship was wrecked off Yarmouth. I had not the sense to go back to Hull, but continued on to London, where I went on a vessel bound for the coast of Africa.

My next voyage was on a Guinea trader, and I fell into terrible misfortunes on this voyage—namely, our ship, making her course towards the Canary Islands, was surprised in the grey of the morning by a Turkish rover of Sallee, and we were all carried prisoners into Sallee, a port belonging to the Moors. I was held as a slave for several years, until I eventually made my escape on board a ship bound for the Brazils.

In Brazil I prospered, and became a successful planter and merchant for many years; however I was not content, but must go to sea again, and boarded

a ship bound for Guinea. The same day I went on board we set sail, standing away to the northward upon our own coast, with design to stretch over for the African coast. We had very good weather all the way upon our own coast, till we came to the height of Cape St. Augustino; from whence, keeping farther off at sea, we lost sight of land. Then a violent tornado or hurricane took us quite out of our knowledge. It began from the south-east, came about to the north-west, and then settled in the north-east; from whence it blew in such a terrible manner that for twelve days together we could do nothing but drive, and, scudding away before it, let it carry us whither ever fate and the fury of the winds directed.

Shipwrecked

A second storm then came upon us, and carried us away with the same impetuosity westward, and drove us so out of the way of all human commerce, that, had our lives been saved as to the sea, we were rather in danger of being devoured by savages than ever returning to our own country.

In this distress, the wind still blowing very hard, one of our men early in the morning called out "Land!" and we had no sooner run out of the cabin to look out in hopes of seeing whereabouts in the world we were, but the ship struck upon a sand, and in a moment, her motion being so stopped, the sea broke over her in such a manner, that we expected we should all have perished immediately, and we were driven into our close quarters to shelter us from the very foam and spray of the sea.

It is not easy for anyone who has not been in the like condition to describe or conceive the consternation of men in such circumstances; we were in a dreadful condition indeed, and had nothing to do but to think of saving our lives as best we could. We had a boat on board, but how to get her off into the sea was a doubtful thing. However, there was no room for debate, for we fancied the ship would break in pieces every minute. The mate of the vessel lays hold of the boat, and with the help of the rest of the men, they got her slung over the ship's side, and, getting all into her, we let go, and committed ourselves to God's mercy and the wild sea.

And now our case was very dismal indeed, for we all saw plainly that the sea went so high, that the boat could not live, and that we should be inevitably drowned. After we had rowed, or rather driven, about a league and a half, a raging wave, mountain-like, came rolling astern of us, and plainly bade us expect the *coup-de-grâce*. In a word, it took us with such a fury, that it overset the boat at once, and we were all swallowed up in a moment.

Nothing can describe the confusion of thought which I felt when I sunk into the water; for though I swam very well, yet I could not deliver myself from the waves so as to draw breath, till that a wave, having driven me, or rather carried me, a vast way on towards the shore, and having spent itself, went back, and left me upon the land almost dry, but half-dead with the water I took in. I had so much presence of mind as well as breath left that, seeing myself nearer the mainland than I expected, I got upon my feet, and endeavoured to make on towards the land as fast as I could, before another wave should return and take me up again. But I soon found it was impossible to avoid it; for I saw the sea come after me as high as a great hill and as furious as an enemy which I had no means or strength to contend with. My business was to hold my breath and rise myself upon the water if I could, and so by swimming pilot myself towards the shore.

The wave that came upon me again buried me at once twenty or thirty feet deep in its own body; and I could feel myself carried with a mighty force and swiftness towards the shore a very great way. I was covered again with water for a good while, but I held out until it had spent itself and begun to subside. Then I struck forward against the return of the waves, and felt ground again with my feet. I stood still a few moments to recover breath, till the water went from me, and then took to my heels and ran with what strength I had farther towards the shore.

The final surge, though it broke over me, did not swallow me up, and the next run I took got me to the mainland, where I sat down upon the grass, quite out of reach of the wild sea.

I Am Saved

I was now landed, and safe on shore, but as for my comrades, I never saw them afterwards, or any sign of them, except three of their hats, one cap, and two shoes that were not fellows.

After a while I began to look round me to see what kind of place I was in, and what was next to be done. I soon found my comforts abate, and that in a word I had a dreadful deliverance: for I was wet, had no clothes to shift me, nor anything either to eat or drink to comfort me, neither did I see any prospect before me, but that of perishing with hunger, or being devoured by wild beasts; and that which was particularly afflicting to me was that I had no weapon either to hunt and kill any creature for my sustenance, or to defend myself against any other creature that might desire to kill me for theirs. In a word, I had nothing about me but a knife, a tobacco-pipe, and a little tobacco in a box; this was all my provision, and this threw me into terrible agonies of mind, that for a while I ran about like a madman. Night coming upon me, I began with a heavy heart to consider what would be my lot if there were any ravenous beasts in that country, seeing at night they always come abroad for their prey.

All the remedy that offered to my thoughts at the
time was to get up into a thick bushy tree like a fir,
but thorny, which grew near me, and where I
resolved to sit all night, and consider the next day
what death I should die, for as yet I saw no prospect
of life. I walked about a furlong from the shore, to
see if I could find any fresh water to drink, which I
did, to my great joy; and having drunk and put a
little tobacco in my mouth to prevent hunger, I went
to the tree, and getting up into it, endeavoured to
place myself so as that if I should sleep I might not
fall; and having cut me a short stick, like a
truncheon, for my defence, I took up my lodging,
and having been excessively fatigued, I fell fast
asleep and slept as comfortably as, I believe, few

could have done in my condition, and found myself the most refreshed with it that I think I ever was on such an occasion.

I Visit the Wreck

When I waked it was broad day, the weather clear, and the storm abated, so that the sea did not rage and swell as before; but that which surprised me most was that the ship was lifted off in the night from the sand where she lay, by the swelling of the tide, and was driven up almost as far as the rocks; this being within about a mile from where I was, and the ship seeming to stand upright still, I wished myself on board, that, at least, I might save some necessary things for my use.

When I came down from my apartment in the tree, I looked about me again, and the first thing I found was the boat, which lay as the wind and the sea had tossed her up upon the land, about two miles on my right hand. I walked as far as I could upon the shore to have got to her, but found a neck or inlet of water between me and the boat, which was about half a mile broad, so I came back for the present, being more intent upon getting at the ship, where I hoped to find something for my present subsistence.

A little after noon I found the sea very calm, and the tide ebbed so far out that I could come within a quarter of a mile of the ship; and here I found a fresh renewing of my grief, for I saw evidently that if we had kept on board, we had been all safe, that is to say, we had all got safe on shore, and I had not been so miserable as to be left entirely destitute of all comfort and company, as I now was; this forced tears from my eyes again, but as there was little relief, I resolved, if possible, to get to the ship, so I pulled off my shirt, for the weather was hot to extremity, and took the water, but when I came to the ship, my difficulty was still greater to know how to get on board, for as she lay aground, and high out of the water, there was nothing within my reach to lay hold of. I swam round her twice, and the second time I spied a small piece of a rope, which I wondered I did not see at first, hang down by the forechains so low as that with great difficulty I got hold of it, and by the help of that rope, got up into the forecastle of the ship. My first work was to search and to see what was spoiled and what was free; and first I found that all the ship's provisions were dry and untouched by the water, and being very well disposed to eat, I went to the bread-room and filled my pockets with biscuit, and ate it as I went about other things, for I had no time to lose. I also found some rum in the great cabin, of which I took a large dram, and which I had indeed need enough of to spirit me for what was before me. Now I wanted nothing but a boat to furnish myself with many things which I foresaw would be very necessary to me.

I Build a Raft

It was in vain to sit still and wish for what was not to be had, and this extremity roused my application. We had several spare yards, and two or three large spars of wood, and a spare top-mast or two in the ship; I resolved to fall to work with these, and I flung as many of them over board as I could manage for their weight, tying every one with a rope that they might not drive away; when this was done I went down the ship's side, and pulling them to me, I tied four of them fast together at both ends as well as I could, in the form of a raft, and laying two or three short pieces of plank upon them crossways, I found I could walk upon it very well, but that it was not able to bear any great weight, the pieces being too light; so I went to work, and with the carpenter's saw I cut a spare top-mast into three lengths, and added them to my raft, with a great deal of labour and pains; but hope of furnishing myself with necessaries encouraged me to go beyond what I should have been able to have done upon another occasion.

My raft was now strong enough to bear any reasonable weight. I first laid all the planks or boards upon it that I could get. I then got three of the seamen's chests, which I had broken open and emptied, and lowered them down upon my raft; the first of these I filled with provision, viz. bread, rice, three Dutch cheeses, five pieces of dried goat's flesh,

which we lived much upon, and a little remainder of European corn which had been laid by for some fowls which we brought to sea with us, but the fowls were killed; there had been some barley and wheat together, but, to my great disappointment, I found afterwards that the rats had eaten or spoiled it all; as for liquors, I found several cases of bottles belonging to our skipper, in which were some cordial waters, and in all about five or six gallons of rack; these I stored by themselves, there being no need to put them into the chest, nor room for them.

Next, tools to work with on shore, and it was after

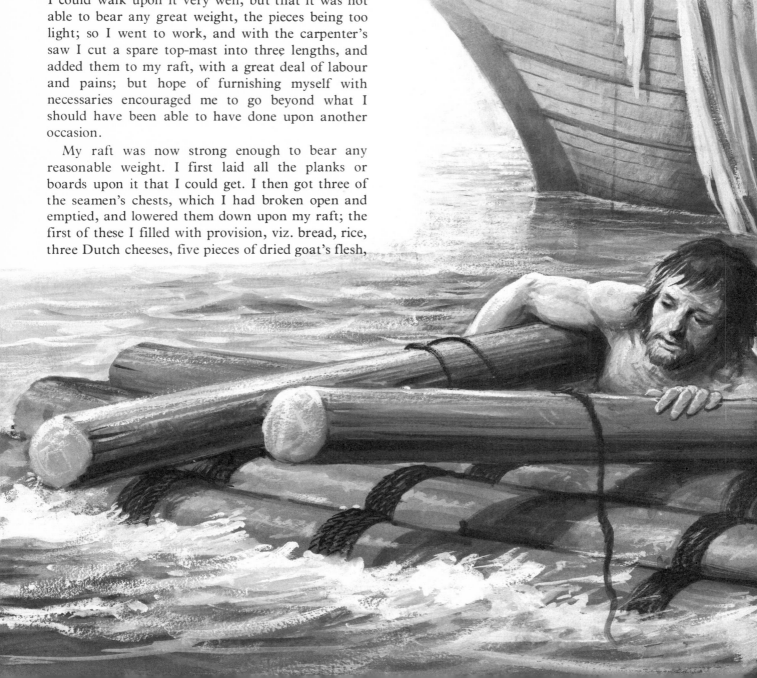

long searching that I found out the carpenter's chest, which was indeed a very useful prize to me; I got it down to my raft, even whole as it was, without losing time to look into it, for I knew in general what it contained.

My next care was for some ammunition and arms; there were two very good fowling-pieces in the great cabin, and two pistols; these I secured first, with some powder-horns, and a small bag of shot, and two old rusty swords; I knew there were three barrels of powder in the ship, but knew not where our gunner had stowed them, but with much search I found

them, two of them dry and good, the third had taken water; those two I got to my raft, with the arms; and now I thought myself pretty well freighted, and began to think how I should get to shore with them, having neither sail, oar, nor rudder, and the least cap full of wind would have overset all my navigation.

I had three encouragements: first, a smooth calm sea; second, the tide rising, and setting in to the shore; third, what little wind there was blew me towards the land; and thus, having found two or three broken oars belonging to the boat, and besides the tools which were in the chest, I found two saws, an axe, and a hammer, and with this cargo I put to sea.

At length I spied a little cove on the right shore of the creek, to which with great pain and difficulty I guided my raft, and at last got so near, that, reaching ground with my oar, I could thrust her directly in; but here I had like to have dipped all my cargo in the sea again; for that shore lying pretty steep, that is to say sloping, there was no place to land but where one end of my float, if it ran on shore, would lie so high, and the other sink lower as before, that it would endanger my cargo again. All that I could do was to wait till the tide was at the highest, keeping the raft with my oar like an anchor to hold this side of it fast to the shore; near a flat piece of ground, which I expected the water would flow over; and so it did. As soon as I found water enough, for my raft drew about a foot of water, I thrust her on upon that flat piece of ground, and there fastened or moored her by sticking my two broken oars into the ground, one on one side near one end, and one on the other side near the other end; and thus I lay till the water ebbed away and left my raft and all my cargo safe on shore.

My next work was to view the country, and seek a proper place for my habitation. There was a hill not above a mile from me, which rose up very steep and high, and which seemed to over-top some other hills, which lay as in a ridge from it northward. I took out one of the fowling-pieces, and one of the pistols, and a horn of powder, and thus armed I travelled for discovery up to the top of that hill, where, after I had with great labour and difficulty got to the top, I saw that I was in an island environed every way with the sea, no land to be seen, except some rocks which lay a great way off, and two small islands less than this, which lay about three leagues to the west.

My Second Voyage to the Wreck

Contented with this discovery, I came back to my raft, and fell to work to bring my cargo on shore, which took me up the rest of that day. At night, I barricaded myself round with the chests and boards I had brought on shore, and made a kind of hut for my lodging.

I now began to consider that I might yet get a great many things out of the ship, which would be useful to me, and particularly some of the rigging, and sails, and such other things as might come to land, and I resolved to make another voyage on board the vessel.

Having got my second cargo on shore, I went to work to make me a little tent with the sail and some poles which I cut for that purpose, and into this tent I brought everything that I knew would spoil, either with rain or sun, and I piled all the empty chests and casks up in a circle round the tent, to fortify it from any sudden attempt, either from man or beast.

When I had this done I blocked up the door of the tent with some boards within, and an empty chest set up on end without, and spreading one of the beds upon the ground, laying my two pistols just at my head, and my gun at length by me, I went to bed for the first time, and slept very quietly all night, for I was very weary and heavy; for the night before I had slept little, and had laboured very hard all day, as well to fetch all those things from the ship, as to get them on shore.

After this I went every day on board, and brought away what I could get.

I had now been thirteen days on shore, and had been eleven times on board the ship. But preparing the twelfth time to go on board, I found the wind begin to rise; however, at low water I went on board, and though I thought I had rummaged the cabin so effectually as that nothing more could be found, yet I discovered a chest with drawers in it, in one of which I found two or three razors, and one pair of scissors, with some ten or a dozen of good knives and forks; in another I found about thirty-six pounds' value in money, some European coin, some Brazilian, some pieces of eight, some gold, some silver. I smiled to myself at the sight of this money. "O drug!" said I aloud, "what art thou good for? Thou art not worth to me, no, not the taking off of the ground; one of those knives is worth all this

heap; I have no manner of use for thee, even remain where thou art, and go to the bottom as a creature whose life is not worth saving."

However, upon second thoughts, I took it away, and wrapping all this in a piece of canvas, I began to think of making another raft, but while I was preparing this, I found the sky overcast, and the wind began to rise, and in a quarter of an hour it blew a fresh gale from the shore; it presently occurred to me that it was in vain to pretend to make a raft with the wind off shore, and that it was my business to be gone before the tide of flood began, otherwise I might not be able to reach the shore at all. Accordingly I let myself down into the water, and swam across the channel which lay between the ship and the sand, and was at last gotten home to my little tent, where I lay with all my wealth about me very secure.

My New Home

In search of a place to settle permanently, I found a little plain on the side of a rising hill, whose front towards this little plain was steep as a house-side, so that nothing could come down upon me from the top; on the side of this rock there was a hollow place worn a little way in like the entrance or door of a cave, but there was not really any cave or way into the rock at all.

On the flat of the green, just before this hollow place, I resolved to pitch my tent. This plain was not above an hundred yards broad, and about twice as long, and lay like a green before my door, and at the end of it descended irregularly every way down into the low grounds by the sea side. It was on the N.N.W. side of the hill, so that I was sheltered from the heat every day, till it came to a west and by south sun, or thereabouts, which in those countries is near the setting.

Before I set up my tent, I drew a half circle before the cave entrance which took in about ten yards in its semi-diameter from the rock, and twenty yards in its diameter, from its beginning and ending.

In this half circle I pitched two rows of strong stakes, driving them into the ground till they stood very firm like piles, the biggest end being out of the ground about five foot and a half, and sharpened on the top. The two rows did not stand above six inches from one another.

Then I took some pieces of cable which I had brought from the ship, and I laid them in rows one

upon another, within the circle, between these two rows of stakes, up to the top, placing other stakes in the inside, leaning against them, about two foot and a half high, like a spur to a post, and this fence was so strong that neither man or beast could scale it.

The entrance into this place I made by a short ladder to go over the top. Into this fence or fortress, with infinite labour, I carried all my riches, all my provisions, ammunition, and stores.

After I had been there about ten or twelve days, it came into my thoughts that I should lose my reckoning of time for want of books and pen and ink, and should even forget the sabbath days from the working days; but to prevent this I cut it with my knife upon a large post, in capital letters, and making it into a great cross I set it up on the shore where I first landed, viz. "I came on shore here on the 30th of Sept. 1659." Upon the sides of this

square post I cut every day a notch with my knife, and every seventh notch was as long again as the rest, and every first day of the month as long again as that long one, and thus I kept my calendar, or weekly, monthly, and yearly reckoning of time. And I must not forget that we had in the ship a dog and two cats, of whose eminent history I may have occasion to say something in its place; for I carried both the cats with me, and as for the dog, he jumped out of the ship of himself, and swam on shore to me the day after I went on shore with my first cargo, and was a trusty servant to me many years; I wanted nothing that he could fetch me, nor any company that he could make up to me, I only wanted to have him talk to me, but that would not do. As I observed before, I found pen, ink, and paper, and I husbanded them to the utmost, and I shall show, that while my ink lasted, I kept things very exact, but after that was gone I could not, for I could not make any ink by any means that I could devise.

And now I began to apply myself to make such necessary things as I found I most wanted, as particularly a chair and a table; for without these I was not able to enjoy the few comforts I had in the world; I could not write, or eat, or do several things with so much pleasure without a table. So I went to work; I made abundance of things, even without tools, and some with no more tools than an adze and a hatchet. But having gotten over these things in some measure, and having settled my household stuff and habitation, made me a table and a chair, and all as handsome about me as I could, I began to keep my journal, of which I shall here give you the copy (though in it will be told all these particulars over again) as long as it lasted, for having no more ink I was forced to leave it off.

My Journal

September 30, 1659

I, poor, miserable Robinson Crusoe, being ship-wrecked during a dreadful storm in the offing, came on shore on this dismal unfortunate island, which I called the Island of Despair, all the rest of the ship's company being drowned, and myself almost dead.

All the rest of that day I spent in afflicting myself at the dismal circumstances I was brought to, viz. I had neither food, house, clothes, weapon, nor place to fly to, and in despair of any relief, saw nothing but death before me, either that I should be devoured by wild beasts, murdered by savages, or starved to death for want of food. At the approach of night, I slept in a tree for fear of wild creatures, but slept soundly though it rained all night.

October 1

In the morning I saw to my great surprise the ship had floated with the high tide, and was driven on shore again much nearer the island. I went upon the sand as near as I could, and then swam on board; this day also it continued raining, though with no wind at all.

From the 1st October to the 24th

All these days entirely spent in many several voyages to get all I could out of the ship.

Oct. 20

I overset my raft and all the goods I had got upon it, but being in shoal water, and the things being chiefly heavy, I recovered many of them when the tide was out.

Oct. 25

It rained all night and all day, with some gusts of wind, during which time the ship broke in pieces.

Oct. 26

I walked about the shore almost all day to find out a place to fix my habitation, greatly concerned to secure myself from an attack in the night, either from wild beasts or men. Towards night, I fixed upon a proper place under a rock, and marked out my encampment, which I resolved to strengthen with a wall.

From the 26th to the 30th

I worked very hard in carrying all my goods to my new habitation, though some part of the time it rained exceeding hard.

The 31st

In the morning I went out into the island with my gun to see for some food, and discover the country, when I killed a she-goat, and her kid followed me

home, which I afterwards killed also because it would not feed.

November 1

I set up my tent under a rock, and lay there for the first night, making it as large as I could with stakes driven in to swing my hammock upon.

Nov. 2

I set up all my chests and boards, and the pieces of timber which made my rafts, and with them formed a fence round me, a little within the place I had marked out for my fortification.

Nov. 3

I went out with my gun and killed two fowls like ducks, which were very good food. In the afternoon went to work to make me a table.

Nov. 4

This morning I began to order my times of work, of going out with my gun, time of sleep, and time of diversion.

Nov. 5

This day went abroad with my gun and my dog, and killed a wild cat.

Nov. 6

After my morning walk I went to work with my table again, and finished it, though not to my liking; nor was it long before I learned to mend it.

Nov. 7

Now it began to be settled fair weather. The 7th, 8th, 9th, 10th, and part of the 12th (for the 11th was Sunday) I took wholly up to make me a chair.

Nov. 13

This day it rained, which refreshed me exceedingly, and cooled the earth, but it was accompanied with terrible thunder and lightning, which frighted me dreadfully for fear of my powder.

Nov. 14, 15, 16

These three days I spent in making little square chests or boxes, which might hold a pound or two pound, at most, of powder, and so putting the powder in, I stowed it in places as secure and remote from one another as possible. On one of these three days I killed a large bird that was good to eat, but I know not what to call it.

Nov. 17

This day I began to dig behind my tent into the rock to make room for my farther conveniency. Note: Three things I wanted exceedingly for this work, viz. a pick-axe, a shovel, and a wheel-barrow or basket; so I desisted from my work, and began to consider how to supply that want and make me some tools; as for a pick-axe, I made use of the iron crows, which were proper enough, though heavy; but the next thing was a shovel or spade; this was so absolutely necessary, that indeed I could do nothing effectually without it, but what kind of one to make I knew not.

Nov. 18

The next day in searching the woods I found a tree of that wood, or like it, which in the Brazils they call the iron tree, for its exceeding hardness; of this, with great labour and almost spoiling my axe, I cut a piece, and brought it home too with difficulty enough, for it was exceedingly heavy.

The excessive hardness of the wood, and having no other way, made me a long while upon this machine, for I worked it effectually by little and little into the form of a shovel or spade, the handle exactly shaped like ours in England, only that the broad part having no iron shod upon it at bottom, it would not last me so long; however, it served well enough for the uses which I had occasion to put it to; but never was a shovel, I believe, made after that fashion, or so long a making.

I was still deficient, for I wanted a basket or a wheel-barrow; a basket I could not make by any means, having no such things as twigs that would bend to make wicker ware, at least none yet found out; and as to a wheel-barrow, I fancied I could make all but the wheel, but that I had no notion of, neither did I know how to go about it; besides, I had no possible way to make the iron gudgeons for the spindle or axis of the wheel to run in, so I gave it over, and so for carrying away the earth which I dug out of the cave, I made me a thing like a hod, which the labourers carry mortar in when they serve the bricklayers.

This was not so difficult to me as the making the shovel; and yet this, and the shovel, and the attempt which I made in vain to make a wheel-barrow, took me up no less than four days; I mean, always excepting my morning walk with my gun, which I seldom failed, and very seldom failed also bringing home something fit to eat.

Nov. 23

My other work having now stood still, because of my making these tools, when they were finished I went on, and working every day, as my strength and time allowed, I spent eighteen days entirely in widening and deepening my cave, that it might hold my goods commodiously.

20

Note:

During all this time, I worked to make this room or cave spacious enough to accommodate me as a warehouse or magazine, a kitchen, a dining-room, and a cellar; as for my lodging, I kept to the tent, except that sometimes in the wet season of the year, it rained so hard that I could not keep myself dry, which caused me afterwards to cover all my place within my pale with long poles in the form of rafters leaning against the rock, and load them with flags and large leaves of trees like a thatch.

A Lucky Escape

December 10

I began now to think my cave or vault finished, when on a sudden (it seems I had made it too large) a great quantity of earth fell down from the top and one side, so much that in short it frightened me, and not without reason too; for I had had a lucky escape. Upon this disaster I had great deal of work to do over again; for I had the loose earth to carry out, and which was of more importance, I had the ceiling to prop up, so that I might be sure no more would come down.

Dec. 11

This day I went to work with it accordingly, and got two shores or posts pitched upright to the top, with two pieces of boards across over each post; this I finished the next day; and setting more posts up with boards, in about a week more I had the roof secured, and the posts, standing in rows, served me for partitions to part of my house.

Dec. 17

From this day to the twentieth I placed shelves, and knocked up nails on the posts to hang everything

up that could be hung up, and now I began to be in some order within doors.

Dec. 20

Now I carried everything into the cave, and began to furnish my house, and set up some pieces of boards, like a dresser, to order my victuals upon, but boards began to be very scarce with me; also I made me another table.

Dec. 24

Much rain all night and all day; no stirring out.

Dec. 25

Rain all day.

Dec. 26

No rain, and the earth much cooler than before, and pleasanter.

Dec. 27

Killed a young goat, and lamed another so that I caught it, and led it home in a string; when I had it home, I bound and splintered up its leg, which was broken. N.B. I took such care of it that it lived, and the leg grew well, and as strong as ever; but by my nursing it so long it grew tame, and fed upon the little green at my door, and would not go away. This was the first time that I entertained a thought of breeding up some tame creatures, that I might have food when my powder and shot was all spent.

Dec. 28, 29, 30

Great heats and no breeze; so that there was no stirring abroad, except in the evening for food; this time I spent in putting all my things in order within doors.

January 1

Very hot still, but I went abroad early and late with my gun, and lay still in the middle of the day. This evening, going farther into the valleys which lay towards the centre of the island, I found there was plenty of goats, though exceeding shy and hard to come at; however I resolved to try if I could not bring my dog to hunt them down.

Jan. 2

Accordingly, the next day I went out with my dog, and set him upon the goats; but I was mistaken, for they all faced about upon the dog, and he knew his danger too well, for he would not come near them.

Jan. 3

I began my fence or wall; which, being still jealous of my being attacked by somebody, I resolved to make very thick and strong.

All this time I worked very hard on my wall, and made my rounds in the woods every day.

Exploring the Island

It was the 15th of July that I began to take a more particular survey of the island itself. I went up the creek first, where, as I hinted, I brought my rafts on shore. I found, after I came about two miles up, that the tide did not flow any higher, and that it was no more than a little brook of running water, and very fresh and good; but this being the dry season, there was hardly any water in some parts of it, at least not enough to run in any stream so as it could be perceived.

I spent all that evening there, and went not back to my habitation, which, by the way, was the first night, as I might say, I had lain from home. In the night I took my first contrivance, and got up into a tree, where I slept well, and the next morning proceeded upon my discovery, travelling nearly four miles, as I might judge by the length of the valley, keeping still due north, with a ridge of hills on the south and northside of me.

At the end of this march I came to an opening, where the country seemed to descend to the west, and a little spring of fresh water, which issued out of the side of the hill by me, ran the other way, that is, due east; and the country appeared so fresh, so green, so flourishing, everything being in a constant verdure or flourish of spring, that it looked like a planted garden.

I descended a little on the side of that delicious vale, surveying it with a secret kind of pleasure (though mixed with my other afflicting thoughts)— to think that this was all my own, that I was king and lord of all this country indefeasibly, and had a right of possession; and if I could convey it, I might have it in inheritance as completely as any lord of a manor in England. I saw here abundance of cocoa trees, orange, and lemon, and citron trees; but all wild, and very few bearing any fruit, at least not then. However, the green limes that I gathered were not only pleasant to eat, but very wholesome; and I mixed their juice afterwards with water, which made it very wholesome, and very cool and refreshing. I found melons on the ground in great abundance, and grapes upon the trees.

I found now I had business enough to gather and carry home; and I resolved to lay up a store, as well of grapes as limes and lemons, to furnish myself for the wet season, which I knew was approaching. Having spent three days on this journey, I then returned home; for so I must now call my cave.

24

My First Boat

Two more seasons passed, then, one day, travelling about my island, I came within view of the sea to the west, and it being a very clear day I fairly descried land; whether an island or continent I could not tell.

This at length put me upon thinking whether it was not possible to make myself a canoe, or periagua, such as the natives of those climates make, even without tools, or, as I might say, without hands, viz. of the trunk of a great tree. I was so intent upon my voyage over the sea in it, that I never once considered how I should get it off the land; and it was really in its own nature more easy for me to guide it over forty five miles of sea, than about forty fathoms of land, where it lay, to set it afloat in the water.

I went to work upon this boat the most like a fool that ever man did, who had any of his senses awake. I pleased myself with the design, without determining whether I was ever able to undertake it; not but that the difficulty of launching my boat came often into my head; but I put a stop to my own enquiries into it, by this foolish answer which I gave myself, "Let's first make it, I'll warrant I'll find some way or other to get it along when 'tis done."

This was a most preposterous method; but the eagerness of my fancy prevailed, and to work I went. I felled a cedar tree: I question much whether Solomon ever had such a one for the building of the temple of Jerusalem. It was five foot ten inches diameter at the end of twenty two foot, after which it lessened for a while, and then parted into branches. It was not without infinite labour that I felled this tree; I was twenty days hacking and hewing at it at the bottom; I was fourteen more getting the branches and limbs and the vast spreading head of it cut off, which I hacked and hewed through with axe and hatchet, and inexpressible labour; after this, it cost me a month to shape it, and dub it to a proportion, and to something like the bottom of a boat, that it might swim upright as it ought to do. It cost me near three months more to clear the inside, and work it out so as to make an exact boat of it. This I did indeed without fire, by mere mallet and chisel, and by the dint of hard labour, till I had brought it to be a very handsome periagua, and big enough to have carried me and all my cargo.

When I had gone through this work, I was extremely delighted with it. The boat was really

much bigger than I ever saw a canoe that was made of one tree, in my life. Many a weary stroke it had cost, you may be sure; and there remained nothing but to get it into the water; and had I gotten it into the water, I make no question but I should have begun the maddest voyage, and the most unlikely to be performed, that ever was undertaken.

But all my devices to get it into the water failed me; though they cost me infinite labour too. This grieved me heartily, and now I saw, though too late, the folly of beginning a work before we count the cost, and before we judge rightly of our own strength to go through with it.

Making an Umbrella

I saved the skins of all the creatures that I killed, I mean four-footed ones, and I had hung them up stretched out with sticks in the sun, by which means some of them were so dry and hard that they were fit for little, but others, it seems, were very useful. The first thing I made of these was a great cap for my head, with the hair on the outside to shoot off the rain; and this I performed so well, that after this I made me a suit of clothes wholly of these skins, that is to say, a waistcoat, and breeches open at knees, and both loose, for they were rather wanting to keep me cool than to keep me warm. I must not omit to acknowledge that they were wretchedly made; for if I was a bad carpenter, I was a worse tailor. However, they were such as I made good shift with; and when I was abroad, if it happened to rain, the hair of my waistcoat and cap being outermost, I was kept very dry.

After this I spent a great deal of time and pains to make me an umbrella; I was indeed in great want of one, and had a great mind to make one; I had seen them made in Brazil, where they are very useful in the great heats which are there; and I felt the heats every jot as great here, and greater too, being nearer the equinox; besides, as I was obliged to be much abroad, it was a most useful thing to me, as well for the rains as the heats. I took a world of pains at it, and was a great while before I could make anything likely to hold; nay, after I thought I had hit the way, I spoiled two or three before I made one to my mind; but at last I made one that answered indifferently well. The main difficulty I found was to make it to let down. I could make it to spread, but if it did not let down too, and draw in, it was not portable for me any way but just over my head, which would not do. However, at last, as I said, I made one to answer, and covered it with skins, the hair upwards, so that it cast off the rains like a penthouse, and kept off the sun so effectually, that I could walk out in the hottest of the weather with greater advantage than I could before in the coolest, and when I had no need of it, could close it and carry it under my arm.

My "Family"

Being now in the eleventh year of my residence, I set myself to study some art to trap and snare the wild goats that roamed the island, to see whether I could not catch some of them alive. To this purpose, I made snares to hamper them, and at length resolved to try a pit-fall, in which I trapped three kids. In about a year and a half I had a flock of about twelve goats, and in two years more I had three and forty. Now I not only had goats flesh to feed on when I pleased, but milk too, sometimes a gallon or two in a day.

It would have made a Stoic smile to have seen me and my little family sit down to dinner, and to see how like a king I dined, too, all alone, attended by my servants. Poll, the young parrot I had caught and tamed, was the only person permitted to talk to me, like a favourite. My dog—who was now grown very old and crazy—sat always at my right hand; and two cats, the descendants of those I had brought on shore, one on one side the table and one on the other, expecting now and then a bit from my hand, as a mark of special favour.

Had anyone in England been to meet such a man as I was, it must either have frighted them, or raised a great deal of laughter. Be pleased to take a sketch of my figure as follows.

I had my great, high shapeless cap, made of a goat's skin, with a flap hanging down behind, as well to keep the sun from me as to shoot the rain off from running into my neck. I had a short jacket of goat-skin, the skirts coming down to about the middle of my thighs; and a pair of open-kneed breeches of the same—the breeches were made of the skin of an old he-goat, whose hair hung down such a length on either side, that like pantaloons it reached to the middle of my legs; stockings and shoes I had none, but had made me a pair of somethings, I scarce know what to call them, but they were of a most barbarous shape— as indeed were the rest of my clothes.

I had on a broad belt of goatskin dried, and in a kind of a frog on either side of this, instead of a sword and dagger hung a little saw and hatchet. I had another belt not so broad, which hung over my shoulder; and at the end of it hung two pouches, both made of goat's skin too—in one of which hung my powder, in the other my shot. I carried my gun on my shoulder; and over my head a great clumsy, ugly, goatskin umbrella—but which, after all, was the most necessary thing I had about me, next to my gun.

As for my face, the colour of it was not really so Mulatto-like as one might expect. My beard I had once suffered to grow until it hung about a quarter of a yard long, but as I had both scissors and razors sufficient, I had cut it pretty short, except what grew on my upper lip, which I had trimmed into a large pair of whiskers, of a length and shape monstrous enough, and such as in England would have passed for frightful.

During this time I built another canoe, which I fitted with mast and sail, and used for short voyages on the sea, but I never ventured far from my little creek.

A Mysterious Footprint

It happened one day about noon going towards my boat, I was exceedingly surprised with the print of a man's naked foot on the shore, which was very plain to be seen in the sand. I stood like one thunderstruck, or as if I had seen an apparition; I listened, I looked round me, I could hear nothing, nor see anything; I went up to a rising ground to look farther; I went up the shore and down the shore, but it was all one, I could see no other impression but that one.

I slept none that night; the farther I was from the occasion of my fright, the greater my apprehensions were, which is something contrary to the nature of such things, and especially to the usual practice of all creatures in fear: but I was so embarrassed with my own frightful ideas of the thing, that I formed nothing but dismal imaginations to myself, even though I was now a great way off it. Some times I fancied it must be the devil; and reason joined in with me upon this supposition; for how should any other thing in human shape come into the place? Where was the vessel that brought them? What marks were there of any other footsteps? And how was it possible a man should come there?

Now I began to take courage, and to peep abroad again, for I had not stirred out of my castle for three days and nights, so that I began to starve for provision; for I had little or nothing within doors but some barley cakes and water. Then I knew that my goats wanted to be milked too, which usually was my evening diversion; and the poor creatures were in great pain and inconvenience for want of it; and indeed, it almost spoiled some of them, and almost dried up their milk.

For several years, I went about the whole island, then, one day, when wandering more to the west point of the island than I had ever gone yet, and looking out to sea, I thought I saw a boat upon the sea, at a great distance; I had found a perspective glass or two in one of the seamen's chests, which I saved out of our ship; but I had it not about me, and this was so remote that I could not tell what to make of it, though I looked at it till my eyes were not able to hold to look any longer. Whether it was a boat or not, I do not know; but as I descended from the hill, I could see no more of it, so I gave it over; only I resolved to go no more out without a perspective glass in my pocket.

When I was come down the hill to the end of the

bones of human bodies; and particularly I observed a place where there had been a fire made, and a circle dug in the earth, like a cockpit, where it is supposed the savage wretches had sat down to their inhuman feastings upon the bodies of their fellow creatures.

I was so astonished with the sight of these things, that I entertained no notions of any danger to myself from it for a long while; all my apprehensions were buried in the thoughts of such a pitch of inhuman, hellish brutality, and the horror of the degeneracy of human nature. I could not bear to stay in the place a moment longer so I got me up the hill again with all the speed I could, and walked on towards my own habitation.

Thus the years passed, and I began to be very well contented with the life I led, if it might but have been secured from the dread of the savages.

But it was otherwise directed; and it may not be amiss for all people who shall meet with my story, to make this just observation from it, namely, how frequently, in the course of our lives, the evil which in itself we seek most to shun and which when we are fallen into it is the most dreadful to us, is often times the very means or door of our deliverance, by which alone we can be raised again from the affliction we are fallen into.

A Cannibal Feast

It was now the month of December, in my twenty-third year; and this being the southern solstice, for winter I cannot call it, was the particular time of my harvest, and required my being pretty much abroad in the fields; when going out pretty early in the morning, even before it was thorough day-light, I was surprised with seeing a light of some fire upon the shore, at a distance from me of about two miles towards the end of the island, where I had observed some savages had been as before; but not on the other side; but to my great affliction, it was on my side of the island.

I was indeed terribly surprised at the sight, and stopped short within my grove, not daring to go out, lest I might be surprised; and yet I had no more peace within, from the apprehensions I had that if these savages, in rambling over the island, should find my corn standing or cut, or any of my works and improvements, they would immediately con-clude that there were people in the place, and would

island, where indeed I had never been before, I was presently convinced that the seeing the print of a man's foot was not such a strange thing in the islands as I imagined; and but that it was a special providence that I was cast upon the side of the island where the savages never came, I should easily have known that nothing was more frequent than for the canoes from the main, when they happened to be a little too far out at sea, to shoot over to that side of the island for harbour; likewise as they often met and fought in their canoes, the victors, having taken any prisoners, would bring them over to this shore, where according to their dreadful customs, being all cannibals, they would kill and eat them; of which hereafter.

When I was come down the hill to the shore, as I said above, being the S.W. point of the island, I was perfectly confounded and amazed; nor is it possible for me to express the horror of my mind, at seeing the shore spread with skulls, hands, feet, and other

then never give over till they had found me out. In this extremity I went back directly to my castle, pulled up the ladder after me, and made all things without look as wild and natural as I could.

Then I prepared myself within, putting myself in a posture of defence; I loaded all my cannon, as I called them, that is to say, my muskets which were mounted upon my new fortification, and all my pistols, and resolved to defend myself to the last gasp, not forgetting seriously to commend myself to the divine protection, and earnestly to pray to God to deliver me out of the hands of the barbarians; and in this posture I continued about two hours; but began to be mighty impatient for intelligence abroad, for I had no spies to send out.

After sitting a while, and musing what I should do, I was not able to bear sitting in ignorance any longer; so setting up my ladder to the side of the hill, I mounted to the top. Pulling out my perspective glass, which I had taken on purpose, I laid me down flat on my belly on the ground, and began to look for the place; I presently found there was no less than nine naked savages, sitting round a small fire they had made, not to warm them, for they had no need of that, the weather being extremely hot; but, as I supposed, to dress some of their barbarous diet of human flesh, which they had brought with them, whether alive or dead I could not know.

They had two canoes with them, which they had hauled up upon the shore; and as it was then tide of ebb, they seemed to me to wait for the return of the flood, to go away again; it is not easy to imagine what confusion this sight put me into, especially seeing them come on my side the island, and so near me too; but when I observed their coming must always be with the current of the ebb, I began afterwards to be more sedate in my mind, being satisfied that I might go abroad with safety all the time of the tide of flood, if they were not on shore before: and having made this observation, I went abroad about my harvest work with the more composure.

As I expected, as soon as the tide made to the westward, I saw them all take the boat, and row (or paddle as we call it) all away. I should have observed, that for an hour and more before they went off, they went to dancing, and I could easily discern their postures and gestures by my glasses: I could not perceive, by my nicest observation, but that they were stark naked, and had not the least

covering upon them; but whether they were men or women, that I could not distinguish.

As soon as I saw them shipped and gone, I took two guns upon my shoulders, and two pistols at my girdle, and my great sword by my side, without a scabbard, and with all the speed I was able to make, I went away to the hill, where I had discovered the first appearance of all; and as soon as I got thither, which was not less than two hours (for I could not go apace, being so loaded with arms as I was), I perceived there had been three canoes more of savages on that place; and looking out farther, I saw they were all at sea together, making over for the main.

This was a dreadful sight to me, especially when going down to the shore, I could see the marks of horror which the dismal work they had been about had left behind it, viz. the blood, the bones, and part of the flesh of human bodies, eaten and devoured by those wretches, with merriment and sport. I was so filled with indignation at the sight, that I began now to premeditate the destruction of the next that I saw there, let them be who or how many soever.

I Rescue a Savage

Two more years passed, then I was surprised one morning early, with seeing no less than five canoes all on shore together on my side the island; and the people who belonged to them all landed and out of my sight. The number of them broke all my measures, for seeing so many, and knowing that they always came four or six, or sometimes more in a boat, I could not tell what to think of it, or how to take my measures to attack twenty or thirty men single handed; so I lay still in my castle, perplexed and discomforted; however, I put myself into all the same postures for an attack that I had formerly provided, and was just ready for action, if anything had presented. Having waited a good while, listening to hear if they made any noise, at length, being very impatient, I set my guns at the foot of my ladder, and clambered up to the top of the hill; standing so, however, that my head did not appear above the hill, so that they could not perceive me by any means; here I observed, by the help of my perspective glass, that they were no less than thirty in number, they

had a fire kindled, and they had meat dressed. How they had cooked it, that I knew not, or what it was; but they were all dancing in I know not how many barbarous gestures and figures, their own way round the fire.

While I was thus looking on them, I perceived by my perspective two miserable wretches dragged from the boats, where it seems they were laid by, and were now brought out for the slaughter: I perceived one of them immediately fell, being knocked down, I suppose with a club or wooden sword, for that was their way, and two or three others were at work immediately cutting him open for their cookery, while the other victim was left standing by himself, till they should be ready for him. In that very moment this poor wretch seeing himself a little at liberty, nature inspired him with hopes of life, and he started away from them, and ran with incredible swiftness along the sands directly towards me, I mean that part of the coast where my habitation was.

I was dreadfully frightened when I perceived him to run my way, and especially when, as I thought, I saw him pursued by the whole body. However, I kept my station, and my spirits began to recover when I found that there were not above three men that followed him, and still more was I encouraged when I found that he outstripped them exceedingly in running, and gained ground on them, so that if he could but hold it for half an hour, I saw easily he would fairly get away from them all.

There was between them and my castle the creek; and this, I saw plainly, he must necessarily swim over, or the poor wretch would be taken there. But when the savage escaping came thither, he made nothing of it, though the tide was then up, but plunging in, swam through in about thirty strokes or thereabouts, landed, and ran on with exceeding strength and swiftness; when the three persons came to the creek, I found that two of them could swim, but the third could not, and that standing on the other side, he looked at the others, but went no further, and soon after went softly back again, which, as it happened, was very well for him in the main.

I observed that the two who swam were yet more

than twice as long swimming over the creek as the fellow was that fled from them. It came now very warmly upon my thoughts, and indeed irresistibly, that now was my time to get me a servant, and perhaps a companion or assistant; and that I was called plainly by providence to save the poor creature's life, I fetched my two guns, and getting up again with the same haste to the top of the hill, I crossed towards the sea; and making a short cut, nearly all downhill, placed myself between pursuers and pursued. I shouted to him that fled, and beckoned for him to come back. I then turned to the two who followed. The first I knocked down with the stock of my piece, the second, while he was raising his bow and arrow, I killed with my first shot.

The poor savage who fled had now stopped and though he saw his enemies lying dead, was so fearful that he stood stock still, and made no effort to come forward. I smiled and made signs to him to advance. At length he did so, then kneeling down placed his face upon the ground. He then took my foot and placed it upon his head, thus displaying that he was my slave forever.

My Man Friday

I took him up and made much of him, and encouraged him all I could. I gave him bread and raisins to eat, and a draught of water. And, having refreshed him, I made signs for him to lie down and sleep, which he did straight away.

He was a comely, handsome fellow, perfectly well made, with straight strong limbs, not too large, tall and well shaped, and about twenty-six years of age. He had a very good countenance, especially when he smiled. His hair was long and black, his forehead very high and large, and a great vivacity and sparkling sharpness in his eyes. The colour of his skin was not quite black, but very tawny.

After he had slumbered, I made him know his name should be Friday, which was the day I saved his life. I likewise taught him that my name was to be Master and taught him to say it, and Yes and No, and to know the meaning of them.

The night came and we slept. In the morning I beckoned him to come with me. I gave him the sword, bows and arrows, and one of the guns to carry, and we marched away to the place these creatures had been.

When we arrived there, my blood chilled in my

veins, for all variety of human remnants littered the ground; skulls, bones, flesh and blood covered half eaten morsels. In short, evidence of a wild cannibal feast. Friday conveyed to me that they had brought over four prisoners to feast upon. Three had been eaten, and he was to be the fourth.

I told Friday to gather all the remains and burn them. When this had been done we returned to my castle. There I kitted out my savage. I gave him a pair of linen drawers which I had out of the poor gunner's chest I mentioned, and which I found in the wreck; and which with a little alteration fitted him very well.

Never man had a more faithful, loving, sincere servant, than Friday was to me; without passions, sullenness, or designs, perfectly obliged and engaged; his very affections were tied to me, like those of a child to a father; and I dare say he would have sacrificed his life for the saving mine upon any occasion whatsoever. The many testimonies he gave me of this put it out of doubt, and soon convinced me that I needed to use no precautions as to my safety on his account.

I was greatly delighted with him, and made it my business to teach him everything that was proper to make him useful, handy, and helpful; but especially to make him speak, and understand me when I spake; and he was the aptest scholar that ever was, and particularly was so merry, so constantly diligent, and so pleased, when he could but understand me, or make me understand him, that it was very pleasant to talk to him.

This was the pleasantest year of all the life that I led in this place. Friday began to talk pretty well, and understand the names of almost everything I had occasion to call for, and of every place I had to send him to, and to talk a great deal to me. Besides the pleasure of talking to him, I had a singular satisfaction in the fellow himself. His simple, unfeigned honesty appeared to me more and more every day.

He told me that up a great way beyond the moon, that was, beyond the setting of the moon, which must be W. from their country, there dwelt white men, like me; and that they had killed "much mans", that was his word. By all which I understood, he meant the Spaniards, whose cruelties in America had been spread over the whole country, and was remembered by all the nations from father to son.

I enquired if he could tell me how I might come from this island, and get among those white men; he told me, "yes, yes", I might go "in two canoe"; I could not understand what he meant, or make him describe to me what he meant by "two canoe", till at last, with great difficulty, I found he meant it must be in a large great boat, as big as two canoes.

This part of Friday's discourse began to relish with me very well, and from this time I entertained some hopes, that one time or other, I might find an opportunity to make my escape from this place; and that this poor savage might be a means to help me to do it.

After Friday and I became more intimately acquainted, and that he could understand almost all I said to him and speak fluently, though in broken English, to me, I acquainted him with my own story, or at least so much of it is as related to my coming into the place, how I had lived there, and how long. I let him into the mystery, for such it was to him, of gunpowder and bullet, and taught him how to shoot; I gave him a knife, which he was wonderfully delighted with, and I made him a belt, with a frog hanging to it, such as in England we wear hangers in; and in the frog, instead of a hanger, I gave him a hatchet, which was not only as good a weapon in some cases, but much more useful upon other occasions.

Preparing for a Voyage

I showed him the ruins of our boat, which we lost when we were wrecked, and which I could not stir with my whole strength then, but was now fallen almost to pieces. Upon seeing this boat, Friday stood musing a great while, and said nothing. I asked him what it was he studied upon; at last said he, "Me see such boat like come to place at my nation."

I did not understand him a good while; but at last, when I had examined farther into it, I understood by him that a boat, such as that had been, came on shore upon the country where he lived; that is, as he explained it, was driven thither by stress of weather. I presently imagined that some European ship must have been cast away upon their coast, and the boat might get loose, and drive ashore; but was so dull, that I never once thought of men making escape from a wreck thither, much less whence they might come; so I only inquired after a description of the boat.

Friday described the boat to me well enough; but brought me better to understand him, when he added with some warmth, "We save the white mans from drown." Then I presently asked him if there was any white mans, as he called them, in the boat. 'Yes," he said, "the boat full white mans." I asked him how many; he told upon his fingers seventeen. I asked him then what became of them; he told me, "They live, they dwell at my nation."

From this time I confess I had a mind to venture over, and see if I could possibly join with these men, who I had no doubt were Spaniards or Portuguese;

not doubting but if I could, we might find some method to escape from thence, being upon the continent, and a good company together; better than I could from an island forty miles off the shore, and alone without help. So after some days I took Friday to work again, by way of discourse, and told him I would give him a boat to go back to his own nation; and accordingly I carried him to my frigate which lay on the other side of the island, and having cleared it of water, for I always kept it sunk in the water, I brought it out, showed it him, and we both went into it.

I found he was a most dexterous fellow at managing it, he would make it go almost as swift and fast again as I could; so when he was in, I said to him, "Well now, Friday, shall we go to your nation?" He looked very dull at my saying so, which it seems was because he thought the boat too small to go so far.

Upon the whole, I was by this time so fixed upon my design of going over with him to the continent, that I told him we would go and make one as big as required and he should go home in it. There were trees enough in the island to have built a little fleet, not of periaguas and canoes, but even of good large vessels. But the main thing I looked at, was to get one so near the water that we might launch it when it was made, to avoid the mistake I committed at first.

At last, Friday pitched upon a tree, for I found he knew much better than I what kind of wood was fittest for it, nor can I tell to this day what wood to call the tree we cut down, except that it was very like the tree we call fustic, or between that and the Nicaragua wood, for it was much the same colour and smell. Friday was for burning the hollow or cavity of the tree out to make it for a boat. But I showed him how rather to cut it out with tools, which, after I had showed him how to use, he did very handily, and in about a month's hard labour we finished it, and made it very handsome, especially when with our axes, which I showed him how to handle, we cut and hewed the outside into the true shape of a boat; after this, however, it cost us near a fortnight's time to get her along as it were inch by inch upon great rollers into the water. But when she was in, she would have carried twenty men with great ease.

I had a farther design, and that was to make a mast and sail and to fit her with an anchor and cable. As to a mast, that was easy enough to get; so I pitched upon a straight young cedar–tree, which I found near the place, and which there was great plenty of in the island, and I set Friday to work to cut it down, and gave him directions how to shape and order it. I was preparing daily for the voyage; and the first thing I did was to lay by a certain quantity of provisions, being the stores for our voyage; and intended, in a week or a fortnight's time, to open the dock and launch our boat.

More Cannibals

I was busy one morning upon some thing of this kind, when I called to Friday, and bade him go to the sea shore, and see if he could find a turtle or tortoise, a thing which we generally got once a week, for the sake of the eggs as well as the flesh. Friday had not been long gone, when he came running back, and flew over my outer wall, or fence, like one that felt not the ground, or the steps he set his feet on; and

before I had time to speak to him, he cried out to me, "O master; O master! O sorrow! O bad!" "What's the matter, Friday?" said I. "O yonder, there," said he, "one, two, three canoe! One, two, three!" By his way of speaking, I concluded there were six; but on inquiry, I found it was but three. "Well, Friday," said I, "do not be frightened"; so I heartened him up as well as I could. However, I saw the poor fellow was most terribly scared.

I entered the wood, with all possible wariness and silence, Friday following close at my heels. I marched till I came to the skirt of the wood, on the side which was next to them; only that one corner of the wood lay between me and them; here I called softly to Friday, and showing him a great tree, which was just at the corner of the wood, I bade him go to the tree, and bring me word if he could see there plainly what they were doing. He did so, and came immediately back to me, and told me they might be plainly viewed there; that they were all about their fire, eating the flesh of one of their prisoners; and that another lay bound upon the sand, a little from them, which he said they would kill next, and which fired all the very soul within me; he told me it was

not one of their nation, but one of the bearded men, who he had told me of, that came to their country in the boat. I was filled with horror at the very naming of the white bearded man, and going to the tree, I saw plainly by my glass a white man who lay upon the beach of the sea, with his hands and his feet tied, with flags, or things like rushes; and that he was a European, and had clothes on.

We Rescue a Spaniard

There was another tree, and a little thicket beyond it, about fifty yards nearer to them than the place where I was, which by going a little way about, I saw I might come at undiscovered, and that then I should be within half shot of them; so I withheld my passion, though I was indeed enraged to the highest degree, and going back about twenty paces, I got behind some bushes, which held all the way till I came to the other tree; and then I came to a little rising ground, which gave me a full view of them, at the distance of about eighty yards.

I had not a moment to lose; for nineteen of the dreadful wretches sat upon the ground, all close

huddled together, and had just sent the other two to butcher the poor Christian, and bring him, perhaps limb by limb to their fire, and they were stooped down to untie the bands at his feet; I turned to Friday. "Now, Friday," said I, "do as I bid thee;" Friday said he would. "Then, Friday," said I, "do exactly as you see me do, fail in nothing." So I set down one of the muskets and the fowling-piece upon the ground, and Friday did the like by his; and with the other musket I took my aim at the savages, bidding him do the like. Then asking him if he was ready, he said "Yes." "Then fire at them," said I; and the same moment I fired also.

Friday took his aim so much better than I, that on the side that he shot, he killed two of them and wounded three more; and on my side, I killed one, and wounded two. They were, you may be sure, in a dreadful consternation; and all of them who were not hurt, jumped up upon their feet, but did not immediately know which way to run, or which way to look; for they knew not from whence their destruction came. Friday kept his eyes close upon me, that, as I had bid him, he might observe what I did. So as soon as the first shot was made, I threw down the piece and took up the fowling-piece, and Friday did the like; he saw me cock and present; he did the same again. "Are you ready, Friday?" said I. "Yes," said he. "Let fly then," said I, "in the name of God," and with that I fired again among the amazed wretches, and so did Friday; and as our pieces were now loaded with what I called swan-shot, or small pistol bullets, we found only two drop; but so many were wounded, that they ran about yelling and screaming like mad creatures, all bloody and miserably wounded, most of them; whereof three more fell quickly after, though not quite dead.

"Now, Friday," said I, laying down the discharged pieces, and taking up the musket which was yet loaded, "follow me," which he did, with a great deal of courage. Upon which I rushed out of the wood, and showed myself, and Friday close at my foot. As soon as I perceived they saw me, I shouted as loud as I could, and bade Friday do so too; and running as fast as I could, which by the way, was not very fast, being loaded with arms as I was, I made directly towards the poor victim, who was, as I said, lying upon the beach or shore, between the place where they sat and the sea. The two butchers who were just going to work with him had left him at the surprise of our first fire, and fled in a terrible fright to the

48

seaside and had jumped into a canoe, and three more of the rest made the same way. I turned to Friday, and bade him step forwards and fire at them. He understood me immediately, and running about forty yards to be near them, he shot at them, and I thought he had killed them all; for I saw them fall of a heap into the boat; though I saw two of them up again quickly. However, he killed two of them and wounded the third; so that he lay down in the bottom of the boat, as if he had been dead.

While my man Friday fired at them, I pulled out my knife and cut the flags that bound the poor victim, and loosing his hands and feet, I lifted him up, and asked him in the Portuguese tongue what he was. He answered in Latin, "Christianus"; but was so weak and faint that he could scarce stand or speak. I took my bottle out of my pocket and gave it him, making signs that he should drink, which he did; and I gave him a piece of bread, which he ate. Then I asked him what countryman he was, and he said, "Espagniole"; and being a little recovered, let me know by all the signs he could possibly make, how much he was in my debt for his deliverance. "Seignior," said I, with as much Spanish as I could make up, "we will talk afterwards, but we must fight now. If you have any strength left, take this pistol and sword, and lay about you." He took them very thankfully, and no sooner had he the arms in his

hands, but, as if they had put new vigour into him, he flew upon his murderers like a fury, and had cut two of them in pieces in an instant. For the truth is, as the whole was a surprise to them, so the poor creatures were so much frighted with the noise of our pieces, that they fell down for mere amazement and fear; and had no more power to attempt their own escape, than their flesh had to resist our shot. And that was the case of those five that Friday shot at in the boat; for as three of them fell with the hurt they received, so the other two fell with the fright.

I kept my piece in my hand still, without firing, being willing to keep my charge ready, because I had given the Spaniard my pistol and sword; so I called to Friday, and bade him run up to the tree from whence we first fired, and fetch the arms which lay there, which he did with great swiftness; and then giving him my musket, I sat down myself to load all the rest again, and bade them come to me when they wanted. While I was loading these pieces, there happened a fierce engagement between the Spaniard and one of the savages, who made at him with one of their great wooden swords, the same weapon that

was to have killed him before, if I had not prevented it. The Spaniard, who was as bold and as brave as could be imagined, though weak, had fought this Indian a good while, and had cut him two great wounds on his head; but the savage being a stout lusty fellow, closing in with him, had thrown him down (being faint) and was wringing my sword out of his hand, when the Spaniard, though undermost,

wisely quitting the sword, drew the pistol from his girdle, shot the savage through the body, and killed him upon the spot, before I, who was running to help him, could come near him.

Friday, being now left to his liberty, pursued the flying wretches with no weapon in his hand but his hatchet; and with that he dispatched those three who, as I said before, were wounded at first and fallen, and all the rest he could come up with; and the Spaniard coming to me for a gun, I gave him one of the fowling-pieces, with which he pursued two of the savages, and wounded them both; but as he was not able to run, they both got from him into the wood, where Friday pursued them and killed one of them; but the other was too nimble for him, and though he was wounded, yet had plunged himself into the sea, and swam with all his might off to those two who were left in the canoe, which three in the canoe, with one wounded, who we know not whether he died or no, were all that escaped our hands of one and twenty.

My thoughts were a little suspended when I had a serious discourse with the Spaniard, and when I understood that there were sixteen more of his countrymen and Portuguese, who having been cast away, and made their escape to that side, lived there at peace indeed with the savages, but were very sore put to it for necessaries and indeed for life.

He told me they were all of them very civil, honest men, and they were under the greatest distress imaginable, having neither weapons nor clothes, nor any food, but at the mercy and discretion of the savages; out of all hopes of ever returning to their own country; and that he was sure, if I would undertake their relief, they would live and die by me.

Having prepared a larger stock of food for all the guests I expected, I gave the Spaniard leave to go over to the main, to see what he could do with those he had left behind him there. I gave him a strict charge in writing not to bring any man with him who would not first swear that he would in no way injure, fight with, or attack the person he should find in the island.

A Ship Approaches

It was no less than eight days I had waited for them, when a strange and unforeseen accident intervened. I was fast asleep in my hutch one morning, when my man Friday came running in to me, and called aloud, "Master, master, they are come, they are come!"

I jumped up, and regardless of danger, I went out, as soon as I could get my clothes on, through my little grove, which by the way was by this time grown to be a very thick wood. I say, regardless of danger, I went without my arms, which was not my custom

to do; but I was surprised, when turning my eyes to the sea, I presently saw a boat at about a league and a half's distance, standing in for the shore, with a shoulder of mutton sail, as they call it; and the wind blowing pretty fair to bring them in; also I observed presently that they did not come from that side which the shore lay on, but from the southernmost end of the island. Upon this I called Friday in, and bade him lie close, for these were not the people we looked for, and that we might not know yet whether they were friends or enemies.

Having fetched my perspective glass, and climbed

the ladder, I had scarce set my foot on the hill when my eye plainly discovered a ship lying at anchor, at about two leagues and a half's distance from me south-south-east, but not above a league and a half from the shore. By my observation it appeared plainly to be an English ship, and the boat appeared to be an English long-boat.

I cannot express the confusion I was in, though the joy of seeing a ship, and one who I had reason to believe was manned by my own countrymen, and consequently friends, was such as I cannot describe; but yet I had some secret doubts hung about me, I cannot tell from whence they came, bidding me keep upon my guard.

In the first place, it occurred to me to consider what business an English ship could have in that part of the world, since it was not the way to or from any part of the world where the English had any traffic; and I knew there had been no storms to drive them in there. It was most probable that they were here upon no good design.

I saw the boat draw near the shore, as if they looked for a creek to thrust in at for the convenience of landing; however, as they did not come quite far enough, they did not see the little inlet where I formerly landed my rafts; but ran their boat on shore upon the beach, at about half a mile from me, which was very happy for me; for otherwise they would have landed just, as I may say, at my door, and would soon have beaten me out of my castle, and perhaps have plundered me of all I had. When they were on shore, I was fully satisfied that they were Englishmen; at least most of them. One or two I thought were Dutch, but it did not prove so.

Mutiny

There were in all eleven men, whereof three of them I found were unarmed, and, as I thought, bound; and when the first four or five of them were jumped on shore, they took those three out of the boat as prisoners. One of the three I could perceive using passionate gestures of entreaty, affliction, and despair, even to a kind of extravagance; the other two I could perceive lifted up their hands sometimes, and appeared concerned indeed, but not to such a degree as the first.

I was perfectly confounded at the sight, and knew not what the meaning of it should be. Friday called out to me in English, as well as he could. "O master!

you see English mans eat prisoner as well as savage mans." "Why," said I, "Friday, do you think they are a-going to eat them then?" "Yes," said Friday, "they will eat them." "No, no, Friday, I am afraid they will murder them indeed, but you may be sure they will not eat them."

All this while I had no thought of what the matter really was, but stood trembling with the horror of the sight, expecting every moment when the three prisoners should be killed; nay, once I saw one of the villains lift up his arm with a great cutlass, to strike one of the poor men; and I expected to see him fall every moment, at which all the blood in my body seemed to run chill in my veins.

After I had observed the outrageous usage of the three men by the insolent seamen, I observed the fellows run scattering about the land, as if they wanted to see the country. I observed that the three other men had liberty to go also where they pleased; but they sat down all three upon the ground, very pensive, and looked like men in despair.

This put me in mind of the first time when I came on shore, and began to look about me; how I gave myself over for lost; how wildly I looked round me; what dreadful apprehensions I had; and how I lodged in the tree all night for fear of being devoured by wild beasts.

As I knew nothing that night of the supply I was to receive by the providential driving of the ship nearer the land, by the storms and tide, by which I have since been so long nourished and supported; so these three poor desolate men knew nothing how certain of deliverance and supply they were, how near it was to them, and how effectually and really they were in a condition of safety, at the same time that they thought themselves lost, and their case desperate.

It was just at the top of high-water when these people came on shore, and while partly they stood parleying with the prisoners they brought, and partly while they rambled about to see what kind of a place they were in, they had carelessly stayed till the tide was spent and the water was ebbed considerably away, leaving their boat aground.

They had left two men in the boat, who, as I found afterwards, having drank a little too much brandy, fell asleep; however, one of them waking sooner than the other, and finding the boat too fast aground for him to stir it, hallooed for the rest who were straggling about, upon which they all soon came to the boat; but it was past all their strength to launch her, the boat being very heavy, and the shore on that side being a soft oozy sand, almost like a quick-sand.

In this condition, like true seamen who are perhaps the least of all mankind given to fore-thought, they gave it over, and away they strolled about the country again; and I heard one of them say aloud to another, calling them off from the boat, "Why, let her alone, Jack, can't ye, she will float next tide," by which I was fully confirmed in the main enquiry, of what countrymen they were.

I knew it was no less than ten hours before the boat could be afloat again, and by that time it would be dark, and I might be at more liberty to see their motions, and to hear their discourse, if they had any. In the meantime, I fitted myself up for a battle, as before, though with more caution, knowing I had to do with another kind of enemy than I had at first. I ordered Friday also, who I had made an excellent marksman with his gun, to load himself with arms. I took myself two fowling-pieces, and I gave him three muskets. My figure indeed was very fierce; I had my formidable goat-skin coat on, with the great cap I have mentioned, a naked sword by my side, two pistols in my belt, and a gun upon each shoulder.

The English Captain

It was my design, as I said above, not to have made any attempt till it was dark; but about two o'clock, being the heat of the day, I found that in short they were all gone straggling into the woods, and, as I thought, were laid down to sleep. The three poor distressed men, too anxious for their condition to get any sleep, were however sat down under the shelter of a great tree, at about a quarter of a mile from me, and, as I thought, out of sight of any of the rest.

Upon this I resolved to discover myself to them, and learn something of their condition. Immediately

I marched in the figure as above, my man Friday at a good distance behind me, as formidable for his arms, as I, but not making quite so staring a spectre-like figure as I did.

I came as near them undiscovered as I could and then, before any of them saw me, I called aloud to them in Spanish, "What are ye, gentlemen?"

They started up at the noise, but were ten times more confounded when they saw me, and the uncouth figure that I made. They made no answer at all, but I thought I perceived them just going to fly from me, when I spoke to them in English.

"Gentlemen," said I, "do not be surprised at me; perhaps you may have a friend near you when you did not expect it." "He must be sent directly from heaven then," said one of them gravely to me, and pulling off his hat at the same time to me, "for our condition is past the help of man." "All help is from heaven, sir," said I. "But can you put a stranger in the way how to help you, for you seem to me to be in some great distress? I saw you when you landed, and when you seemed to make applications to the brutes that came with you, I saw one of them lift up his sword to kill you."

The poor man, with tears running down his face, and trembling, looked like one astonished, returned, "Am I talking to God or man?" "I am an Englishman and disposed to assist you," I said. "I have one servant only; we have arms and ammunition; tell us freely, can we serve you? What is your case?"

"Our case," said he, 'sir, is too long to tell you, while our murderers are so near; but in short, sir, I was commander of that ship; my men have mutinied against me; they have been hardly prevailed upon not to murder me, and at last have set me on shore in this desolate place, with these two men with me, one my mate, the other a passenger, where we expected to perish, believing the place to be uninhabited, and know not yet what to think of it."

"Have they any fire-arms?" said I. He answered they had only two pieces, and one which they left in the boat. "Well then," said I, "leave the rest to me; I see they are all asleep, it is an easy thing to kill them all; but shall we rather take them prisoners?" He told me there were two desperate villains among them, that it was scarce safe to show any mercy to; but if they were secured, he believed all the rest would return to their duty. I asked him which they were. He told me he could not at that distance describe them; but he would obey my orders in anything I would direct. "Well," said I, "let us retreat out of their view or hearing, lest they awake, and we will resolve further"; so they willingly went back with me, till the woods covered us from them.

"Now," said I, "here are three muskets for you, with powder and ball; tell me next what you think is proper to be done." He showed me all the testimony of his gratitude that he was able; but offered to be wholly guided by me. I told him I thought it was hard venturing anything; but the best method I could think of was to fire upon them at once, as they lay; and if any was not killed at the first volley, and offered to submit, we might save them, and so put it wholly upon God's providence to direct the shot.

He said very modestly, that he was loth to kill them, if he could help it, but that those two were incorrigible villains, and had been the authors of all the mutiny in the ship, and if they escaped, we should be undone still; for they would go on board and bring the whole ship's company, and destroy us all. "Well then," says I, "necessity legitimates my advice, for it is the only way to save our lives." However, seeing him still cautious of shedding blood, I told him they should go themselves, and manage as they found convenient.

We Make Our Plans

In the middle of this discourse, we heard some of them awake, and soon after we saw two of them on their feet. I asked him if either of them were of the men who he had said were the heads of the mutiny. He said, "No." "Well then," said I, "you may let them escape, and Providence seems to have wakened them on purpose to save themselves. Now," says I, "if the rest escape you, it is your fault."

Animated with this, he took the musket I had given him in his hand and a pistol in his belt, and his two comrades with him, with each man a piece in his hand. The two men who were with him, going first, made some noise, at which one of the seamen who was awake turned about, and seeing them coming, cried out to the rest; but it was too late then; for the moment he cried out, they fired—I mean the two men, the captain wisely reserving his own piece.

They had so well aimed their shot at the men they knew, that one of them was killed on the spot, and the other very much wounded; but not being dead, he started up upon his feet, and called eagerly for help to the other; but the captain, stepping to him, told him, 'twas too late to cry for help, he should call upon God to forgive his villainy, and with that word knocked him down with the stock of his musket, so that he never spoke more. There were three more in the company, and one of them was also slightly wounded. By this time I was come, and when they saw their danger, and that it was in vain to resist, they begged for mercy. The captain told them he would spare their lives, if they would give him an assurance of their abhorrence of the treachery they had been guilty of, and would swear to be faithful to him in recovering the ship, and afterwards in carrying her back to Jamaica, from whence they came. They gave him all the protestations of their sincerity that could be desired, and he was willing to believe them, and spare their lives, which I was not against; only I obliged him to keep them bound hand and foot while they were upon the island.

While this was doing, I sent Friday with the captain's mate to the boat, with orders to secure her and bring away the oars and sail, which they did; and by and by, three straggling men that were (happily for them) parted from the rest, came back upon hearing the guns fired, and seeing their captain, who before was their prisoner, now their conqueror, they submitted to be bound also; and so our victory was complete. Our business now was to consider how to recover the ship.

He agreed with me as to that, but told me he was perfectly at a loss what measures to take; for that there were still six and twenty hands on board, who having entered into a cursed conspiracy, by which they had all forfeited their lives to the law, would be hardened in it now by desperation; and would carry it on, knowing that if they were reduced, they should be brought to the gallows as soon as they came to England, or to any of the English colonies; and that therefore there would be no attacking them with so small a number as we were.

I mused for some time upon what he had said, and found it was a very rational conclusion; and that therefore some thing was to be resolved on very speedily, as well to draw the men on board into some snare for their surprise, as to prevent their landing upon us, and destroying us. Upon this it presently

occurred to me that in a little while the ship's crew, wondering what was become of their comrades and of the boat, would certainly come on shore in their other boat to seek for them, and that then perhaps they might come armed, and be too strong for us. This he allowed was rational.

Upon this, I told him the first thing we had to do was to stave the boat which lay upon the beach, so that they might not carry her off; and taking everything out of her, leave her so far useless as not to be fit to swim. Accordingly we went on board, took the arms which were left on board out of her, and whatever else we found there, which was a bottle of brandy and another of rum, a few biscuit cakes, a horn of powder, and a great lump of sugar in a piece of canvas; the sugar was five or six pounds; all which was very welcome to me, especially the brandy and sugar, of which I had had none left for many years.

We knocked a great hole in her bottom, so that if they had come strong enough to master us, yet they could not carry off the boat. Indeed, it was not much in my thoughts that we could be able to recover the ship; but my view was that if they went away without the boat, I did not much question to make her fit again, to carry us away to the Leeward Islands, and call upon our friends, the Spaniards, in my way, for I had them still in my thoughts.

While we were thus preparing our designs, and had first by main strength heaved the boat up upon the beach, so high that the tide would not float her

off at high-water-mark, we heard the ship fire a gun, and saw her make a waft with her antient, as a signal for the boat to come on board; but no boat stirred; and they fired several times, making other signals for the boat. At last, when all their signals and firing proved fruitless, and they found the boat did not stir, we saw them, by the help of my glasses, hoist another boat out, and row towards the shore; and we found as they approached that there were no less than ten men in her, and that they had fire-arms with them.

As the ship lay almost two leagues from the shore,

we had a full view of them as they came, and a plain sight of the men, even of their faces, because the tide having set them a little to the east of the other boat, they rowed up under shore, to come to the same place where the other had landed, and where the boat lay.

The captain knew the persons and characters of all the men in the boat, of whom he said that there were three very honest fellows, but that as for the boatswain, who it seems was the chief officer among them and all the rest, they were as outrageous as any of the ship's crew.

We had, upon the first appearance of the boat's coming from the ship, considered of separating our prisoners, and had indeed secured them effectually.

Two of them, of whom the captain was less assured than ordinary, I sent with Friday and one of the three (delivered men) to my cave, and he stood sentinel over them at the entrance. The other prisoners had better usage; two of them were kept pinioned indeed, because the captain was not free to trust them; but the other two were taken into my service upon their captain's recommendation, and upon their solemnly engaging to live and die with us;

so with them and the three honest men, we were seven men, well armed; and I made no doubt we should be able to deal well enough with the ten that were coming, considering that the captain had said there were three or four honest men among them also. As soon as they got to the place where their other boat lay, they run their boat in to the beach and came all on shore, hauling the boat up after them, which I was glad to see; for I was afraid they would rather have left the boat at anchor, some distance from the shore, with some hands in her to guard her; and so we should not be able to seize the boat.

Being on shore, the first thing they did, they ran all to their other boat, and it was easy to see that they were under a great surprise, to find her stripped as above, of all that was in her, and a great hole in her bottom.

After they had mused a while upon this, they set up two or three great shouts, hallooing with all their might, to try if they could make their companions hear; but all was to no purpose. Then they came all close in a ring, and fired a volley of their small arms, which indeed we heard, and the echoes made the woods ring, but it was all one; those in the cave we were sure could not hear, and those in our keeping, though they heard it well enough, yet dared give no answer to them.

They were so astonished at the surprise of this, that as they told us afterwards, they resolved to go all on board again to their ship, and let them know that the men were all murdered, and the long-boat staved; accordingly they immediately launched their boat again, and got all of them on board.

The captain was terribly amazed and even confounded at this, believing they would go on board the ship again, and set sail, giving their comrades for lost, and so he should still lose the ship, which he was in hopes we should have recovered; but he was quickly as much frighted the other way.

They had not been long put off with the boat, but we perceived them all coming on shore again; but with this new measure in their conduct, which it seems they consulted together upon, namely to leave three men in the boat, and the rest to go on shore, and go up into the country to look for their fellows.

This was a great disappointment to us; for now we were at a loss what to do; for our seizing those seven men on shore would be of no advantage to us if we let the boat escape; because they would then row away to the ship, and then the rest of them would be sure to weigh and set sail, and so our recovering the ship would be lost.

However, we had no remedy but to wait and see what the issue of things might present. The seven men came on shore, and the three who remained in

the boat put her off to a good distance from the shore, and came to anchor to wait for them; so that it was impossible for us to come at them in the boat.

Those that came on shore kept close together, marching towards the top of the little hill under which my habitation lay; and we could see them plainly, though they could not perceive us. We could have been very glad they would have come nearer to us, so that we might have fired at them, or that they would have gone farther off, that we might have come abroad.

We waited a great while though very impatient for their removing; and were very uneasy, when after long consultations, we saw them start all up and march down towards the sea.

It seems they had such dreadful apprehensions upon them of the danger of the place, that they resolved to go on board the ship again, give their companions over for lost, and so go on with their intended voyage with the ship.

Battle with the Mutineers

As soon as I perceived them go towards the shore, I imagined it to be as it really was, that they had given over their search, and were for going back again; and the captain, as soon as I told him my thoughts, was ready to sink at the apprehensions of it; but I presently thought of a stratagem to fetch them back again, and which answered my end to a tittle.

I ordered Friday and the captain's mate to go over the little creek westward, and bade them halloo as loud as they could, and wait till the seamen heard them; that as soon as they heard the seamen answer them, they could return it again, and then keeping out of sight, take a round, always answering when the other hallooed, to draw them as far into the island, and among the woods, as possible, and then wheel about again to me, by such ways as I directed them.

They were just going into the boat, when Friday and the mate hallooed, and they presently heard them, and answering, ran along the shore westward, towards the voice they heard, where they were presently stopped by the creek, where the water being up, they could not get over, and called for the boat to come up and set them over, as indeed I expected.

When they had set themselves over, I observed that the boat being gone up a good way into the creek, and as it were, in a harbour within the land, they took one of the three men out of her to go along with them, and left only two in the boat, having fastened her to the stump of a little tree on the shore.

This is what I wished for, and immediately leaving Friday and the captain's mate to their business, I took the rest with me, and crossing the creek out of their sight, we surprised the two men before they were aware; one of them lying on shore, and the other being in the boat. The fellow on shore was between sleeping and waking, and going to start up; the captain, who was foremost, ran in upon him and knocked him down, and then called out to him in the boat to yield, or he was a dead man.

There needed very few arguments to persuade a single man to yield, when he saw five men upon him, and his comrade knocked down; besides, this was, it seems, one of the three who were not so hearty in the mutiny as the rest of the crew, and therefore was easily persuaded, not only to yield, but afterwards to join sincerely with us.

In the meantime, Friday and the captain's mate so well managed their business with the rest, that they drew them by hallooing and answering, from one hill to another, and from one wood to another, till they not only heartily tired them, but left them where they were very sure they could not reach back to the boat before it was dark; and indeed they were heartily tired themselves also by the time they came back to us.

We had nothing now to do but to watch for them in the dark, and to fall upon them, so as to make sure work with them.

It was several hours after Friday came back to me, before they came back to their boat; and we could hear the foremost of them long before they came quite up, calling to those behind to come along, and could also hear them answer and complain how lame and tired they were, and not able to come any faster, which was very welcome news to us.

At length they came up to the boat; but 'tis impossible to express their confusion when they found the boat fast aground in the creek, the tide ebbed out, and their two men gone. We could hear them call to one another in a most lamentable manner, telling one another they were gotten into an enchanted island; that either there were inhabitants in it, and they should all be murdered, or else there were devils and spirits in it, and they should be all carried away and devoured.

They hallooed again, and called their two comrades by their names a great many times, but no answer. After some time, we could see them, by the little light there was, run about wringing their hands like men in despair; and that sometimes they would go and sit down in the boat to rest themselves, then come ashore again, and walk about again, and so over the same thing again.

My men would fain have me give them leave to fall upon them at once in the dark; but I was willing to take them at some advantage, so to spare them, and kill as few of them as I could; and especially I was unwilling to hazard the killing any of our own men, knowing the others were very well armed. I resolved to wait to see if they did not separate; and therefore to make sure of them, I drew my ambuscade nearer, and ordered Friday and the captain to creep upon their hands and feet as close to the ground as they could, that they might not be discovered, and get as near them as they could possibly, before they offered to fire.

They had not been long in that posture, but that the boatswain, who was the principal ringleader of the mutiny, and had now shown himself the most dejected and dispirited of all the rest, came walking towards them with two more of their crew; when they came nearer, the captain and Friday, starting up on their feet, let fly at them.

The boatswain was killed upon the spot, the next man was shot in the body and fell just by him, though he did not die till an hour or two after; and the third ran for it.

At the noise of the fire, I immediately advanced with my whole army, which was now eight men, namely myself generalissimo, Friday my lieutenant-general, the captain and his two men, and the three prisoners-of-war, who we had trusted with arms.

We came upon them indeed in the dark, so that they could not see our number; we bade them surrender, as the captain was standing by with fifty men ready to attack. This, of course, was a fiction as our active force numbered only eight in all. But it worked.

My army came up and seized them all, upon their boat, only I kept myself and one more out of sight, for reasons of state.

Our next work was to repair the boat, and think of seizing the ship.

Upon the captain's coming to me, I told him my project for doing this, which he liked wonderfully well, and resolved to put it in execution the next morning, but first it would be necessary to divide the prisoners.

I told him that he should go and take three of the worst of them, and send them pinioned to the cave where the others lay. This was committed to Friday and the two men who came on shore with the captain.

Retaking the Ship

With the coming of morning the captain had no difficulty before him, but to furnish his two boats, stop the breach of one, and man them. He made his passenger captain of one, with four other men; and himself, and his mate, and five more, went in the other; and they contrived their business very well, for they came up to the ship about midnight. As soon as they came within call of the ship, he hailed them and told them they had brought off the men and the boat, but that it was a long time before they had found them, and the like; holding them in a chat till

they came to the ship's side; when the captain and the mate, entering first with their arms, immediately knocked down the second mate and carpenter with the butt-end of their muskets. Being very faithfully seconded by their men, they secured all the rest that were upon the main and quarter decks, and began to fasten the hatches to keep them down who were below, when the other boat and their men, entering at the fore chains, secured the fore-castle of the ship, and the scuttle which went down into the cook-room, making three men they found there prisoners.

When this was done, and all safe upon deck, the captain ordered the mate with three men to break into the round-house where the new rebel captain lay, and having taken the alarm, was gotten up, and with two men and a boy had gotten fire-arms in their hands; and when the mate with a crow split open the door, the new captain and his men fired boldly among them, and wounded the mate with a musket ball, which broke his arm, and wounded two more of the men but killed nobody.

The mate, calling for help, rushed however into the round-house, wounded as he was, and with his pistol shot the new captain through the head, the bullet entering at his mouth, and came out again behind one of his ears, so that he never spoke a word; upon which the rest yielded, and the ship was taken effectually, without any more lives lost.

As soon as the ship was thus secured, the captain ordered seven guns to be fired, which was the signal agreed upon with me, to give me notice of his success, which you may be sure I was very glad to hear, having sat watching upon the shore for it till near two o'clock in the morning.

I Am Rescued at Last

Having thus heard the signal plainly, I laid me down; and it having been a day of great fatigue to me, I slept very sound, till I was something surprised with the noise of a gun; and presently starting up, I heard a man call me and presently I knew the captain's voice, when climbing up to the top of the hill, there he stood, and pointing to the ship, he embraced me in his arms. "My dear friend and deliverer," said he, "there's your ship, for she is all yours, and so are we and all that belong to her." I cast my eyes to the ship, and there she rode within little more than half a mile of the shore; for they had weighed her anchor as soon as they were masters of her; and the weather being fair, had brought her to an anchor just against the mouth of the little creek; and the tide being up, the captain had brought the pinnace in near the place where I at first landed my rafts, and so landed just at my door.

I was at first ready to sink down with surprise; for I saw my deliverance indeed visibly put into my hands, all things easy, and a large ship just ready to carry me away whither I pleased to go.

At first, for some time, I was not able to answer him one word; but as he had taken me in his arms, I held fast by him, or I should have fallen to the ground.

He perceived the surprise, and immediately pulled a bottle out of his pocket, and gave me a dram of cordial, which he had brought on purpose for me. After I had drunk it, I sat down upon the ground; and though it brought me to myself, yet it was a good while before I could speak a word to him.

We consulted what was to be done with the prisoners we had, and decided that, since we knew them to be incorrigible and refractory to the last degree, it was best to leave them on the island. I therefore let them into the story of my living there

and put them in the way of making it easy to them. In a word, I gave them every part of my own story.

When we, my man Friday and I, took leave of this island, we carried on board for relics the great goat's-skin cap I had made, my umbrella, and my parrot; also I forgot not to take the money I formerly mentioned, which had lain by me so long useless that it was grown rusty, or tarnished, and could hardly pass for silver till it had been cleaned and polished.

And thus we left the island, the nineteenth of December as I found by the ship's account, in the year 1686, after I had been upon it eight and twenty years, two months, and nineteen days.

In this vessel, after a long voyage, we arrived in England, the eleventh of June, in the year 1687.